CRC Series
in
MARINE SCIENCE

Editor-in-Chief
J. Robert Moore
Director, Marine Science Institute
The University of Texas
Austin, Texas

CRC HANDBOOK OF MARINE SCIENCE
Editors
F. G. Walton Smith
Dean Emeritus, School of Marine and Atmospheric Science
University of Miami
President, International Oceanographic Foundation
Miami, Florida
Frederick A. Kalbler
President, Hydrobiological Services, Inc.
Naranja, Florida

CRC HANDBOOK OF GEOPHYSICAL EXPLORATION AT SEA
Editor
Richard A. Geyer
Professor Emeritus
Department of Oceanography
Texas A & M University
College Station, Texas

MAN AND THE MARINE ENVIRONMENT
Editor
Robert A. Ragotzkie
Director, Sea Grant Institute
University of Wisconsin
Madison, Wisconsin

CRC HANDBOOK OF MARINE SCIENCE: COMPOUNDS FROM MARINE ORGANISMS
Authors
Joseph T. Baker
Vreni Murphy
Roche Research Institute of Marine Pharmacology
Sydney, Australia

CRC HANDBOOK OF COASTAL PROCESSES AND EROSION
Editor
Paul D. Komar
School of Oceanography
Oregon State University
Corvallis, Oregon

CRC HANDBOOK OF MARICULTURE
Editor
James P. McVey
Fishery Biologist (Aquaculture Specialist)
National Marine Fisheries Service
Office of International Fisheries
Jakarta, Indonesia

CRC Handbook of Coastal Processes and Erosion

Editor

Paul D. Komar, Ph.D.
Professor of Oceanography
Oregon State University
Corvallis, Oregon

Editor-in-Chief
CRC Series in Marine Science

J. Robert Moore, Ph.D.
Professor
Marine Science Institute
The University of Texas
Austin, Texas

CRC Press, Inc.
Boca Raton, Florida

Library of Congress Cataloging-in-Publication Data
Main entry under title:

CRC handbook of coastal processes and erosion.

(CRC series in marine science)
Includes bibliographies and index.
1. Coast changes. 2. Erosion. I. Komar, Paul D., 1939- . II. Title: C.R.C. handbook of coastal processes and erosion. III. Series.
GB451.2.C73 1983 551.3′6 83-3773
ISBN 0-8493-0225-0

This book represents information obtained from authentic and highly regarded sources. Reprinted material is quoted with permission, and sources are indicated. A wide variety of references are listed. Every reasonable effort has been made to give reliable data and information, but the author and the publisher cannot assume responsibility for the validity of all materials or for the consequences of their use.

Direct all inquiries to CRC Press, Inc., 2000 Corporate Blvd., N.W., Boca Raton, Florida, 33431.

Second Printing, 1984
Third Printing, 1985
Fourth Printing, 1991

International Standard Book Number 0-8493-0225-0

Library of Congress Card Number 83-3773
Printed in the United States

PREFACE

Every coast has its story to tell of property losses to the sea. Medieval towns in England disappeared as waves cut away at the land. Hurricanes and nor'easters periodically wreak havoc along the eastern seaboard and Gulf Coasts of the United States, and the severe storms of the not so pacific Pacific have resulted in untold instances of property losses.

Such tales of woe have been recounted many times, and it is not the purpose of the present volume to repeat them once again. Instead, the objective is to focus more on the physical processes that produce coastal erosion. During the past quarter century our understanding of coastal-zone processes has increased vastly, thanks to the many scientists and engineers who have focused their research efforts in this environment. These studies have spanned the entire range of physical processes from the waves and currents in the nearshore to the response of the beach, via sand transport, resulting in a changing coastal morphology. Of these many processes, the present volume focusses only on those which directly relate to the generation of coastal erosion. Some chapters deal exclusively with the physical processes, while others provide examples of erosion problems. The examples were not selected to elicit a "gee-whiz" response, but instead to illustrate via case studies the physical processes or combinations of processes. Although most of the chapter topics have clear implications for issues of coastal-zone management, these issues are not belabored as several other books are already available in this area. The objective here is to provide a state-of-the-art presentation of the science of coastal erosion processes.

Paul D. Komar
February 1983

EDITOR-IN-CHIEF

The Editor-in-Chief of the CRC Marine Science Series, **Professor J. Robert Moore,** received his B.S. degree (honors) from the University of Houston, in geology, his M.A. from Harvard University, also in geology, and his Ph.D. from the University of Wales (Aberystwyth), in geology and oceanography. From 1956 to 1966, he was a Senior Scientist at the Texaco Research Laboratories, Houston, Texas, engaged in research on marine sedimentation and geo-resources. From 1966 to 1977, Dr. Moore was Professor of Geology at the University of Wisconsin, Madison, where he was also Director of the Marine Research Laboratory and the Marine Mining Program. In 1977, he became Director, Institute of Marine Science, University of Alaska, and in 1979, assumed his present affiliation as Professor of Marine Studies, at the University of Texas, at Austin. Professor Moore's principal research interests are in seafloor mineral resources, marine geology, and global oceanographic processes and resources. His research has involved studies in the Bering Sea, Irish Sea, Western Pacific, Gulf of Mexico, Gulf of Alaska, Atlantic shelf and several coastal areas in the United States and abroad. He is a member of the Marine Board of the National Academy, the A.A.P.G. Marine Geology Committee, the N.A.C.O.A. Marine Mining Panel, Organizing Chairman of the CHEMRAWN IV Conference, and member of The Geochemical Society, Soc. of Economic Paleontologists and Mineralogists, Marine Technology Society, Am. Assoc. of Petroleum Geologists, Challenger Society, and a Fellow of the Geological Society of London. He is Editor of the *Marine Mining Journal,* and Exec. Secretary of the International Marine Mining Association. Since 1966, he has been an active consultant to major international corporations, chiefly on matters related to ocean resources, utilization, exploration, and development. He is currently conducting two major marine research programs, one in Palauan waters of the Western Pacific and the other in the Bering Sea off Alaska. He resides in Austin, Texas.

THE EDITOR

Paul D. Komar, Ph.D., is a Professor of Oceanography at Oregon State University. He received his training in mathematics and geology at the University of Michigan, Ann Arbor, where he obtained A.B. and M.S. degrees. In 1969, Dr. Komar received his Ph.D. degree in oceanography from the Scripps Institution of Oceanography, The University of California, San Diego. This was followed by a postdoctoral year of research at the Wallingford Hydraulics Research Station, England, and at St. Andrews University, Scotland, supported by a NATO fellowship. He assumed his present position in 1970.

Dr. Komar is a member of the American Association for the Advancement of Science, the American Geophysical Union, the Geological Society of America, and the Society of Economic Paleontologists and Mineralogists. He has served as associate editor for several scientific journals including *Sedimentology*, *Marine Geology*, and the *Bulletin* of the Geological Society of America.

Dr. Komar is the author of more than 80 papers and one textbook. His research interests are broad, including coastal processes, deep-sea sedimentary processes, the geomorphic evolution of the planet Mars, and the fluid mechanics of igneous dike formation.

CONTRIBUTORS

Robert G. Dean, Sc.D.
Graduate Research Professor
Department of Coastal and Oceanographic Engineering
University of Florida
Gainesville, Florida

Robert Dolan, Ph.D.
Professor
Department of Environmental Sciences
University of Virginia
Charlottesville, Virginia

Edward Hands
U.S. Army Waterway Experiment Station
Coastal Engineering Research Center
Vicksburg, Mississippi

Bruce Hayden, Ph.D.
Associate Professor
Department of Environmental Sciences
University of Virginia
Charlottesville, Virginia

Robert A. Holman, Ph.D.
Assistant Professor
School of Oceanography
Oregon State University
Corvallis, Oregon

Gerald G. Kuhn, Ph.D.
Research Associate
Scripps Institution of Oceanography
La Jolla, California

E. M. Maurmeyer, Ph.D.
College of Marine Studies
University of Delaware
Lewes, Delaware

Suzette Kimball May, Ph.D.
Research Associate
University of Virginia
Charlottesville, Virginia

Dag Nummedal, Ph.D.
Associate Professor
Department of Geology
Louisiana State University
Baton Rouge, Louisiana

Francis P. Shepard, Ph.D.
Professor Emeritus of Submarine Geology
University of California, San Diego
Scripps Institution of Oceanography
La Jolla, California

A. D. Short, Ph.D.
Senior Lecturer
Coastal Studies Unit
Department of Geography
University of Sydney
Sydney, Australia

Tsuguo Sunamura, Ph.D.
Associate Professor
Institute of Geoscience
University of Tsukuba
Ibaraki, Japan

L. D. Wright, Ph.D.
Professor
Virginia Institute of Marine Science
School of Marine Science
College of William and Mary
Gloucester Point, Virginia

DEDICATION

To Francis P. Shepard

Numbered among his many accomplishments within a distinguished career has been a concern for the coastal zone, both as a scientist and as a citizen.

TABLE OF CONTENTS

Chapter 1

BEACH PROCESSES AND EROSION — AN INTRODUCTION

Paul D. Komar

Whether it be on a barrier island under the onslaught of a hurricane or the long-term wearing away of rocky sea cliffs, coastal erosion is the result of a complex interaction of physical processes. Depending on the circumstances, the water motions of waves, tides, storm surge, and nearshore currents combine and interact with the coastal land which in turn has considerable inherent variability.

The purpose of this chapter is to briefly summarize these several agents that play a role in coastal erosion. Many of these processes will be dealt with at greater length in subsequent chapters, with more details on the resulting erosion. This chapter will serve only as an abstract for this later material, and to present an introduction to waves and currents that will serve as background material needed in later chapters. More complete reviews on coastal processes in general can be found in Wiegel,[54] King,[28] Silvester,[48] Komar,[30] and Horikawa.[24]

This brief summary will begin with waves — certainly the primary agent in coastal erosion — examining their generation, measurement, travel, and eventual breaking at the coastline. These waves generate nearshore currents which in turn transport the beach sediments, a movement which results in a rearrangement of the shoreline with possible sites of erosion.

WAVES — THEIR GENERATION, TRAVEL, AND ARRIVAL AT THE COAST

The waves one sees every day on the ocean, and those primarily responsible for coastal erosion, are generated by winds blowing over the water surface. There are, of course, waves such as the tsunami that are not generated by winds, and although they can be extremely destructive their occurrence is too rare for them to be a significant factor in coastal erosion and so will not be considered here.

Wind-generated waves are important as energy-transfer agents; first obtaining their energy from the winds, transfering it across the expanses of the oceans, and then delivering it to the coastal zone where it can be the primary cause of erosion or can generate a variety of nearshore currents and sand transport patterns. This transferral of energy is depicted schematically in Figure 1. The generation of the waves is primarily dependent upon three storm factors: the speed of the wind, the duration of the storm, and the fetch area over which the storm occurs. The duration is important in that the longer the winds have been blowing, the greater the amount of energy that can be transferred to the growing waves. The fetch area has a similar effect; once the waves travel out of the storm area they no longer acquire additional energy, so that the larger the fetch area, the more energy the waves can potentially obtain.

The simplest type of wave is shown in Figure 2, characterized by a height, H, length, L, and period, T, the time it takes for successive crests to pass a fixed position such as the piling shown. However, waves in the storm area, termed *sea* (Figure 1), do not fit this ideal picture but instead are a complex pattern of wave crests and troughs, with no two waves seeming to have the same height or period. If an individual wave crest is followed it often is observed to progressively decrease in height and eventually disappear. This complex pattern of waves results because a storm does not simply generate waves of one fixed height and period, but instead, a whole range or spectrum is generated. When the wind first blows across an initially quiet body of water only small ripples are formed with periods less than

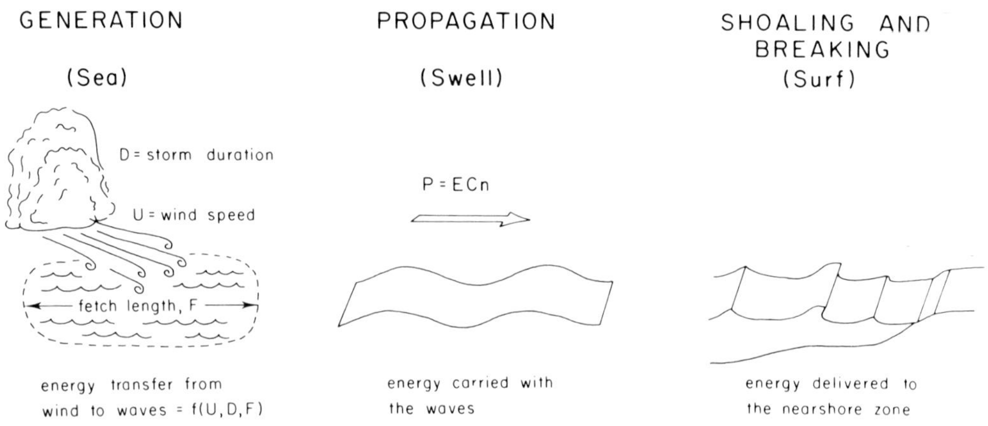

FIGURE 1. The generation of waves by storm winds. The energy transfer depending on the factors illustrated, followed by the transfer of energy across the sea as swell waves, finally arriving at the coastline where the energy is important to many nearshore processes.

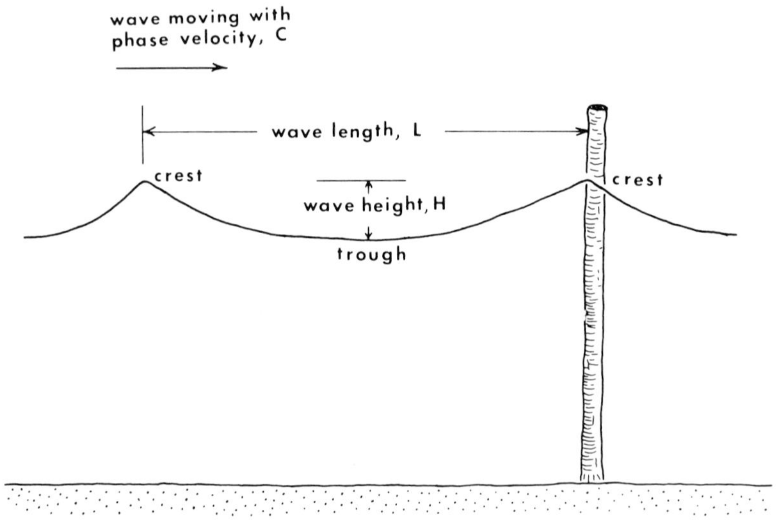

FIGURE 2. The parameters that, along with the wave period T, describe simple oscillatory waves. (From Komar, P. D., *Beach Processes and Sedimentation,* Prentice-Hall, Englewood Cliffs, N.J., 1976. With permission.)

1 second and heights of only a couple of centimeters. As time passes, waves with longer and longer periods will be formed, but small ripples will continue to be present and waves with a range of periods will now exist. The longer-period waves have longer wave lengths, and this also permits them to achieve greater heights without breaking and thereby losing energy. Concomitant with the progressive increase in wave periods present in the area of generation is an increase in wave heights.

Characterizing the waves in the area of generation is obviously more difficult than in the case of the simple sinusoidal wave of Figure 2 which could be defined by one height and period. Once a record is obtained of the water-surface elevations, determined by a variety of techniques including water-pressure sensors, wave staffs, surface "glitter" and other remote sensing methods, the analysis usually takes one of two possible paths (Figure 3). A statistical analysis of wave heights can be performed, noting the maximum wave height in

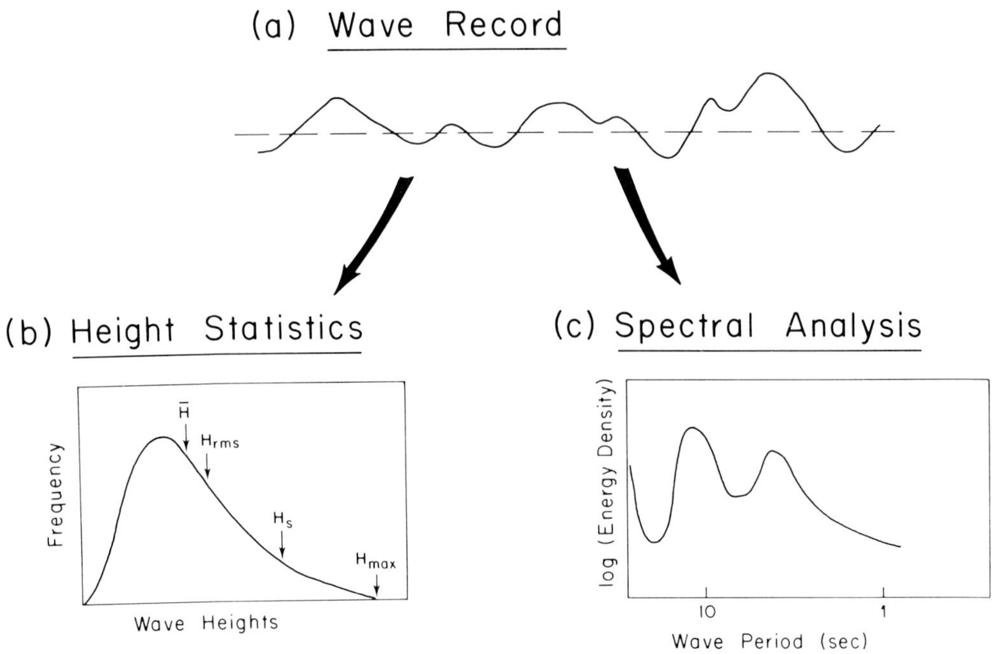

FIGURE 3. The analysis of a wave record (a) usually involves a consideration of the statistics of the wave heights (b) or a spectral analysis approach (c).

the record, the average height, or a root-mean-square wave height. A commonly used statistical wave height is the *significant wave height*, H_s, defined as the average of the highest one third of the waves. Its choice was based on the impression that in many applications the larger waves are more important than the small waves, and H_s thereby provides a more representative wave measure than, for example, the average wave height. It has also been shown that H_s roughly corresponds to a visual estimate of a representative wave height in that the observer naturally tends to weight his observation toward the larger waves. It has been demonstrated theoretically[38] and by measurements[18,44] that under fully developed storm waves these statistical wave heights form well-defined ratios. For example, the ratio of H_s to the average wave height is approximately 1.56, and its ratio to the root-mean-square wave height is 1.42. The maximum wave height observed in a wave record depends on the length of that record, so there is no fixed value for its ratio with the other wave statistics. One can also define a significant wave period or average wave period, but these have somewhat less physical reality than the corresponding wave heights, and as we shall see, their use can lead to mistaken results.

An alternate and more realistic approach to analyzing waves is through *spectral analysis* (Figure 3). This approach views the complex wave patterns as the sum of many simple sinusoidal waves of differing periods and heights (energies). A summation of even a few sinusoidal waves quickly leads to complex wave patterns such as those observed in the area of generation. For example, the highest waves found under a storm are usually produced by the chance summation of several of the simpler waves, a summation which may occur for only a few seconds — after which the simple waves move apart and no longer sum to form the exceptionally large wave. Spectral analysis is the reverse of this summation, starting with the observed complex pattern as in Figure 3(a) and breaking it down into the individual sinusoidal waves. Figure 3(c) depicts schematically what is typically obtained with such a technique. The spectrum shows the amount of energy contained at each wave period or frequency, the energy being proportional to the wave height squared. Therefore, the spectrum indicates the heights of all the sinusoidal waves that when summed would yield the complex waves measured. In the spectrum illustrated most of the energy is seen to occur at two periods, 12 and 4 seconds indicating that there were two principal sets of waves present, although it is seen that a wide range of periods from less than 1 second to greater than 12 seconds was present. The rapid upturn at very high periods would be caused by water-level changes due to tides, the analysis evaluating this change as a long-period wave (which, of course, the tide is). This illustrative spectrum is typical in that at a coastal site one commonly records waves arriving from a distant storm, the 12 second peak, as well as waves being locally generated by the coastal winds (the 4 second peak).

A discussion of the methods employed for predicting the wave conditions that result from a certain combination of storm conditions is beyond the scope of this introductory chapter. Two general approaches are available, one which yields a representative significant wave height and period, or approaches which yield the complete spectrum. The several text books available discuss these various approaches, and the U.S. Army Corps of Engineers *Shore Protection Manual*[8] provides a useful cookbook methodology. In general the spectrum methods provide a conceptual advance over the significant wave approach[42] and yield more realistic descriptions of the waves, a factor that is especially important in following the movement of the waves across the oceans. Spectral analysis techniques are now sufficiently available that significant wave height approaches should be employed only in the simplest of applications.

Derivations of the equations that describe the wave motions can be found in several texts. In deep water individual waves travel at the phase velocity, C, related to the wave period by

$$C = \frac{g}{2\pi} T \tag{1}$$

However, in deep water groups of waves and their energy travel at only half this rate. One important consequence of the relationship of Equation 1 is that the waves sort themselves according to their periods once they leave the storm area of generation. By a process called *wave dispersion,* the highest-period waves in the spectrum move out most rapidly and would be the first to arrive and be measured at some distant coastal site. This sorting by the wave period also narrows the spectrum so that the complex sea in the storm area is progressively transformed into regular *swell waves* (Figure 1) which can be better characterized by a dominant wave height and period such as the simple waves of Figure 2. It is apparent that when the waves are characterized by one significant wave height and period it is not possible to satisfactorily examine these subsequent changes brought about by wave dispersion.

Swell waves are known to travel over the expanse of the oceans with relatively little loss of energy, the second stage in the energy transfer depicted in Figure 1. Important to coastal erosion, the energy the waves have obtained from the storm winds is thereby efficiently carried to the nearshore zone where it is expended over a relatively narrow area. The storm itself can be thousands of kilometers distant with no direct effect on the coast. For example, Snodgrass et al.[49] followed waves that had been generated in the South Pacific near Antarctica, tracing their progress as far as the shores of Alaska.

The wave energy, E, is related directly to the wave height by

$$E = \frac{1}{8} \rho g H^2 \tag{2}$$

where ρ is the water density. The flux of energy carried by the moving waves is then

$$P = ECn \tag{3}$$

where Cn, together, is the group velocity of the waves, the rate of energy transfer, n being a function of the water depth[30] but with n = 1/2 in deep water and n = 1 in shallow water (as discussed in Reference 30, the terms "deep water" and "shallow water" have precise technical meanings related to simplifying approximations of the more complex general equations). The energy flux P can also be viewed as the power of the waves per unit crest length. This is the power delivered to the coastal zone when the waves reach the nearshore and break. Without significant bottom frictional drag on the waves, and for the moment neglecting wave refraction, P remains constant as the waves move across the ocean. Even when the waves reach shallow water it is P of Equation 3 that remains relatively constant, not the energy E of Equation 2. As a consequence, when the waves leave deep water and begin to shoal, Cn in general decreases so that E and hence the wave height H must increase in order for P to remain constant. Therefore, for the most part the wave heights increase as they approach shore, although there is a brief stage in which they decrease where n increases faster than C decreases.[30]

In shallow water the phase velocity of the waves is no longer determined by their period but instead is governed by the water depth, h, according to the relationship

$$C = \sqrt{gh} \tag{4}$$

or by

$$C = \sqrt{g(h + H)} \tag{5}$$

if the wave height H is significant in comparison with h, as it generally is as the waves approach the shoreline. This dependence on water depth gives rise to *wave refraction,* a rotation or bending of the waves in response to the bathymetry over which they pass. If one part of a particular wave crest is in deeper water than another portion further along the shore, the portion in deeper water will advance more rapidly. As a result, waves arriving obliquely to a coastline will tend to rotate so as to become more nearly parallel with the shoreline. Under simple refraction one has *Snell's Law,*

$$\frac{\sin\alpha}{C} = \text{constant} \tag{6}$$

where α is the angle between the wave crests and the shoreline or depth contour, and Equation 3 is modified to

$$P = (ECn)\cos\alpha \simeq \text{constant} \tag{7}$$

With more complex topography, for example an offshore submarine canyon, the wave refraction becomes more complicated with a focusing and defocusing of the wave energy, much as a lens focuses rays of light. An example is shown in Figure 4. The techniques for constructing wave refraction diagrams such as this are summarized in CERC.[8] In addition to drawing the wave crests, the *wave rays* are also shown: lines that are normal to the crests and are therefore in the direction of wave advance and energy propagation. The wave power P is conserved between adjacent rays, so that a convergence of rays implies a focusing of the wave energy leading to greater wave heights, and conversely, progressive separation of rays represents defocusing. As seen in Figure 4, the greater water depths of the submarine canyons cause a defocusing in their shoreward lees and a focusing to either side. As a result, the waves directly shoreward of the canyons will be smaller than in the shallower sections between the canyons. It is apparent then that wave refraction can be an important factor in coastal erosion, the coastal sites where the waves are focused undergoing greater wave

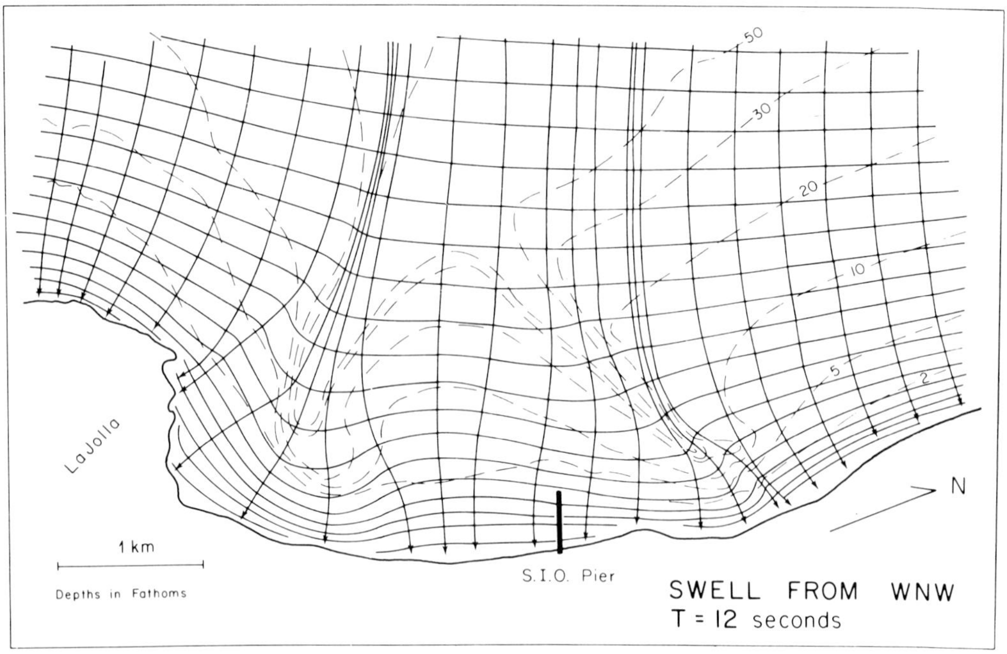

FIGURE 4. Wave refraction at La Jolla, Calif. The patterns are due primarily to the presence of two submarine canyons, La Jolla Canyon to the south (left) and Scripps Canyon to the north (right). Both the wave crests and the orthogonal set of wave rays are shown.[43]

attack. Goldsmith[17] provides an analysis of the refraction patterns of waves from various directions arriving along the shorelines of Delaware south to North Carolina, examining the potential for erosion. In addition, as will be discussed later in this chapter, the longshore variations in wave heights brought about by refraction can give rise to nearshore currents, specifically to rip currents (compare Figures 4 and 10).

It is also seen in Figure 4 that as the waves approach shore adjacent crests become closer together, that is, the wave length L decreases. This follows from the relationship

$$L = CT \tag{8}$$

the decrease in L accompanying the reduction in the wave speed, C, of the shoaling waves, the period T remaining constant during the process. Due to the decrease in L and increase in heights, H, the wave steepness H/L progressively increases as the waves approach the shoreline. The wave crests also become narrower and peaked, the troughs becoming wide and flat.

Eventually the waves oversteepen, become unstable, and break, generally either by plunging or spilling. The nature of their breaking depends on their initial steepness and on the slope of the beach.[10] It is found that the ratio of the breaker height, H_b, to depth is

$$\gamma = \frac{H_b}{h_b} \simeq 0.75 - 1.2 \tag{9}$$

the exact value depending mainly on the beach slope.[27] Equation 9 is a useful criterion for estimating the position of the breaker zone on a particular beach. Several formulas have been proposed for the prediction of the breaker height itself from the known deep-water wave height, H_o, and period, T. That proposed by Komar and Gaughan[35] and further substantiated by Weishar and Byrne[53] is the simple expression

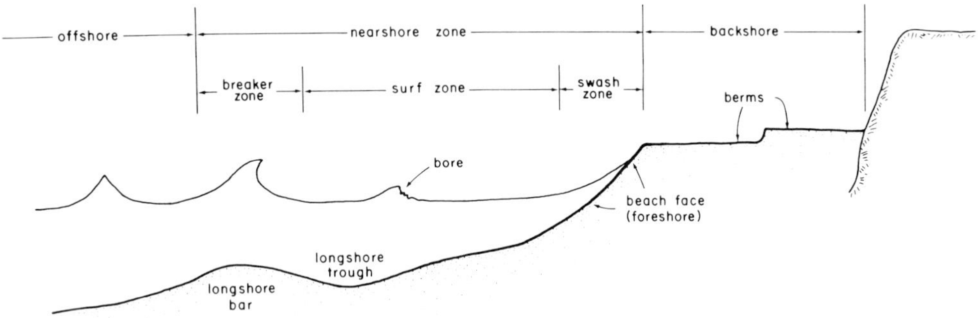

FIGURE 5. Some of the terminology that is used to describe the nearshore zone and beach profiles.

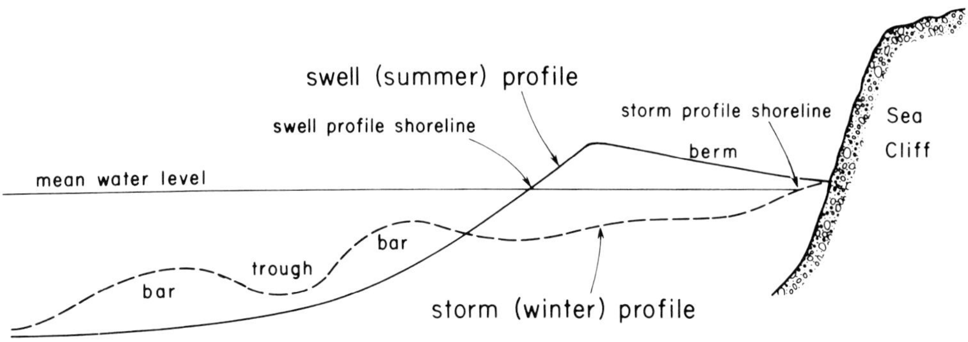

FIGURE 6. Changes in the beach profile brought about by high-energy storm waves vs. the lower-energy swell waves. (From Komar, P. D., *Beach Processes and Sedimentation,* Prentice-Hall, Englewood Cliffs, N.J., 1976. With permission.)

$$H_b = 0.39\ g^{1/5}(TH_0^2)^{2/5} \qquad (10)$$

This formula does not include the effects of wave refraction, but can be altered to do so.

In general, waves with a range of heights are arriving at the coast, and according to Equation 9 the larger of these will break in somewhat greater water depths than the smaller waves. This gives rise to a *breaker zone* (Figure 5) whose width depends on the wave-size range and on the beach slope. On low-sloping beaches the broken waves transform into *bores* that travel across the *surf zone* and finally surge up the *beach face* as *swash* (Figure 5). The nature of the beach profile, and hence the patterns or nearshore waves, depends on the grain size of the beach sediments and on the energy level of the waves. Normally the coarser the sediment grain size the steeper the over-all beach slope. On steep, coarse-sand to cobble beaches the waves usually break directly at the base of the beach face and swash up the slope, a surf zone with bores being absent. The morphology of the beach profile also depends on the wave-energy level, high-wave conditions tending to shift sand offshore, eroding it from the beach *berm* and depositing it into bars in the region of the breaker zone. A return of low waves reverses the process. Along most coastlines there is a seasonality to the wave-energy levels, wave heights tending to be larger during winter storms, producing a seasonal change in the beach profiles as shown schematically in Figure 6. Due to this seasonality, the two profile types are often termed the *summer profile* and *winter profile.* Other terminology has been employed to eliminate the seasonal connotations, which do not apply in places such as Bermuda, the terms *swell profile* and *storm profile* more clearly indicating the dependence on the wave conditions. Such changes are clearly important to coastal erosion, the loss of the berm in the storm (winter) profile permitting the wave swash to more directly attack the coastal property.

THE WATER LEVEL FACTOR — TIDES AND STORM SURGE

Another factor important to the occurrence and severity of coastal erosion is the water level as influenced by the tides and any superimposed storm surge. High water levels can result in flooding of low-lying coastal areas and shift the surf zone landward so that the waves directly attack coastal properties.

The astronomical tides are produced by the combined gravitational attraction of the moon and sun acting upon large bodies of water.[9,11] Twice each month, at new and full moon, the earth, moon, and sun align *(syzygy)* so that the forces of the moon and sun combine to produce increased tidal ranges called *spring tides*. If a storm occurs at a time of spring tides the water levels of the high tides permit the storm waves to break closer to or directly on the coastal property, thereby influencing the degree of the resulting erosion.

Wood[55] demonstrated in a comprehensive fashion that unusually high spring tides have been a significant factor in historic occurrences of coastal flooding and erosion. Termed *perigean spring tides,* these exceptional tides occur when there is both an alignment between the earth, moon, and sun, and the moon is located at its perigee position, i.e., its closest approach to the earth in its elliptical orbit. Since the tide-generating force varies inversely as the cube of the distance between the earth and moon, it becomes a maximum when the moon is at perigee. The combined conditions of perigee-syzygy add about 40% to the tidal range, thereby significantly affecting the water levels.

Wood[55] described the occurrence of a perigean spring tide as "a window for potential flooding." In themselves, the unusual water levels may not produce appreciable coastal flooding and erosion but do increase the potential for those conditions. He also demonstrated that many of our major coastal flooding events in the past occurred when perigean spring tides combined with strong, onshore winds that further raised the water levels as well as generating large waves. He found that over 100 cases of major coastal flooding associated with these conditions occurred on the North American coastline in the past 341 years (1635 to 1976).

The most significant water-level change caused by meteorological factors such as strong winds and abrupt atmospheric pressure reduction is the *storm surge,* sometimes called a meteorological tide. The rise in water level may be rather abrupt, induced by strong onshore winds and low atmospheric pressures accompanying a hurricane or typhoon. In addition to generating large surface waves, the winds of the hurricane can push water toward the coast, holding it there as a set-up of the mean water level. The exceptionally low pressures of the storm system also contribute by "humping" up the water level, the ocean surface acting as an inverse barometer (a drop of 1 in. in the atmospheric pressure produces an approximate 13-in. rise in the water level). Another factor is the heavy rains which commonly accompany the hurricane, the rise in the coastal water levels inhibiting the seaward flow of the runoff so that flooding results.

With respect to coastal erosion, the storm surge may raise the water levels by several meters above the highest high tides normal for the coastal zone, thus covering areas generally not affected by wave attack. Considerable destruction results from the large waves superimposed on the exceptionally high levels since shoreline structures are now in the surf zone. And storm surge is a very important factor in producing barrier-island washovers and breaching (Chapter 5).

Storm surges have their greatest impact in coastal areas that experience such storms and are topographically low lying. They have been exceptionally destructive along the North Sea coasts of the Netherlands and Germany, the Gulf of Mexico and eastern coasts of the U.S., the low coastal areas of Bangladesh and eastern India, and in Japan. Storm-generated water-level rises can be significant in the shallower Great Lakes, especially at the Buffalo and Toledo ends of Lake Erie. Bascom[1] provides descriptions of the destructive effects of storm surges at some of these locations.

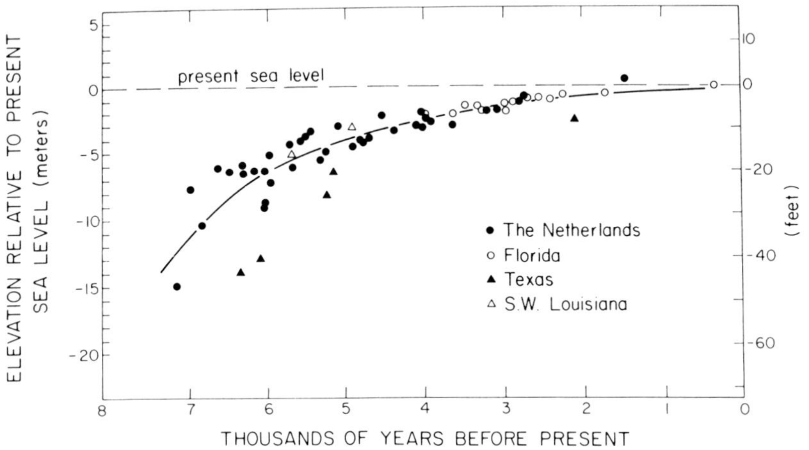

FIGURE 7. Sea-level curve for the past 8000 years based on the data compilation[46] from tectonically "stable" areas of the world.

Silvester,[48] Horikawa,[24] and Sorensen[50] provide good reviews of the state of our understanding of the physical processes of storm surge. We are now in the position of making reasonable predictions of the response of the coastal waters to various storm conditions. Yet little can be done as countermeasures to prevent their erosive destruction. Where the potential economic impact warrants and the geography is suitable, coastal dikes have been constructed. But in many areas the best response is to keep dwellings out of susceptible low-lying areas, to establish an early warning system, and to construct escape routes. These measures are particularly important on barrier islands which may be entirely awash during a storm.

SEA LEVEL

For more than 3 million years water has periodically been locked up within large continental glaciers and then released, producing alternating sea-level lowerings and rises. Within the last 20,000 years the sea level has changed by more than 100 m exerting a considerable influence on our coasts.

A timetable of the changes in sea level has been obtained by dating materials such as submerged peat beds, beach rock, and fossil intertidal animals, material that has a known narrow relationship to past stands of the sea. If these have not been transported subsequent to their formation, then dating provides an indication of the sea level at that particular time. It is apparent, however, that such materials indicate only a relative sea level at that location, not an absolute (eustatic) sea level due to the volume of water in the oceans, since in the meantime the level of that chunk of earth could also have changed.

A number of investigators have developed chronologies of sea level variations in this way, some extending over the past 50,000 years or more (Shepard,[45] Curray,[10] Shepard and Curray,[46] Milliman and Emery,[41] and Kraft et al.[36]). The data employed are restricted to continental shelves of reasonably stable areas so as to reduce the land elevation component due to tectonism, etc. These chronologies indicate that sea level stood some 130 m lower than at present about 15,000 to 20,000 years ago at the time of the last major glacial advance. With melting of the glaciers there was initially a rapid rise in sea level, averaging about 8 mm/year, until approximately 7000 years ago when it slowed down to only a couple of millimeters per year at the present time.

Of particular relevance to the development of our present-day coasts and to coastal erosion is the sea-level rise within the last few thousand years. A compilation of data by Shepard

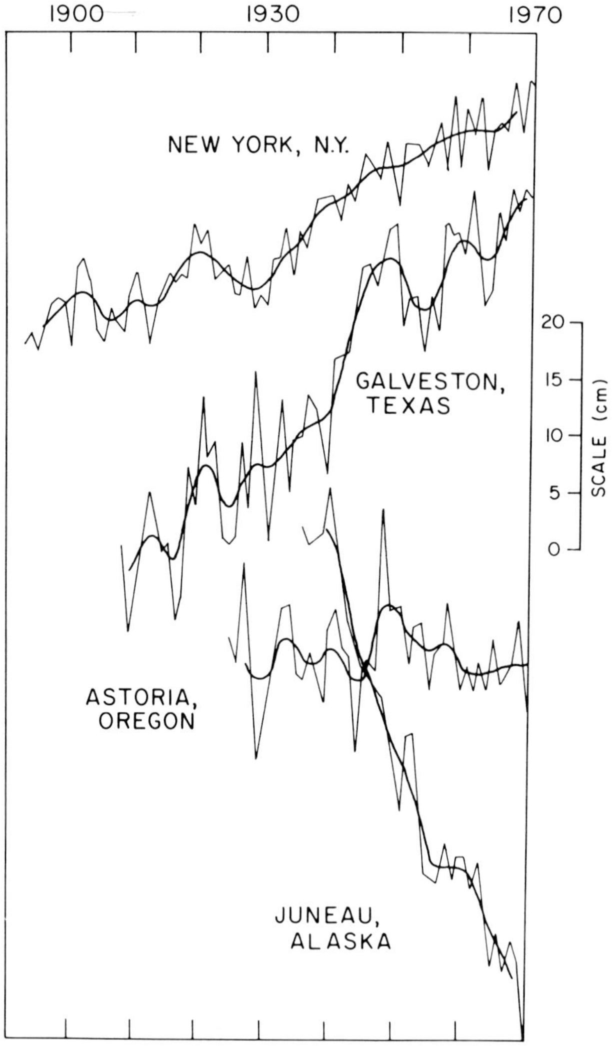

FIGURE 8. Yearly average sea levels as determined from tide-gauge records at various coastal sites, illustrating the effects of a slow rise in the water level of the oceans plus local changes in the levels of the land masses. (From Hicks, S. D., *Shore Beach*, 40(1), 20, 1972. With permission.)

and Curray[46] from tectonically "stable" areas of the world is shown in Figure 7, indicating a progressive rise in the water level to its present position. However, controversy still exists concerning the sea level during this recent period as some data from stable areas indicate that at times the sea was 2 to 3 m higher than at present.[2,15,51]

Long-term tidal gauge records support a present-day rise in sea level averaging on the order of 2 to 3 mm/year.[14,19,21-23,40] The tide-gauge records are particularly useful in examining the net sea-level change at a specific site, the product of the general rise in sea level (eustatic) due to glacial melting plus any local land-level changes. It is, of course, this local net water-level change that is important to erosion at that site. Examples of records up to 80 years in length are shown in Figure 8. The curve from New York is typical of those for much of the east coast of the U.S., indicating an average rise of approximately 3.0 mm/year (see Reference 23, Table 1). This rate is due to the combined effects of a rising sea level with

Table 1
THE BUDGET OF LITTORAL SEDIMENTS

Credit	Debit	Balance
Longshore transport into area	Longshore transport out of area	Beach deposition or erosion
River transport	Wind transport out	
Sea cliff erosion	Offshore transport	
Onshore transport	Deposition in submarine canyons	
Biogenous deposition	Solution and abrasion	
Hydrogenous deposition	Mining	
Wind transport onto beach		
Beach nourishment		

From Bowen, A. J. and Inman, D. L., Coastal Eng. Res. Cent. Tech. Memo No. 19, U.S. Army Corp of Engineers, Washington, D.C., 1966.

a nearly equal contribution from land subsidence. The result from Galveston, Tex. indicates a much higher apparent sea level rise, averaging 6.0 mm/year, produced by the substantial sinking of that portion of the Gulf Coast. Data from the tide gauge at Astoria, Ore., yield a nearly static sea level, apparently caused by the land of this active continental margin rising at the same rate as the eustatic sea level rise due to glacial melting. An extreme case is that of Juneau, Alaska, where the land is rising at such a rate that there is a net lowering of the local water level and a seaward retreat of the shoreline.

Eliminating data that are obviously being greatly affected by anomalous localized subsidence or emergence, Hicks[23] combined the data from 27 tidal stations in an attempt to obtain an average sea level series for the U.S. The resulting average curve for the period 1940 to 1975 indicates that sea level rose at the average rate of 1.5 ± 0.3 mm/year. In a similar analysis, Emery[14] concluded that the rate of sea-level rise has increased by a factor of 2 or 3 during the past decade as compared with pre-1970 rates. As a result, he obtains a median rate of 3 mm/year rise during the past 40 years.

Although these values of sea level rise, amount to a few millimeters per year, seem small and perhaps of negligible importance to coastal erosion problems, the rise is inexorable and cumulative. A rise of 4 mm/year, common on the east coast of the U.S., amounts to 40 cm in 100 years, a significant change, especially in low-lying coastal areas such as the barrier islands. In a recent detailed analysis of sea-level changes, Gornitz et al.[19] predicted that, due to the sharp global warming trend underway since the 1960s, it can be expected that the sea level will rise by about 40 to 60 cm by the year 2050. In areas such as the east coast and Gulf coast of the U.S. where the land is also sinking, the local apparent rise in sea level could approach and even exceed 1 m. The ramifications of this sea level rise to erosion problems on barrier islands will be discussed in Chapter 5.

Attempts have been made to devise models which predict the response of beaches and the amount of shoreline retreat resulting from a specific rise in water level. These models will be examined in Chapter 7. Although of a different cause, there have been substantial "sea-level" changes in the Great Lakes of the U.S., some rises amounting to 34 mm/year over a 4- to 5-year period,[20] resulting in considerable losses of shoreline property. This erosion in the Great Lakes will be discussed in Chapter 8, where the measured shoreline recession will be used to test one simple model such as those presented in Chapter 7.

NEARSHORE CURRENTS AND SEDIMENT TRANSPORT

Waves reaching a beach generate a variety of nearshore currents (Figure 9), the pattern of which depends in large part on the angle of wave breaking. One end-member occurs

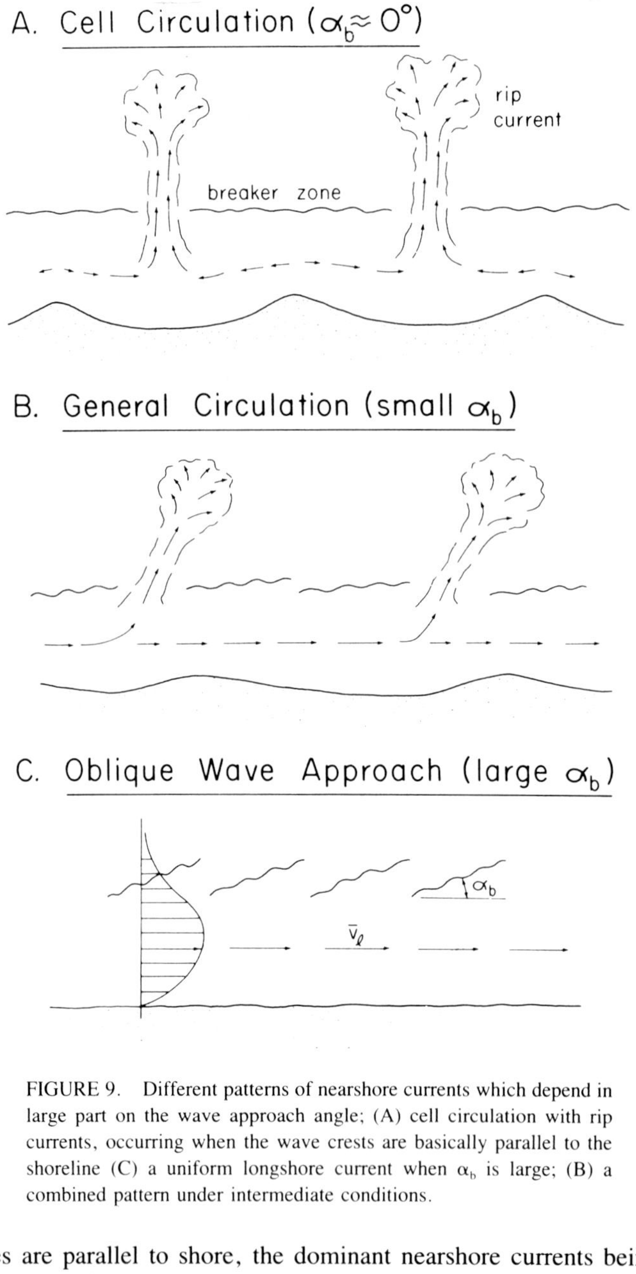

FIGURE 9. Different patterns of nearshore currents which depend in large part on the wave approach angle; (A) cell circulation with rip currents, occurring when the wave crests are basically parallel to the shoreline (C) a uniform longshore current when α_b is large; (B) a combined pattern under intermediate conditions.

when the waves are parallel to shore, the dominant nearshore currents being the cell circulation with seaward flowing rip currents (Figure 9A). The other extreme occurs when waves break at significant angles (roughly, $\alpha_b > 5$ to 10°), producing a longshore current that is continuous along the shoreline (Figure 9C). The general case combines aspects of both systems (Figure 9B).

Bowen[3] and Bowen and Inman[5] have shown that the cell circulation with rip currents is produced by longshore variations in the wave breaker heights. Waves breaking on a sloping beach produce a wave *set-up* in the nearshore, a rise in the water level above the still-water level (the mean sea level in the absence of the waves). Important to the generation of rip currents is that the set-up is higher shoreward of the high breakers than shoreward of the lower breakers. Therefore, if a longshore variation in breaker heights does exist, there will be concomitant longshore gradient in the water level in the surf zone. This will produce a

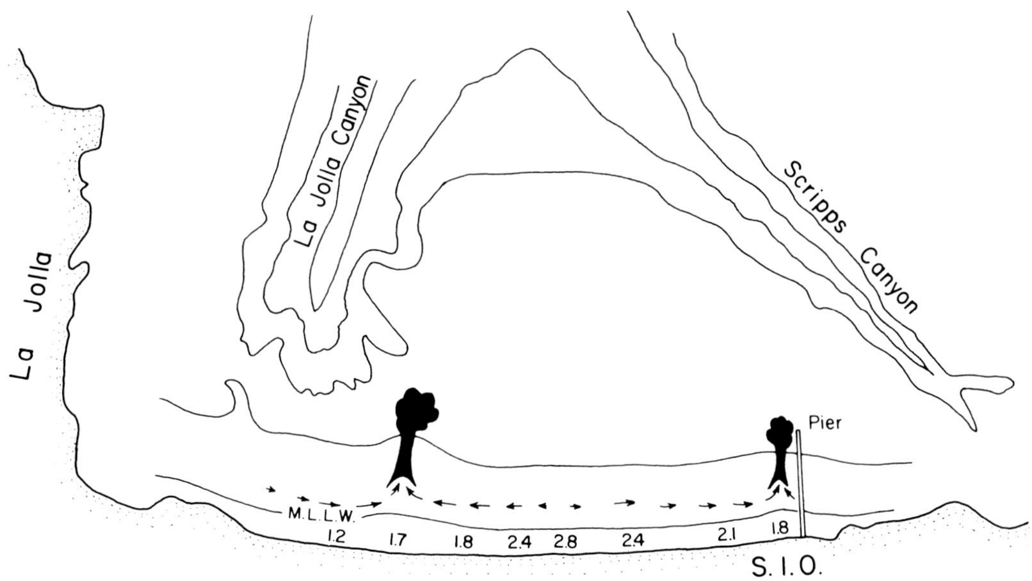

FIGURE 10. Rip currents at La Jolla, Calif. caused by longshore variations in the wave breaker heights (given in meters) according to the hypothesis of Bowen.[3] Compare with Figure 4 which illustrates the wave refraction patterns responsible for the longshore variations. (From Shepard, F. P. and Inman, D. L., *Proc. 1st Conf. Coastal Eng.*, American Society of Civil Engineers, New York, 1951. With permission.)

longshore flow of water toward the low breakers where the water level is lowest, at which point the current turns seaward as a rip current. Rip currents, therefore, tend to be found in positions of lowest breaker heights. The complete analysis of this generation mechanism can be found in Bowen,[3] and a more detailed summary is given in Komar.[30]

The question then becomes: what produces the longshore variations in wave breaker heights which cause the cell circulation? The clearest mechanism is by wave refraction over an irregular offshore topography as in Figure 4, focusing the waves at shoreline locations to produce higher breakers. The rip current systems occurring on the beach corresponding to the wave refraction diagram of Figure 4 have been investigated, and are shown in Figure 10 together with measured breaker heights. It is seen that the rips occur in the expected location of wave-ray divergence and lowest breaker heights. Longshore currents flow away from the positions of highest breakers, reading the rip currents.

However, rip currents are commonly observed on long, straight beaches lacking an irregular offshore topography necessary to produce wave refraction and longshore variations in breaker heights. In cases such as these, a number of rip currents are generally observed with a fairly even spacing (roughly four times the width of the surf zone). Bowen and Inman[5] have explained such occurrences as being due to the presence of edge waves in the nearshore, the edge waves interacting with the incoming swell waves, summing in places to produce high breakers and interfering in other locations to yield low breakers. Edge waves are discussed in greater detail in Chapter 2, including an examination of how they are measured, their role in generating a variety of nearshore sand bars, and their importance to wave run-up on the beach and hence to shoreline erosion.

Given sufficient time, the currents of the cell circulation can rearrange the beach sediments. Most commonly the rip currents transport sand offshore to beyond the breaker zone, hollowing out embayments in the process. A series of rip currents can thereby produce a series of embayments separated by cuspate projections, one type of rhythmic shoreline form. As seen in the example of Figure 11, not all of the rip embayments develop to the same degree, some cutting further into the beach berm than others. At times these embayments can cut

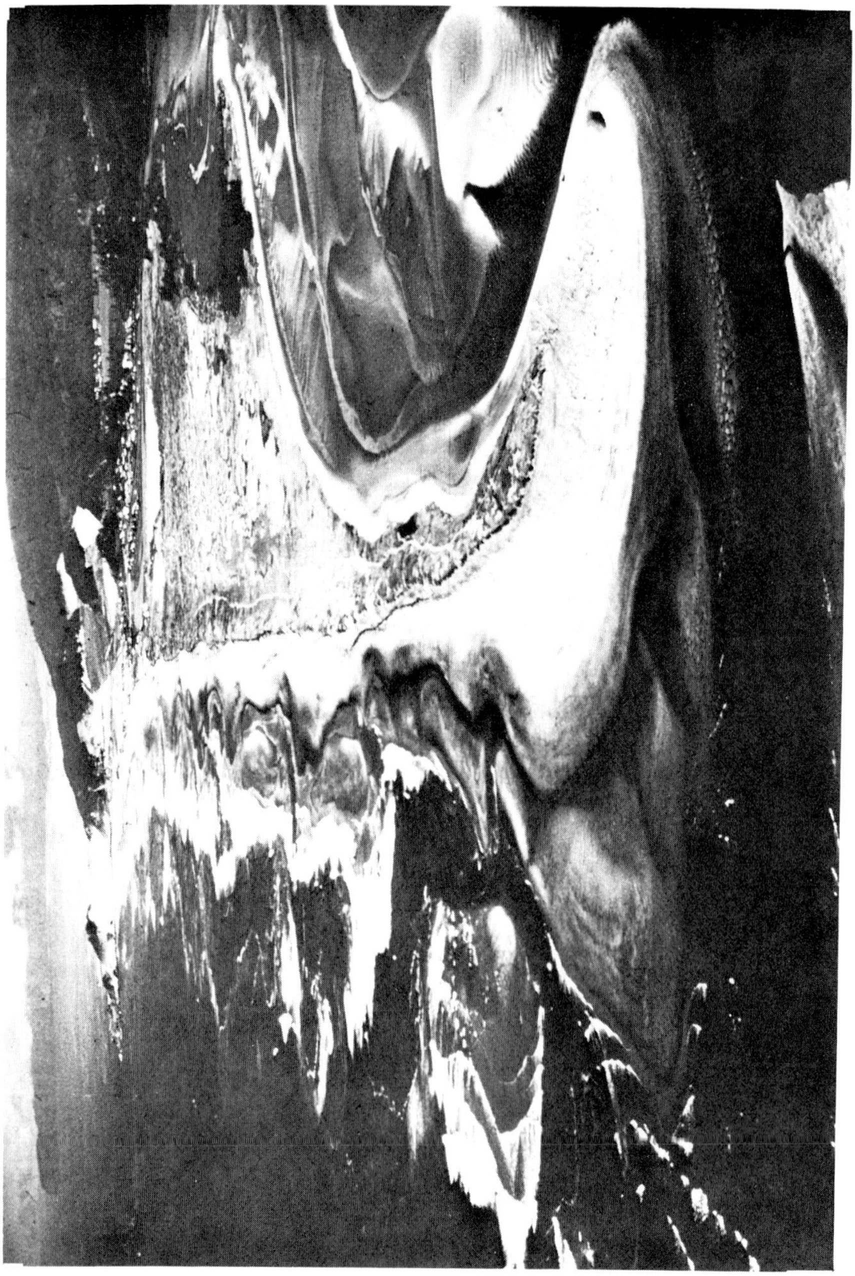

FIGURE 11. An irregular but rhythmic shoreline at Nestucca Spit, Ore., the principal embayments being due to rip currents. Such embayments, together with large storm waves, caused a breaching of the spit at its narrowest section in February 1978.[31]

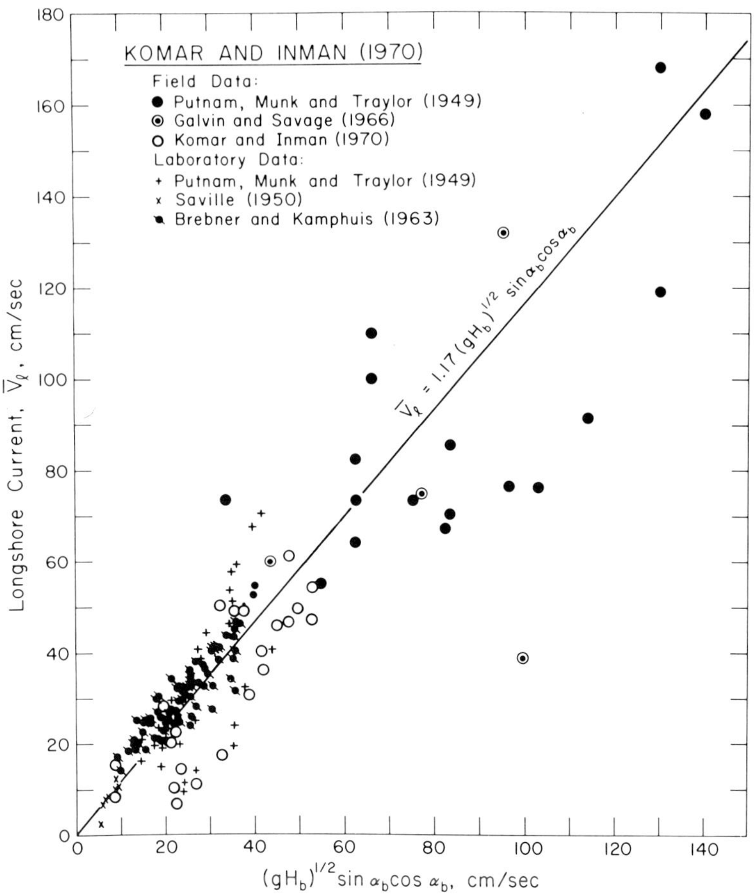

FIGURE 12. Field and laboratory measurements of longshore currents compared with Equation 11. (From Komar, P. D., *J. Waterway Port Coastal Eng.*, WW4, 460, 1979. With permission.)

entirely through the beach berm and begin to impinge on the adjacent coastal property, become sites of erosional losses. Dolan[12] has documented such an erosion process on barrier islands of the eastern U.S., the rip embayments directly threatening man-made structures and often controlling the locations of barrier island washovers. In Chapter 4 it is similarly demonstrated that such rip current embayments are an important factor in erosion on the Oregon coast, the other factors being exceptional storm waves and spring tides. In Chapter 3 Wright and Short show that rip currents and their embayments are important to morphology cycles on Australian beaches.

Longshore currents produced by waves breaking at an angle to the shoreline, Figure 9C, have been shown to be generated by the momentum (or radiation stress) carried by the waves.[39] The formula derived by Longuet-Higgins[39] reduces to the simple expression[32]

$$\bar{v}_\ell = 1.19(gH_b)^{1/2}\sin\alpha_b\cos\alpha_b \qquad (11)$$

so that the magnitude of the current, $\bar{v}_\ell$, can be estimated directly from the measured breaker height, H_b, and breaker angle, α_b. A comparison of Equation 11 with field and laboratory wave-basin measurements is shown in Figure 12. Longshore currents generated by an oblique wave approach vary in magnitude across the width of the neashore, increasing with distance from the shoreline, reaching a maximum usually just beyond the mid-surf position, and decreasing rapidly outside the breaker zone. The magnitude given by Equation 11 is that of

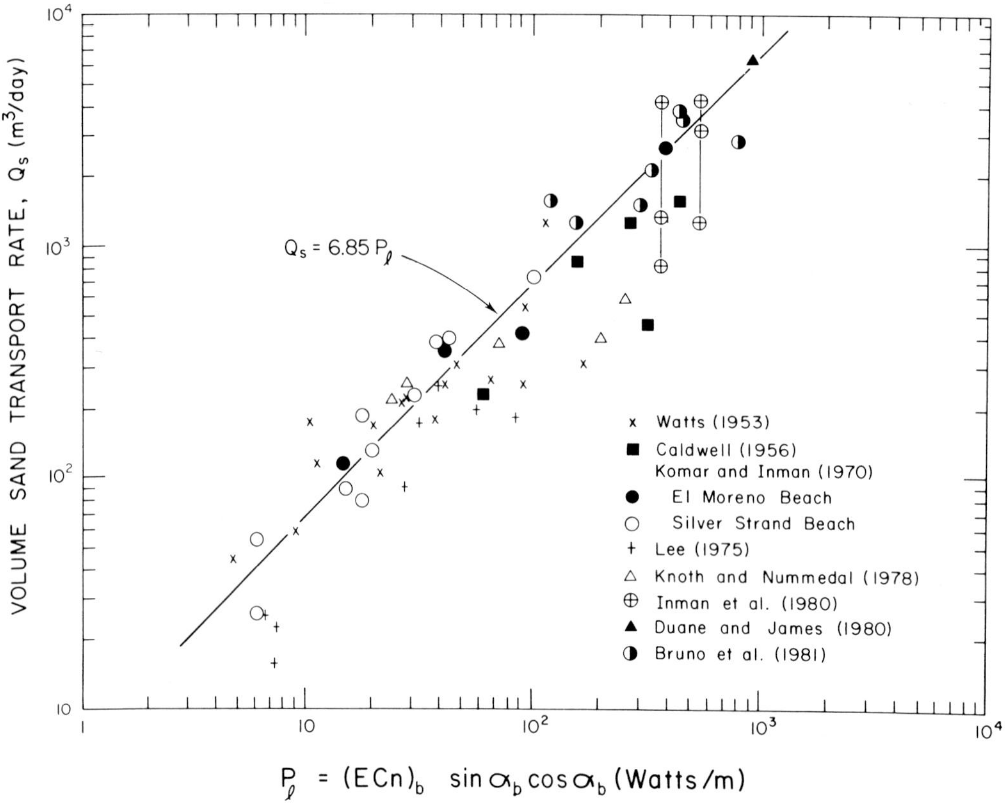

FIGURE 13. The longshore volume transport rate of sand on a beach vs. P_ℓ of Equation 12, leading to the empirical relationship of Equation 13. (From Komar, P. D., in *Coastal and Shelf Dynamical Processes,* Johns, B., Ed., Elsevier, New York, 1983. With permission.)

the current at approximately the mid-surf position, that location being where most measurements of this current have been obtained (only sparse data exist on the complete velocity profile).

The longshore current of Equation 11 transports beach sediments that have been placed into motion by the waves, and being continuous along extensive stretches of the coast, can potentially move the sediments for many kilometers in the longshore direction. This sand movement is termed the *littoral drift,* and is most readily apparent when blocked by a jetty or breakwater built across the nearshore. Such structures act as dams to the littoral drift so that sand accumulates on the updrift side and erosion occurs downdrift. There are many examples along the shorelines of the world of major erosion resulting from this cause. Chapter 9 documents several case studies to illustrate this erosion process.

Watts,[52] Caldwell,[7] and Bruno et al.[6] have examined the quantities of sand trapped by jetties and related these volumes to the causative wave conditions. Additional measurements of sand transport have been obtained using sand tracers[13,26,29,34] and bed-load traps.[37] These measurements are reviewed in Komar[33] and are shown plotted in Figure 13 vs. P_ℓ defined as

$$P_\ell = (ECn)_b \sin\alpha_b \cos\alpha_b \tag{12}$$

where the wave power or energy flux of Equation 3 is now being evaluated at the breaker zone. P_ℓ is often termed "the longshore component of the wave power" due to the presence of the $\sin\alpha_b$ factor, although there are technical objections to such an interpretation. The straight line shown fitted to the data in Figure 13 yields the empirical expression

$$Q_s = 6.8\ P_\ell \qquad (13)$$

where Q_s is the total volume transport rate of the littoral drift given in units m^3/day, and P_ℓ must have the units W/m. Equation 13 is equivalent to the formulas given by Komar and Inman[34] and by CERC,[8] the former being in a dimensionless form and the latter being in the English system of units. There is a factor 2 systematic difference in comparison with the CERC formula, but this is due to reliance of the CERC formula on the significant wave height, H_s, whereas in Equation 13 P_ℓ must be evaluated with the root-mean-square wave height, H_{rms}. With the spectral analysis approach, the energy within the spectrum is summed to yield a direct evaluation of the energy E required in Equation 12, so that direct considerations of wave heights are not involved. This is the correct energy of the waves, and when this is compared with the various wave statistics ($\bar{H}$, H_{rms}, H_s, etc.) it is the root-mean-square wave height that corresponds to this value of E in Equation 2. Since H_s/H_{rms} = 1.418, it is apparent that an evaluation of E based on H_s would be a factor $(1.418)^2 \simeq 2$ higher than the correct wave energy level. However, in the CERC[8] littoral drift formula the proportionality coefficient is reduced by a corresponding factor of 2, so that in practice it yields the same littoral drift evaluation as Equation 13. Of importance is that whenever a relationship such as Equation 13 is employed, one must be careful to evaluate the wave parameters and P_ℓ in the correct manner. And if one wants to use H_s, for example, in the calculations of P_ℓ, the coefficient of Equation 13 must be changed to 6.8/2 = 3.4 so as to yield the correct estimate of the resulting littoral drift.

With relatively simple offshore bathymetry and wave refraction, according to Equation 7 we have

$$(ECn)_b \cos\alpha_b = (ECn)_o \cos\alpha_o = \text{constant} \qquad (14)$$

so that this portion of P_ℓ does not necessarily have to be evaluated at the breaker zone, but instead can be measured or estimated by a wave prediction scheme at some offshore location. This leaves only the $\sin\alpha_b$ factor to be measured at the breaker zone for the complete evaluation of P_ℓ. However, this factor is critical in that a small change in α_b produces a significant variation in $\sin\alpha_b$ and thus in P_ℓ. For example, an increase in α_b from 5 to 6° produce a 20% increase in $\sin\alpha_b$ and P_ℓ. Breaker angles are typically small and difficult to measure precisely, and it follows that errors or uncertainties in their measurement lead to large errors in the evaluation of P_ℓ and thus in the estimated littoral drift, Q_s.

It is seen in Figure 13 that there is considerable scatter to the data upon which Equation 13 is based, and this also contributes to a large uncertainty in the evaluation of Q_s. Therefore, our evaluations of littoral drift rates at some coastal site have a high degree of uncertainty unless based on direct measurements (such as accumulation at a jetty in the area). Particularly difficult is the common condition where the drift direction is reversing, either daily or on a seasonal basis. The net littoral drift evaluated over the duration of a year or longer might be a relatively small value in comparison with the drift existing under any one storm condition or season. From the methods described above, the final uncertainty in the evaluation of this net drift could be much greater than the net drift itself, and quite conceivably the direction of the net drift might be incorrectly evaluated. Such considerations of littoral drift rates and directions must be undertaken with caution, and the effects of engineering structures and any geomorphic factors in the region should be included in the examination.

An important application of littoral drift formulas, such as Equation 13, is in numerical computer models which simulate shoreline changes. These have been used to examine sand blockage by jetties and groins, the impact on the adjacent beaches of sediment dredging on the shallow continental shelf, and the effects of longshore variations in the littoral drift. The techniques of such models are reviewed in Chapter 10 together with examples relevant to coastal erosion problems.

THE BUDGET OF LITTORAL SEDIMENTS

In essence, beach erosion at a particular site is a result of more sand leaving the area than is arriving at the location. There is a resulting deficit in the balance — an erosion. This is the basis of what is commonly termed *the budget of littoral sediments,* the complete analysis of which was first introduced by Bowen and Inman.[4] It is basically an application of the principle of continuity or conservation of mass to the littoral sediments. In practice, the approach consists of evaluating the various sedimentary contributions (credits) and losses (debits) and equating theses to the net gain or loss (balance of sediments) in a given sedimentary compartment or stretch of coast. This balance is thereby reflected in local beach erosion or deposition, depending on whether the balance is in the ''red'' or ''black''.

Table 1 summarizes the most common *sources* and *sinks* (losses) of sand that play a role in the budget. In general, river sand supply, sea-cliff erosion, and littoral drift into the area constitute the major natural sources. Note that unnatural sources such as beach nourishment can also constitute a credit. On the debit side are littoral drift out of the area and offshore transport to deep water, a special case of which is losses to submarine canyons such as along the California coast. Again, there can be unnatural losses such as beach sand mining.

The budget of littoral sediments attempts to quantify these various sources and sinks, evaluating the littoral drift, estimating the contributions from rivers, and so on. These quantities are then balanced to evaluate any resulting erosion (negative balance) or deposition (positive balance). In practice, it is difficult or nearly impossible to make reasonable estimates of some of these quantities. For example, losses to or gains from the offshore present a particular difficulty. Very often the best-known component of the budget is the balance itself, obtained by monitoring the erosion or deposition rate over a number of years. With that knowledge, one can sometimes work backwards to arrive at estimates of what some of the source and sink terms must be to yield that balance.

In some locations the coastal area may be divided into natural compartments to which the budget is applied. Headlands are particularly useful in this regard if they block any possible contribution or escape of sand as a littoral drift. A good example of this is the southern California coast which is divided into four discrete sedimentation cells, each cell containing a complete cycle of littoral transport and sedimentation. Rivers are the principal sources of sediments for the cells and the chief sinks are the series of submarine canyons which bisect the continental shelf and intercept the sand as it moves southward along the coast.[25]

An analysis of the budget of littoral sediments is perhaps most useful in assessing possible impacts by man. For example, once a budget has been developed for the natural conditions found at a certain coastal site, one can make quantitative evaluations of the effects of a proposed river damming that would cut off one of the sources. Other questions that can be examined might be the effects of river or beach sand mining, the placement of riprap along an eroding sea cliff that supplies sand to the beach, or the construction of jetties that will block sand that is normally carried into the area as littoral drift.

In spite of the obvious usefulness of the budget of littoral sediments it has been used far too infrequently in the examination of coastal problems. The study of Bowen and Inman[4] of the Point Arguello, Calif. coastal area remains the best published example of its application. Komar[30] reviews the approach in general, examining how the various sources and sinks are evaluated and provides a hypothetical example for illustrativc purposes. Several chapters of the present volume have direct pertinence to the budget of sediments, dealing with sea-cliff erosion (Chapters 12 and 13) and beach nourishment (Chapter 11). Chapter 4 provides an example of a budget of sediments that was devised to examine the possible role of beach sand mining in causing erosion of a nearby spit.

REFERENCES

1. **Bascom, W.,** *Waves and Beaches*, Anchor Books, Garden City, N.Y., 1980.
2. **Bloch, M. R.,** A hypothesis for the change of ocean levels depending on the albedo of the polar ice caps, *Palaeogeogr. Palaeoclimatol., Palaeoecol.*, 1, 127, 1965.
3. **Bowen, A. J.,** Rip currents. I. Theoretical investigations, *J. Geophys. Res.*, 74, 5467, 1969.
4. **Bowen, A. J. and Inman, D. L.,** Budget of littoral sands in the vicinity of Point Arguello, Calif. *U.S. Army Coastal Eng. Res. Cent. Tech. Memo.*, Tm-19, 1966.
5. **Bowen, A. J. and Inman, D. L.,** Rip currents. II. Laboratory and field observations, *J. Geophys. Res.*, 74, 5479, 1969.
6. **Bruno, R. O., Dean, R. G., Gable, C. G., and Walton, T. L.,** Longshore sand transport study at Channel Island Harbor, Calif., *U.S. Army Coastal Eng. Res. Cent. Tech. Pap.*, No. 81—2, 1981.
7. **Caldwell, J. M.,** Wave action and sand movement near Anaheim Bay, Calif., *U.S. Army Beach Erosion Board Tech. Memo.*, No. 68, 1956.
8. **CERC,** Shore Protection Manual, Coastal Eng. Res. Cent., U.S. Army Corps of Engineers, U.S. Government Printing Office, Washington, D.C., 1973.
9. **Clancy, E. P.,** *The Tides*, Anchor Books, Garden City, N.Y., 1969, 228.
10. **Curray, J. R.,** Late Quaternary history, continental shelfs of the U.S., in *The Quaternary of the United States*, Wright, H. E. and Frey, D. G., Eds., Princeton University Press, Princeton, N.J., 1965.
11. **Defant, A.,** *Ebb and Flow*, University of Michigan Press, Ann Arbor, 1958.
12. **Dolan, R.,** Coastal landforms: crescentic and rhythmic, *Geol. Soc. Am. Bull.*, 82, 177, 1971.
13. **Duane, D. B. and James, W. R.,** Littoral transport in the surf zone elucidated by an Eularian sediment tracer experiment, *J. Sediment. Petrol.*, 50, 929, 1980.
14. **Emery, K. O.,** Relative sea levels from tide-gauge records, *Proc. Natl. Acad. Sci.*, 77(12), 6968, 1980.
15. **Fairbridge, R. W.,** Eustatic changes in sea level, *Phys. Chem. Earth*, 4, 99, 1961.
16. **Galvin, C. J.,** Wave breaking in shallow water, in *Waves and Beaches*, Meyer, R. E., Ed., Academic Press, New York, 1972, 413.
17. **Goldsmith, V.,** Wave climate models for the continental shelf: Critical links between shelf hydraulics and shoreline processes, in *Beach and Nearshore Sedimentation*, Davis, R. A. and Ethington, R. L., Eds., Spec. Publ. No. 24, Society of Economic Paleontologists and Mineralogists, Tulsa, Okla., 1976, 24.
18. **Goodknight, R. C. and Russell, T. L.,** Investigation of the statistics of wave heights, Pap. 3524, *J. Waterways Harbors*, 89, 29, 1963.
19. **Gornitz, V., Lebedeff, S., and Hansen, J.,** Global sea level trend in the past century, *Science*, 215, 1611, 1982.
20. **Hands, E. B.,** Implications of submergence for coastal engineers, *Coastal Sediments*, 77, 149, 1977.
21. **Hicks, S. D.,** On the classification and trends cf long period sea level series, *Shore Beach*, 40(1), 20, 1972.
22. **Hicks, S. D.,** Vertical crustal movements from sea level movements along the east coast of the U.S., *J. Geophys. Res.*, 77(30), 5930, 1972.
23. **Hicks, S. D.,** An average geopotential sea level series for the U.S., *J. Geophys. Res.*, 83, 1377, 1978.
24. **Horikawa, K.,** *Coastal Engineering*, John Wiley & Sons, New York, 1978.
25. **Inman, D. L. and Frautschy, J. D.,** Littoral processes and the development of shorelines, *Coastal Engineering*, American Society of Civil Engineers, New York, 1966, 511.
26. **Inman, D. L., Zampol, J. A., White, T. E., Hanes, D. M., Waldorf, B. W., and Kastens K. A.,** Field measurements of sand motion in the surf zone, *Proc. 17th Conf. Coastal Eng.*, American Society of Civil Engineers, New York, 1980, 1215.
27. **Iverson, H. W.,** Gravity Waves, Circ. No. 521, National Bureau of Standards, Washington, D.C., 1951, 9.
28. **King, C. A. M.,** *Beaches and Coasts*, 2nd ed., St. Martin's Press, New York, 1972.
29. **Knoth, J. S. and Nummedal, D.,** Longshore sediment transport using fluorescent tracer, *Coastal Sediments '77*, American Society of Civil Engineers, New York, 1978, 383.
30. **Komar, P. D.,** *Beach Processes and Sedimentation*, Prentice-Hall, Englewood Cliffs, N.J., 1976.
31. **Komar, P. D.,** Wave conditions on the Oregon coast during the winter of 1977-78 and the resulting erosion of Nestucca Spit, *Shore Beach*, 46(4), 3, 1978.
32. **Komar, P. D.,** Beach-slope dependence of longshore currents, *J. Waterway Port Coastal Eng.*, WW4, 460, 1979.
33. **Komar, P. D.,** Nearshore currents and sand transport on beaches, in *Coastal and Shelf Dynamical Processes*, Johns, B., Ed., Elsevier, New York, 1983.
34. **Komar, P. D. and Inman, D. L.,** Longshore sand transport on beaches, *J. Geophys. Res.*, 75, 5914, 1970.
35. **Komar, P. D. and Gaughan, M. K.,** Airy wave theory and breaker height prediction, *Proc. 13th Conf. Coastal Eng.*, American Society of Civil Engineers, New York, 1972, 405.

36. **Kraft, J. C., Biggs, R. B., and Halsey, S. D.,** Morphology and vertical sedimentary sequence models in Holocene transgressive barrier systems, in *Coastal Geomorphology*, Coates, D. R., Ed., State University of New York, Binghamton, 1973, 321.
37. **Lee, K. K.,** Longshore currents and sediment transport in west shore of Lake Michigan, *Water Resour. Res.*, 11(6), 1029, 1975.
38. **Longuet-Higgins, M. S.,** On the statistical distribution of the height of sea waves, *J. Mar. Res.*, 11, 245, 1952.
39. **Longuet-Higgins, M. S.,** Longshore currents generated by obliquely incident sea waves, parts 1 and 2, *J. Geophys. Res.*, 75, 6778, 1970; 75, 6790, 1970.
40. **Marmer, H. A.,** Changes in sea level determined from tide observations, *Proc. 2nd Conf. Coastal Eng.*, American Society of Civil Engineers, New York, 1952, 62.
41. **Milliman, J. D. and Emery, K. O.,** Sea levels during the past 35,000 years, *Science*, 162, 1121, 1968.
42. **Munk, W. H.,** Comments on review by C. L. Bretschneider of U.S. Navy Dept H. O. Pub. No. 603, 38(5), 118, 1957.
43. **Munk, W. H. and Traylor, M. A.,** Refraction of ocean waves, a process linking underwater topography to beach erosion, *J. Geol.*, 55, 1, 1947.
44. **Putz, R. R.,** Statistical distributions for ocean waves, *Trans. Am. Geophys. Union*, 33, 685, 1952.
45. **Shepard, F. P.,** Thirty-five thousand years of sea level, in *Essays in Marine Geology in Honor of K. O. Emery*, University of Southern California Press, Los Angeles, 1963, 1.
46. **Shepard, F. P. and Curray, J. R.,** Carbon-14 determination of sea level changes in stable areas, *Prog. Oceanogr.*, 283, 1967.
47. **Shepard, F. P. and Inman, D. L.,** Nearshore circulation, *Proc. 1st Conf. Coastal Eng.*, American Society of Civil Engineers, New York, 1951, 50.
48. **Silvester, R.,** *Coastal Engineering*, Elsevier, Amsterdam, 1974.
49. **Snodgrass, D., Groves, G. W., Hasselmann, K. F., Miller, G. R., Munk, W. H., and Powers, W. H.,** Propagation of ocean swell across the Pacific, *Philos. Trans. R. Soc. London, Ser. A*, 259, 431, 1966.
50. **Sorensen, R. M.,** *Basic Coastal Engineering*, John Wiley & Sons, New York, 1978.
51. **van Andel, Tj. H. and Laborel, J.,** Recent high relative sea level stand near Recife, Brazil, *Science*, 145 (3632), 580, 1964.
52. **Watts, G. M.,** A study of sand movement at South Lake Worth Inlet, Florida, *U.S. Army Beach Erosion Board Tech. Memo*, No. 42, 1953.
53. **Weishar, L. L. and Byrne, R. J.,** Field study of breaking wave characteristics, *Proc. 16th Coastal Eng. Conf.*, American Society of Civil Engineers, New York, 1979, 487.
54. **Wiegel, R. L.,** *Oceanographic Engineering*, Prentice-Hall, Englewood Cliffs, N. J., 1964.
55. **Wood, F. J.,** The Strategic Role of Perigean Spring Tides in Nautical History and North American Coastal Flooding, 1635—1976, NOAA, U.S. Department of Commerce, Washington, D.C., 1977.

Chapter 2

EDGE WAVES AND THE CONFIGURATION OF THE SHORELINE

R. A. Holman

Storm waves incident on a beach have their maximum energy at the breakpoint and their minimum at the shoreline, particularly on shallow beaches. It is therefore somewhat paradoxical that storms often result in tremendous erosion and the rearrangement of sediment in the inner surf and swash zones, just where the incident waves have their least expression. Clearly, there must be some other processes acting which may not be as apparent to the eye but whose effect is weighted toward the shoreline. Edge waves are just such a class of motions. Because of their dominance of inner surf zone processes and their remarkable ability to simply explain a number of nearshore morphologies, edge waves have attracted an increasing amount of attention.

The mathematical solution for edge waves was first discovered in 1846 by the mathematician Sir George Stokes.[23] For a long time they were felt to be a mathematical oddity, of no consequence to the real world,[21] a somewhat premature judgment in light of the growing literature on the subject. From the point of view of the mathematicians, edge waves are simply the normal modes of oscillation on a sloping beach, a mathematical way of saying if you kicked the water on a beach you would make edge waves that would progress along the beach (as well as other waves that would escape to sea, usually called leaky modes). The key distinction of edge waves is that their direction of progress is purely alongshore with refraction preventing their escape seaward. This tends to concentrate their energy toward the shoreline.

It is often difficult to visualize a progressive edge wave moving along the shoreline. Figure 1 attempts to provide a clear illustration of this motion. Examining Figure 1a first, we see that the waveform is largest at the shoreline with an exponential offshore decay, and is sinusoidal in the longshore. The wavefcrm is of course progressive in the longshore with Figures 1b, 1c, and 1d showing the gradual progression.

Mathematically, the edge wave is written in the same way as described above; with a shoreline amplitude, a description of the offshore form, and a description of the progressive nature of the wave in the longshore direction. Thus, the waveform, η, can be written

$$\eta(x,y,t) = a\,e^{-kx}\cos(ky - \sigma t) \qquad (1)$$

where a is the amplitude at the shoreline, x and y are, respectively, the offshore and longshore coordinates, and k and σ are the wavenumber and radial frequency of the edge wave ($k = 2\pi/L$ and $\sigma = 2\pi/T$, where L and T are the wavelength and period).

It turns out that the exponential offshore form illustrated in Figure 1 is just the simplest edge wave form and that a whole set of edge wave "modes" exist, each with a slightly more complicated offshore form than the previous. These are denoted by a model number n. For instance, Figure 2 shows a mode 1 ($n = 1$) edge wave. The main difference between this and Figure 1 (which is a mode 0 wave) is the offshore behavior of the wave. The mode 1 wave has an offshore zero crossing such that an elevation high at the shoreline is accompanied by a low offshore and vice versa. The offshore forms for edge wave modes 0 to 3 are shown in Figure 3. They become increasingly complicated, with the model number n determining the number of zero crossings before the eventual offshore decay. The mathematical form for the offshore behavior, denoted by $\phi(x)$, can be determined for certain simple beach profiles. The common assumption is that the beach is plane with a uniform

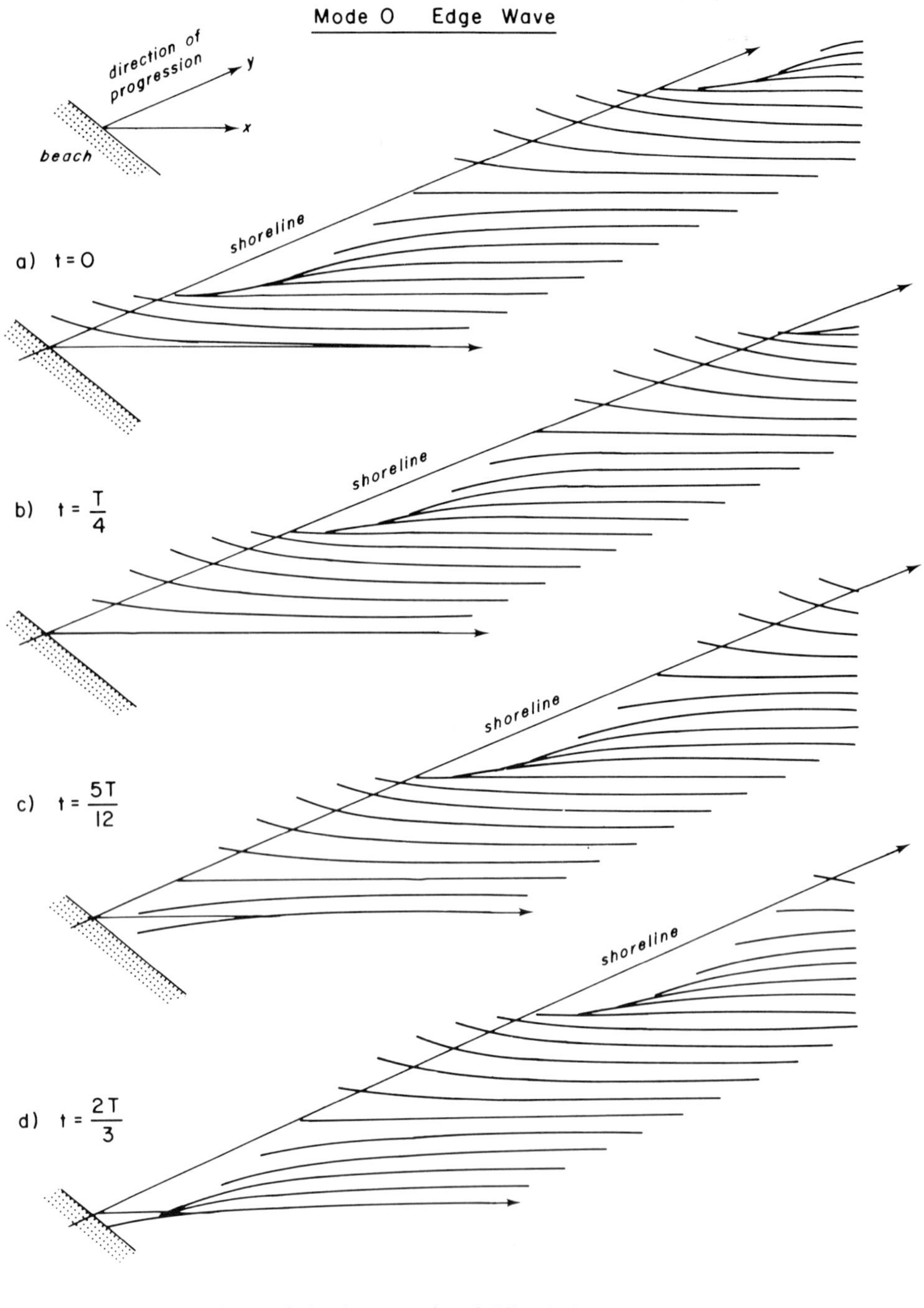

FIGURE 1. A progressive, mode 0, edge wave viewed obliquely from offshore. Seaward is to the right, the shoreline to the left. The shoreline slants away to the upper right. The sinusoidal longshore structure is clearly visible as well as the exponential offshore decay. The time progression of the waves is shown by the four figures, each later in time than the one above it.

slope, tan β.[7,24] In that case the offshore waveform is just a sequence of polynomials times the exponential decay (Table 1). We can actually express this generally as

$$\phi(x) = L_n(2kx)\, e^{-kx} \quad (2)$$

where the polynomials of Table 1 are actually the Laguerre polynomials, L_n, of order n.

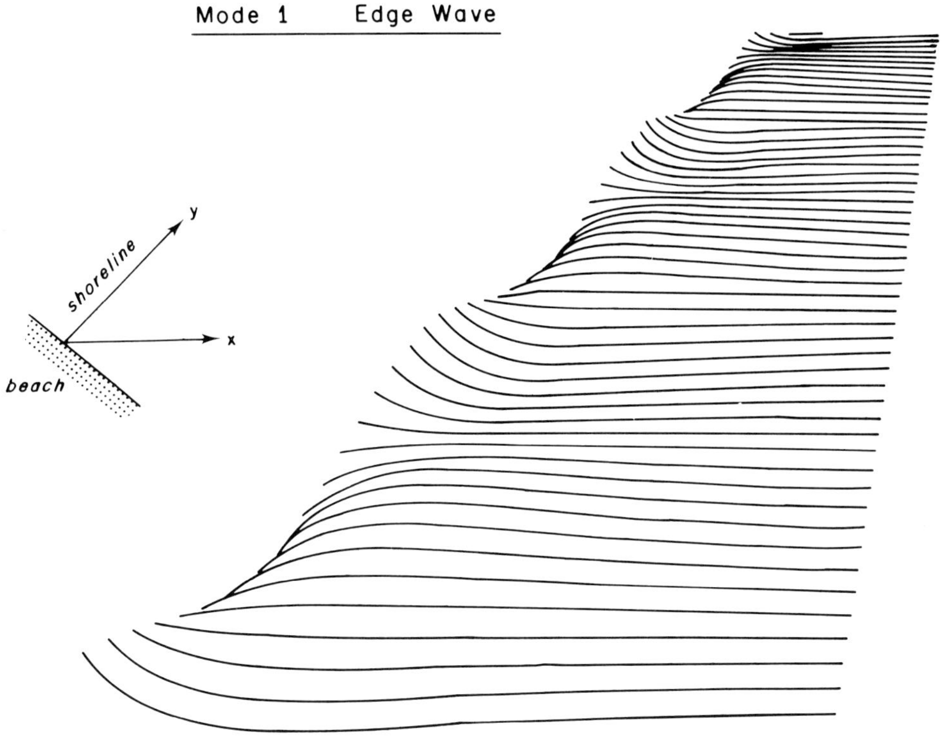

FIGURE 2. Perspective view of a mode 1 edge wave, again viewed obliquely from offshore. The wave is again sinusoidal in the longshore, but this time shows an offshore antinode. This more complicated structure allows the generation of sand bars.

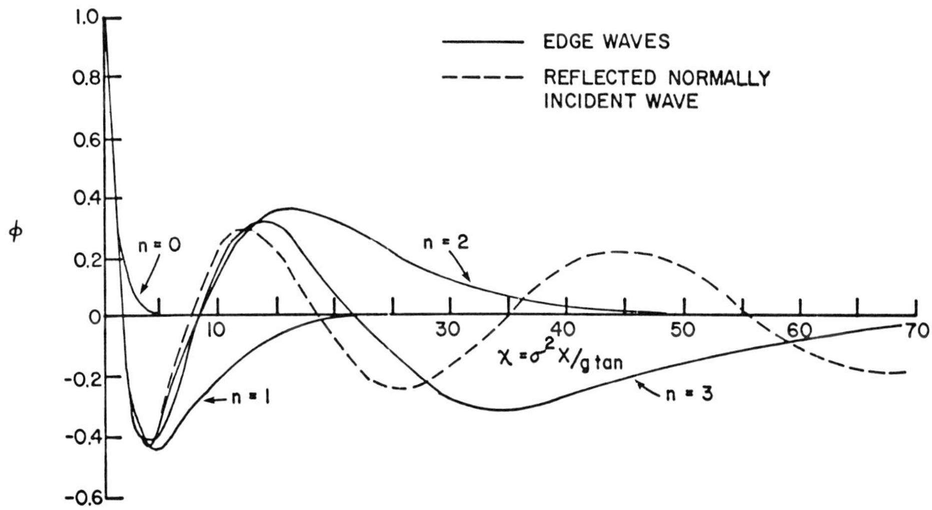

FIGURE 3. The offshore structure of edge wave modes 0 to 3 plotted in terms of nondimensional offshore distance χ. Note the similarity between the modes, particularly near the shore. Note further that the standing incident wave, an example of the leaky modes, also looks similar.

Table 1
OFFSHORE DESCRIPTION OF THE FIRST FOUR EDGE WAVE MODES

n	$\phi(x)$
0	$1 \cdot e^{-kx}$
1	$(1 - 2kx) \cdot e^{-kx}$
2	$(1 - 4kx + 2k^2x^2) \cdot e^{-kx}$
3	$(1 - 6kx + 6k^2x^2 - 4/3k^3x^3) \cdot e^{-kx}$

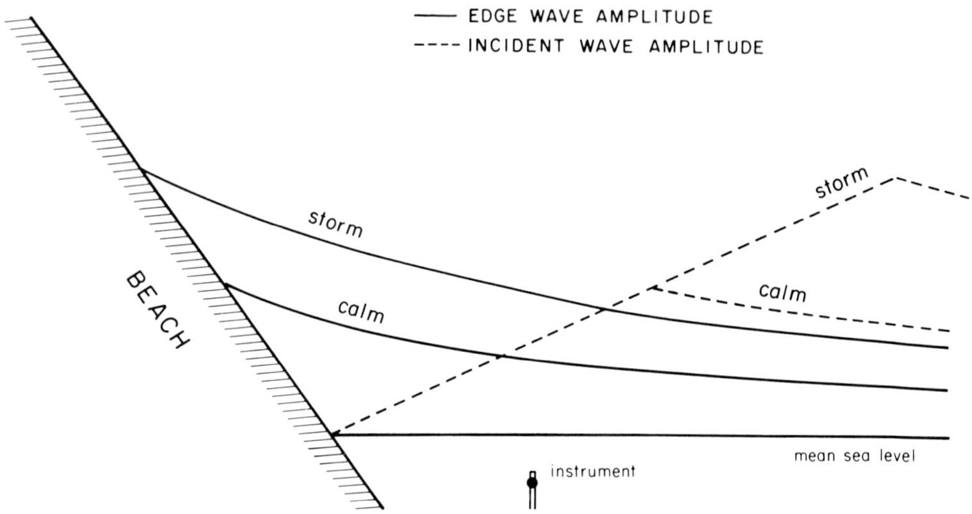

FIGURE 4. Offshore transect showing the response of the incident waves (dashed lines) and edge waves to storm and calm or average conditions. As long as incident dissipation is large, edge waves will always dominate the inner surf zone during storms.

The other piece of information we require is the edge-wave dispersion relation. As with wind waves, the wavelength and period of an edge wave are interdependent. For a plane beach, the dispersion relation was derived by Ursell[24] as

$$L = \frac{gT^2}{2\pi} \sin[(2n + 1)\beta] \qquad (3)$$

Thus, the wavelength depends on the particular mode number as well as the period and beach slope, with higher modes being longer.

Equation 2 shows that the offshore structure is a function of kx ~ x/L. From the dispersion relations we can express this offshore distance as

$$\frac{\sigma^2 x}{g\beta} \frac{1}{(2n + 1)} = \frac{k_o x}{(2n + 1)} = \frac{\chi}{(2n + 1)}$$

providing a convenient nondimensionalization for plots such as Figure 3.

The Importance of Edge Waves During Storms

In the introduction it was pointed out that erosion is often very pronounced at the shoreline despite the fact that, for shallow beaches, incident waves may have lost most of their energy through breaking by the time they reach the shoreline. This erosion was attributed to the growth of edge waves. We can now more easily understand this process.

Figure 4 illustrates what happens to incident and edge waves with the onset of a storm. The dashed line schematically illustrates the shoaling and breaking of incident waves of average wave height. As the waves shoal, wave height increases until the wave crest becomes unstable and breaking begins. Empirical evidence shows that this occurs when the water depth, h, is roughly equal to the wave height H, that is $H_b = \gamma h_b$ where γ is a constant of order 1. Energy is then dissipated through breaking as the waves progress to shallower water so as to maintain this relation. This decrease of wave height is depicted as the dashed straight line in Figure 4 and goes approximately to zero at the shoreline. The change in the incident waves during storms is shown by the dashed line marked "storm". To first order there is no change in the inner surf zone; the larger waves simply break further offshore making the surf zone wider but leaving the wave height in the inner surf zone the same. The edge waves, on the other hand, are not limited and will show an increase in amplitude toward the shoreline, represented graphically by the solid lines, the lower line being the amplitude during times of average incident wave height. We do not know a great deal about the forcing of edge waves by incident waves but it seems reasonable that larger incident waves will force larger edge waves. (Guza and Thornton[12] and Holman[13] show data to demonstrate that the low-frequency swash amplitude, probably edge waves, varies linearly with incident wave amplitude.) The upper solid curve represents this increased edge wave amplitude during storms. If we seek to explain shoreline erosion as a response to a change in wave conditions within the inner surf zone, then the edge waves are clearly a more likely candidate for the erosive mechanism than the incident waves.

Edge Wave Dynamics

The details of edge wave dynamics are not well known and have certainly not been well demonstrated in the field. It has been recognized that different "regimes" exist depending primarily on the beach slope. Steep beaches are generally termed *reflective* while shallow beaches are termed *dissipative*. Reflective beaches are associated with beach cusps and subharmonic edge waves, that is, edge waves having twice the period of the incident waves.[11] The subharmonic forcing mechanism is a nonlinear mechanism whereby a single-incident wave train can force two oppositely progressing edge waves whose sum gives the standing wave pattern which can hollow out rhythmic beach cusps. The period of the edge waves generated under reflective conditions is quite specific: twice the period of the incident waves. Also, both field and lab evidence show that this relatively weak resonance cannot overcome the damping of wave breaking, so the subharmonic regime seems limited to steep reflective beaches. (The wide occurrence of cusps, even on beaches with relatively wide surf zones, may imply the existence of a small reflective regime right at the shoreline in spite of wave-breaking further offshore).

Shallow beaches are associated with wide surf zones of turbulent dissipation. Observations of the wave runup and measurements of water velocities in the inner surf zone show that the dominant wave motions are of much longer period than the incident waves, periods on the order of 100 sec being common. These low-frequency waves have been called infragravity waves, the low-frequency band in the spectrum being the infragravity band. Gallagher[8] proposed a forcing mechanism for the infragravity edge waves which was later expanded and tested in the lab by Bowen and Guza.[4] The forcing for the infragravity edge waves is linked to the modulation, or beating, of the incident waves. This slow variation of breaker heights is a common observation and has led to the surfer's adage that every seventh or ninth wave is the largest. Gallagher demonstrated that this modulation could, under the right conditions, transfer energy to edge waves of the same frequency as the modulation. Since

FIGURE 5. Oblique aerial view of welded sand bars on Cape Cod. The ability of edge waves to quantitatively explain the very rhythmic features has provoked much interest. (From Aubrey, D. G., *Oceans*, 23(4), 4, 1980. With permission.)

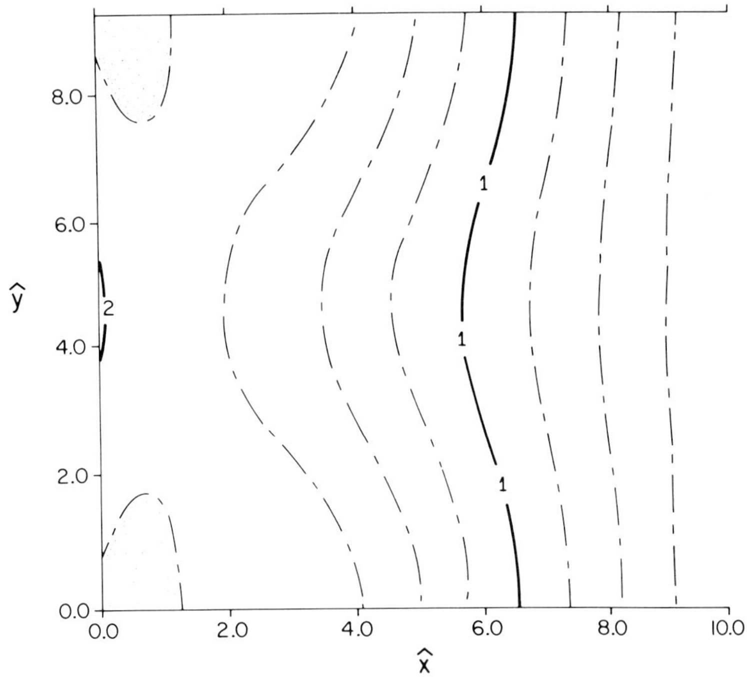

FIGURE 6. Predicted crescentic bar morphology from the numerical model of Holman and Bowen[14] for a standing, mode 1, edge wave. Offshore is to the right, the shoreline to the left. Only one longshore cycle is shown, the pattern repeating along the coast. A profile origin of 0 is chosen offshore with values increasing shoreward. This is just one example of the quantitative beach morphologies easily predictable from edge wave models. ($\hat{x} = k_o x$, $\hat{y} = k_o y$.)

edge wave energy is contained to the nearshore with no leakage, then even a weak forcing mechanism could allow the eventual buildup of large, long edge waves.

There are a number of attractions of infragravity edge waves. First, being of low frequency, they are of long wavelength (Equation 3) and so can be linked to large-scale rhythmic features in the nearshore. We will explore this connection later in this chapter. Second, the forcing mechanism is effective even on a dissipative beach. Another point of interest is that no single frequency is predicted as was the case for the subharmonic; the beating of the incident waves seems broad-banded. This poses somewhat of a problem since the large-scale morphology, such as sand bars, that we hope to explain with edge waves could only have been created if edge waves of a particular wavelength, hence frequency, acted to force the sand into the particular bar formation. Much of the interest in the study of infragravity waves is in finding frequency selection mechanisms.

Edge Waves and Beach Morphology

One of the most attractive features of edge waves is their ability to explain many nearshore morphological features simply and quantitatively without having to invoke a liberal degree of arm waving. This is particularly true for features which are rhythmic in the longshore direction (for example, Figure 5). Guza and Inman,[11] and later Guza and Bowen[9] were able to convincingly link edge waves to beach cusps. Bowen and Inman,[6] in a 1971 paper which stimulated much of the current interest in edge waves, showed that crescentic bars could be simply explained as the result of a standing edge wave motion. Rip currents, which can hollow out embayments into the beach (Chapter 1), were linked to synchronous edge waves by Bowen and Inman[5] in 1969. More recently, Bowen (1980)[3] showed a mechanism whereby

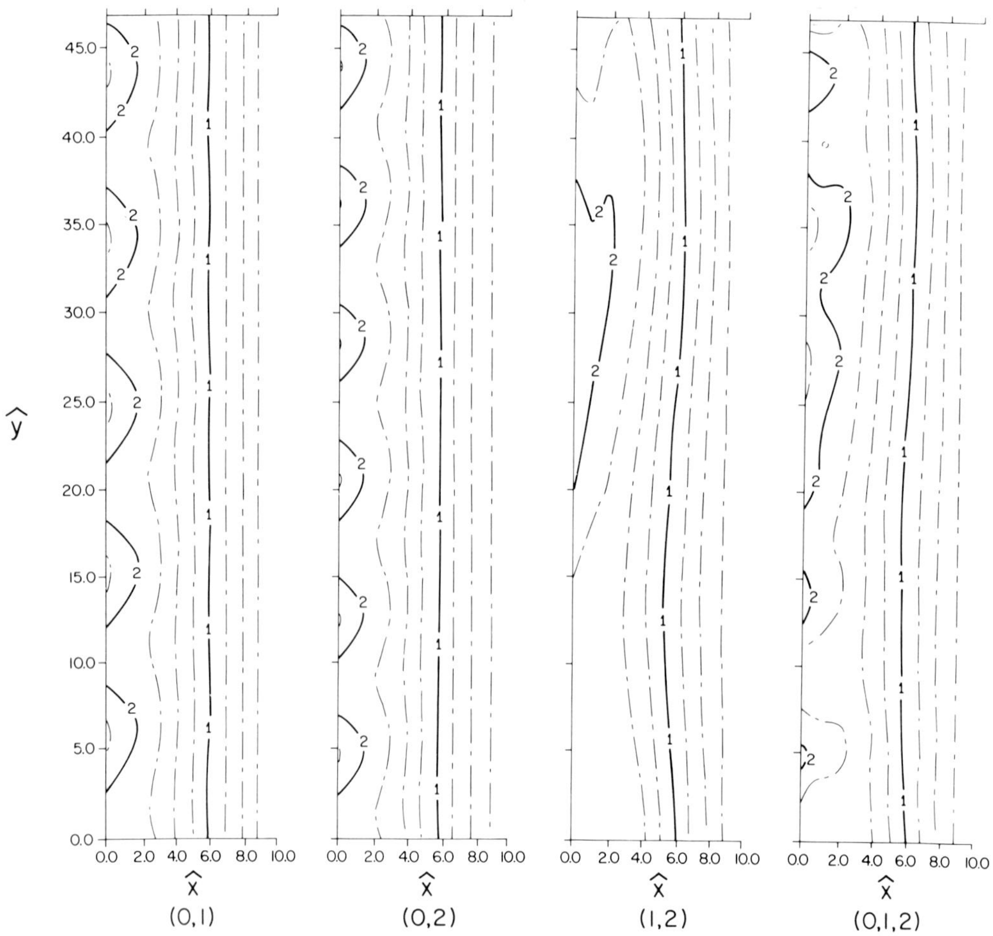

FIGURE 7. Morphology predicted[14] for the simultaneous interaction of three progressive edge waves, in this case modes 0, 1, and 2. The rather complex result on the right is just the sum of the results for each of the possible two wave interactions. Note that each of the four figures will repeat in the longshore. While this makes the complex result rhythmic, it is unlikely that an observer would note the regularity.

multiple bars could be linked to progressive edge waves. Finally, Holman and Bowen[14] showed how welded sand bars (Figure 5) and even apparently complex morphologies can be quantitatively explained through simple edge wave interactions.

The details of a sediment transport mechanism whereby edge waves can change beach topography has been discussed by Bowen[3] and Bailard and Inman.[2] They show that any wave motion can influence a preexisting beach topography through its drift velocity and that the local beach slope will alter so that the tendency of the sediment to be shaken "downhill" just balances the local "push" exerted by the drift velocity. Specifically, the perturbation beach slope pattern β' due to the drift velocity pattern, $\underset{\sim}{u}_1$ of an edge wave is

$$\beta' \cong \frac{-5\underset{\sim}{u}_1 w}{u_0^2} \tag{4}$$

where w is the fall velocity of the sediment, u_0 is the orbital velocity of the incident waves. The drift velocity can be calculated, for example, following Hunt and Johns[15] to give β' which can then be integrated to produce a perturbation beach topography due to the presence of edge waves.

What can be said from this formulation? First, the presence of node and antinodes in the offshore structure of edge waves (Figure 3) leads to zeros and changes of sign of drift velocities. Where drift velocities converge a sand bar is built, where they diverge a trough forms. A standing edge wave (which can be considered the sum of two oppositely traveling progressive waves) can be expressed as

$$\eta = 2a\ \phi(x)\cos(ky)\cos(\sigma t) \tag{5}$$

The drift velocity will similarly be longshore periodic, and, by Equation 4, we expect the bottom topography to also show a periodicity. For example, Figure 6 shows one longshore cycle of the crescentic sand bar which would arise due to a standing mode 1 edge wave. It is this prediction of a regular longshore periodicity that has provided the most compelling evidence of the importance of edge waves on real beaches.

An interesting extension of this work was provided by Holman and Bowen[14] who showed that any two edge waves of the same frequency can interact to form periodic morphology. Even two different edge wave modes, both progressive in the same direction, will give a standing component of flow, in that case generating welded sand bars (Figure 5). The extension to three interacting waves, conceptually very simple, gives rise of an apparently very complex morphology (Figure 7).

What is the relevance of this discussion to problems of erosion? When erosion does occur it commonly shows a longshore dependence (Chapters 1, 3, 4, and 6). The post-storm morphology is often periodic (giant cusps, crescentic or welded sand bars) and can be quantitatively explained in terms of an edge-wave generation mechanism. This implies that the edge-wave mechanism may be dominant in determining the nature of shoreline erosion, an idea which is consistent with the potential for growth of edge-wave energy during storms. Interestingly, the large length scales of many of the features implies that the generating waves must have been long, typically in the infragravity band.

Field Evidence for Edge Waves

An important task has been to find field evidence for the existence and relevance of edge waves in nature. This task has proven exceedingly difficult, largely due to the complex nature of edge waves. It is interesting to examine the lines of evidence that have been pursued in the effort "to catch an edge wave".

The primary characteristics of field data that have been examined for the presence of edge waves have been the offshore structure, the longshore structure, and the phase relationship of the orbital velocities attributable to edge waves. The edge wave structure changes most rapidly in the offshore direction, so most of the studies have concentrated on offshore profiles. Only recently have substantial studies of the longshore structure been carried out.

Perhaps the most successful person in measuring edge waves in the field has been David Huntley at Dalhousie University. In two papers, Huntley and Bowen[17,18] show evidence for subharmonic edge waves. The evidence relies on the offshore decay of the energy in the subharmonic frequency band matching the predicted decay $[\phi^2(x)]$ for a mode zero edge wave, and the zero phase relation between the onshore and longshore velocities, which is only really consistent with a standing edge wave.

Subharmonic edge waves are easier to distinguish than infragravity edge waves for several reasons. First, there are theoretical reasons to expect that only the mode 0 will be forced[10] instead of a possible mix of a number of modes. Second, the subharmonic edge waves are effectively isolated in the spectrum, being the only energy (on reflective beaches) between the incident band and the infragravity band. Thus, the situation is somewhat similar to a lab experiment. No such simplicity exists for the infragravity band. Theory predicts that practically a continuum of frequencies and modes can be forced, so the resulting superposition can be complex.

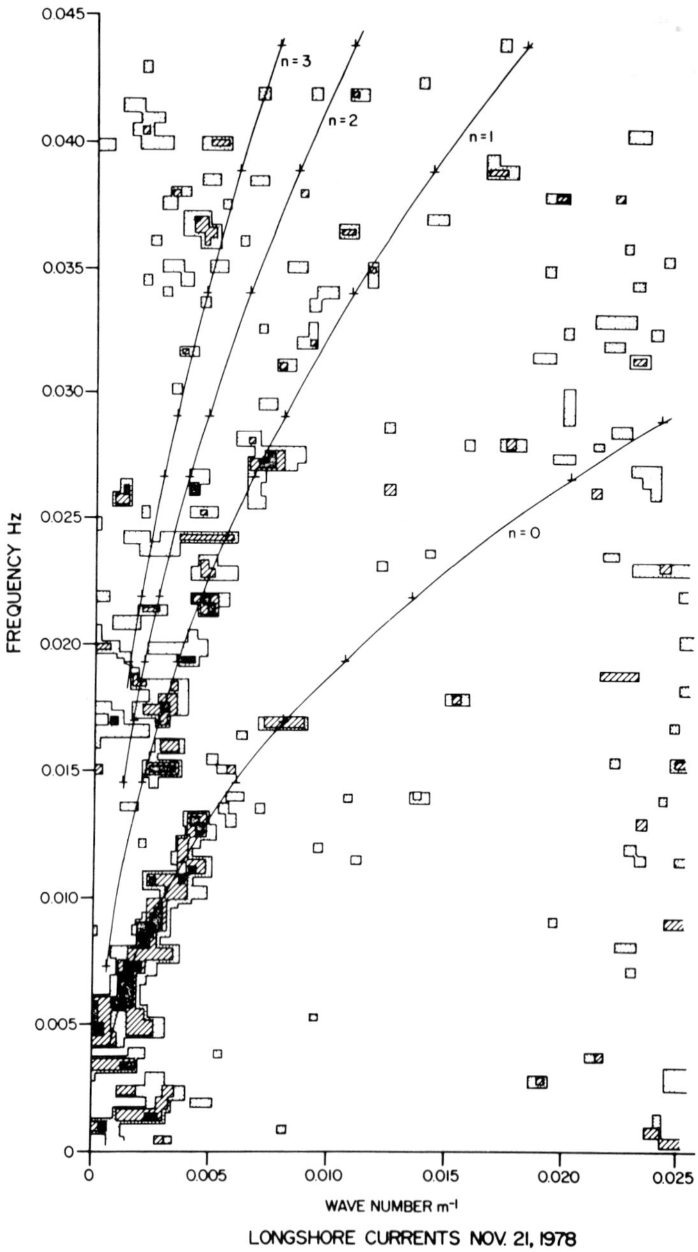

FIGURE 8. Frequency-wavenumber spectrum of longshore currents from Torrey Pines Beach, November 21, 1979. The darkness of the rectangles represents degree of enhancement, a measure of spectral energy. Interestingly, most of the energy lies on the edge wave dispersion curves (solid lines) so is in the form of edge waves. (From Huntley, D. A., Guza, R. T., and Thornton, E. B., *J. Geophys. Res.*, 86(C7), 6451, 1981. With permission.)

A number of investigators have presented data from the infragravity band which are consistent with an edge-wave form,[13,22,25] but are not conclusive. The primary difficulty lies in the similarity, especially close to the shoreline (Figure 3), in offshore structure of the edge wave modes and a standing incident wave produced by wave reflection from the beach (normally called a leaky mode). Unfortunately, a progressive edge wave and a standing

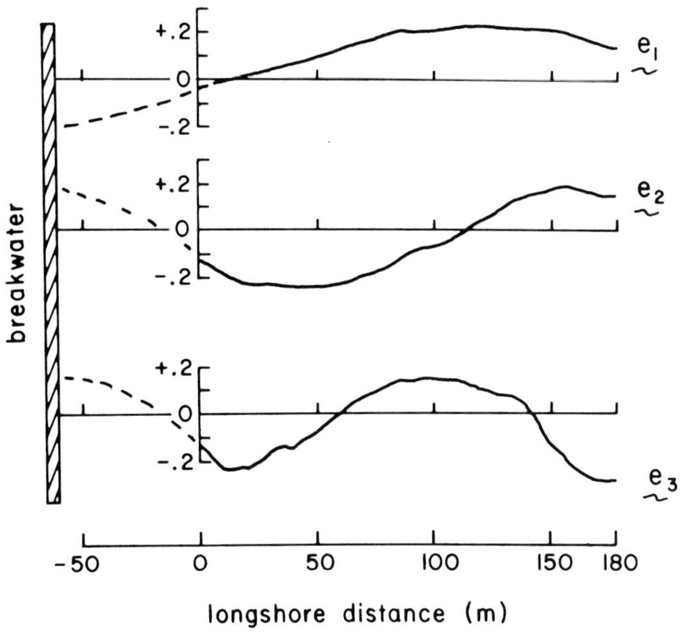

FIGURE 9. Results of empirical eigenfunction analysis of runup variability from a stretch of beach near a breakwater. The first three eigenfunctions are shown. Note that the results of this statistically objective analysis shows sinusoidal longshore modes each of which would have an antinode at the breakwater, consistent with standing edge waves. (From Katoh, K., *Rep. Port Harbor Inst.*, 20(3), 3, 1981. With permission.)

incident wave also show the same phase relationships between onshore and longshore velocities, so these criteria are of little help. The analysis by Huntley[16] is a good example of developing a relatively convincing case for the existence of infragravity edge waves using data from an offshore transect.

Definitive evidence for infragravity edge waves was only found when the longshore variation of the waveform was examined. Huntley et al.[19] accomplished this by using an enormous dataset covering measurements from 19 electromagnetic flowmeters, including a longshore array 500 m in length. Cross-spectra were calculated between all instrument pairs in the longshore array. For each frequency band the cospectral energy was plotted vs. longshore lag. A maximum-likelihood spectrum was then computed to determine the longshore wavelength. The result of this rather tedious process is a two-dimensional wavenumber-frequency spectrum (Figure 8). Interestingly, most of the energy lined up along the theoretical dispersion curves (from Equation 3) proving the wave motion to have been low-mode edge waves.

A second experiment which shows the existence of edge waves through their longshore behavior is that of Katoh.[20] He used a sequence of aerial photographs of wave runup as his dataset, digitizing the runup at a series of longshore locations for each picture. He then performed an empirical orthogonal function analysis to find the longshore modes which best explain the runup variability. Figure 9 shows the first three eigenvectors. These show a clear longshore periodicity which, when extrapolated to a long breakwater just outside the analysis area, shows an antinode at the breakwater for each case. This is compelling evidence for the existence of standing edge waves against the breakwater. Unfortunately, the time between frames was too long to allow spectral analysis of the eigenvector weights to show the frequencies of the modes.

Discussion

The connection of edge waves to shoreline erosion follows from several lines of reasoning. First, unlike incident waves which are limited in the surf zone by breaking, edge waves can grow during storms and may come to dominate the inner surf zone.[13] Second, the large-scale morphology which remains after a storm is commonly in a form consistent with edge wave generation. This implies that the dominant erosive mechanism during the storm and storm decay may be edge waves.

There has been a nagging resistance to acknowledging the existence of edge waves as an important mode of motion in the nearshore. The studies of Huntley and Katoh have now unequivocally proven their existence. However much work remains to be done. Theoretical ideas on edge-wave dynamics must be tested as well as the link between edge-wave motions and resulting rhythmic morphology. To add to the complications of field work in the nearshore zone, there is a need to carry out these experiments during storms, when erosion and the reworking of morphology occur. The work ahead, while necessary and potentially very rewarding, is enough to make any experimentalist shudder.

ACKNOWLEDGMENTS

This research was supported by the Office of Naval Research under contract N00014-81-K-0008. I would like to thank Paul Komar, Peter Howd, and Holly Garrow for helpful comments and suggestions, and Lynn Dickson for typing the several drafts of the manuscript.

REFERENCES

1. **Aubrey, D. G.,** Our dynamic coastline, *Oceans,* 23(4), 4, 1980.
2. **Bailard, J. A. and Inman, D. L.,** An energetics bedload transport model for a plane sloping beach; local transport, *J. Geophys. Res.,* 86, 2035, 1981.
3. **Bowen, A. J.,** Simple Models of Nearshore Sedimentation; Beach Profiles and Longshore Bars, in *The Coastline of Canada,* Pub. 80-10, McCann, S. B., Ed., Geological Survey of Canada, Ottawa, 1980, 1.
4. **Bowen, A. J. and Guza, R. T.,** Edge waves and surf beat, *J. Geophys. Res.,* 83(C4), 1913, 1978.
5. **Bowen, A. J. and Inman, D. L.,** Rip current. 1. Theoretical investigations, *J. Geophys. Res.,* 74(23), 5467, 1969.
6. **Bowen, A. J. and Inman, D. L.,** Edge waves and crescentic bars, *J. Geophys. Res.,* 76(36), 8662, 1971.
7. **Eckart C.,** *Surface Waves on Water of Variable Depth,* Ref. No. 51-12, Scripps Institute of Oceanography, La Jolla, Calif., 1951.
8. **Gallagher, B.,** Generation of surf beat by non-linear wave interaction, *J. Fluid Mech.,* 49(1), 1, 1971.
9. **Guza, R. T. and Bowen, A. J.,** On the amplitude of beach cusps, *J. Geophys. Res.,* 86(C5), 4125, 1981.
10. **Guza, R. T. and Davis, R. E.,** Excitation of edge waves by waves incident on a beach, *J. Geophys. Res.,* 79(9), 1285, 1974.
11. **Guza, R. T. and Inman, D. L.,** Edge waves and beach cusps, *J. Geophys. Res.,* 80(21), 2997, 1975.
12. **Guza, R. T. and Thornton, E. B.,** Swash oscillations on a natural beach, *J. Geophys. Res.,* 87(C1), 483, 1982.
13. **Holman, R. A.,** Infragravity energy in the surf zone, *J. Geophys. Res.,* 86(C7), 6442, 1981.
14. **Holman, R. A. and Bowen, A. J.,** Bars, bump and holes: models for the generation of complex beach topography, *J. Geophys. Res.,* 87(C1), 457, 1982.
15. **Hunt, J. N. and Johns, B.,** Currents induced by tides and gravity waves, *Tellus,* 15, 343, 1963.
16. **Huntley, D. A.,** Long period waves on a natural beach, *J. Geophys. Res.,* 81(36), 6441, 1976.
17. **Huntley, D. A. and Bowen, A. J.,** Field observations of edge waves, *Nature (London),* 243(5403), 160, 1973.
18. **Huntley, D. A. and Bowen, A. J.,** Field observations of edge waves and their effect on beach material, *J. Geol. Soc.,* 131, 68, 1975.
19. **Huntley, D. A., Guza, R. T., and Thornton, E. B.,** Field observations of surf beats. 1. Progressive edge waves, *J. Geophys. Res.,* 86(C7), 6451, 1981.

20. **Katoh, K.,** Analysis of edge waves by means of empirical eigenfunctions, *Rep. Port Harbor Inst.*, 20(3), 3, 1981.
21. **Lamb, H.,** *Hydrodynamics*, 6th Ed., Dover Publications, New York, 1932.
22. **Sasaki, T., Horikawa, K., and Hotta, S.,** Nearshore currents on a gently sloping beach, *Proc. 15th Conf. Coastal Eng.*, American Society of Civil Engineers, New York, 1976, 626.
23. **Stokes, G. G.,** Report on recent research in hydrodynamics, *Br. Assoc. Adv. Sci. Rep.*, pap. i 167, 1846.
24. **Ursell, F.,** Edge waves on a sloping beach, *Proc. R. Soc. A*, 214, 79, 1952.
25. **Wright, L. D., Chappell, J., Thom, B. G., Bradshaw, M. P., and Cowell, P.,** Morphodynamics of reflective and dissipative beach and inshore systems: southeastern Australia, *Mar. Geol.*, 32, 105, 1979.

Chapter 3

MORPHODYNAMICS OF BEACHES AND SURF ZONES IN AUSTRALIA

L. D. Wright and A. D. Short

INTRODUCTION

Beach erosion need not necessarily involve net long-term shoreline retreat in order to present problems for coastal zone inhabitants and developers. Shoreline oscillations involving alternating erosion and accretion can be equally or even more destructive in their effects when the amplitude of change is large and the oscillation is aperiodic, episodic, or only weakly periodic. It is small consolation for a coastal resident or manager whose home or recreational development has fallen into the sea to know that the beach will someday return. Understanding the mechanisms and environmental conditions that cause short-term changes and favor different amplitudes of beach mobility is thus as important to coastal planning and defense as is understanding the causes of chronic beach recession.

Most studies of onshore-offshore sediment migration and associated alternations between erosion and accretion have been strongly influenced by the classic model of "cyclic" or seasonal cycles in beach profile from the accreted "swell" (or "summer") profile to the eroded "storm" (or "winter") profile.[19,21] An abundance of engineering literature exists dealing with criteria for predicting the probability of erosion or accretion occurring over simple beach topographies (e.g., Dean[3]). The critical conditions for erosion have been defined largely in terms of wave steepness, sediment fall velocity, and beach (or bed) slope. Owing to the complexity of the problem, the roles played by preexisting natural beach topographies in altering the forms of the sediment-transporting fluid motions within the surf zone (for example to produce various low-frequency standing oscillations, rips, etc.) have received less attention. In order to understand beach erosion processes more fully, it is necessary to appreciate the relationship that exists between the scales, frequencies, and intensities of various modes of surf-zone motion and the topographic state that often exists as an inheritance from antecedent conditions.

Our field studies of different Australian beaches and surf zones encompass a wide range of temporarily and spatially varying environmental conditions, sediment types, and morphodynamic states.[12,13,16,22-25] These studies point to several important generalizations that should bear heavily on considerations of the processes of beach erosion:

1. Depending on local environmental conditions, sediments, and antecedent wave conditions, beaches and surf zones may be dissipative, reflective, or in any of at least four intermediate states.
2. The relative contributions of incident waves, subharmonic and infragravity edge waves, and net surf zone circulations to near-bottom currents and resultant sand transport vary with beach state.
3. The actual mechanisms that cause beach cut and the wave energy required to induce beach cut are dependent on beach state.
4. As beach state changes with time, for example in response to changing breaker height, the signature of hydrodynamic processes mutates, permitting the evolution of morphodynamic regimes that are free, to varying degrees, from complete forcing by deepwater wave conditions.
5. Beach and surf zone mobility (as expressed by the normal range of temporal change) shows a close relationship to modal (or most recurrent) beach state.

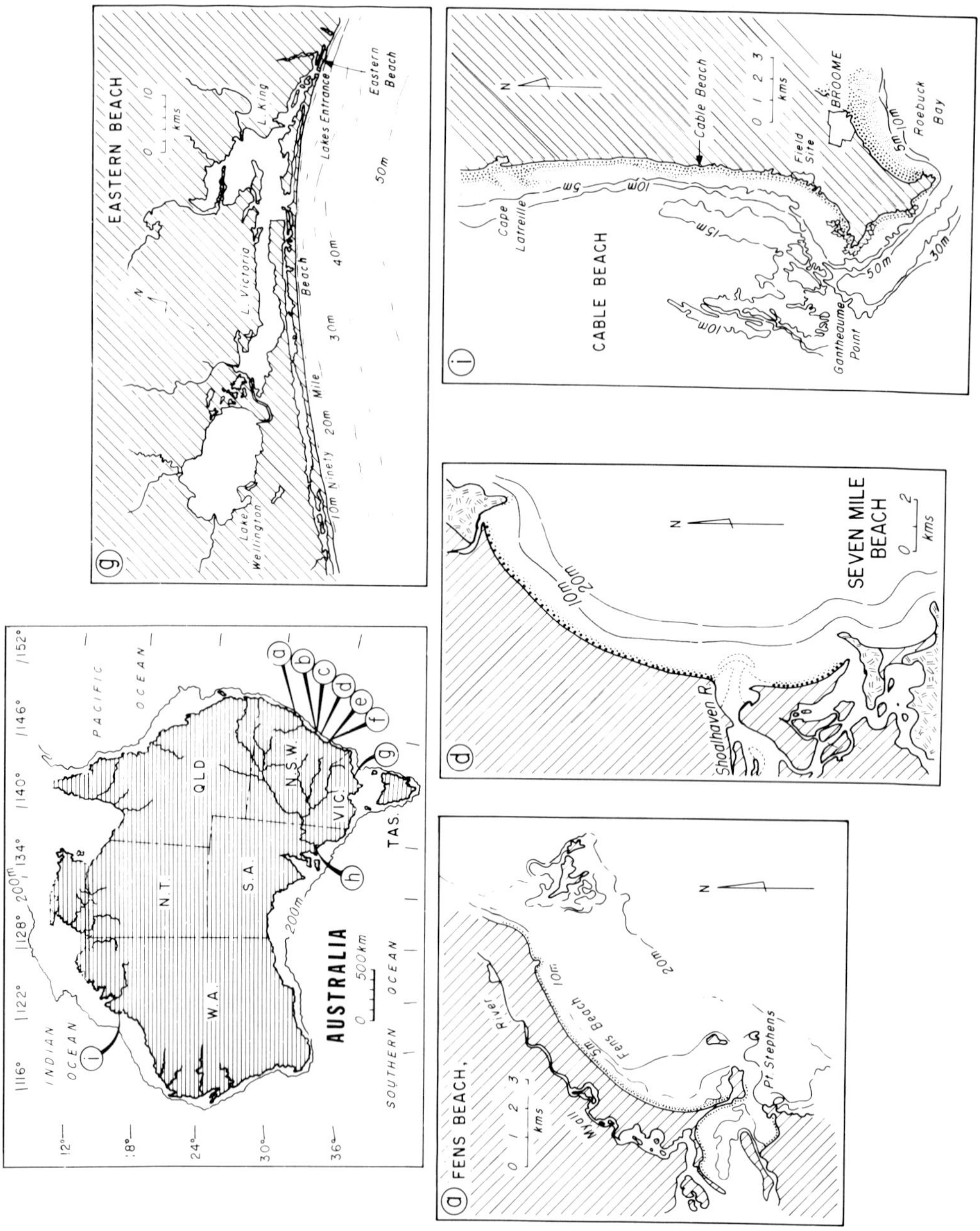
AUSTRALIA
0 500km
INDIAN OCEAN
PACIFIC OCEAN
SOUTHERN OCEAN
W.A.
N.T.
S.A.
QLD
N.S.W.
VIC.
TAS.
200m
116°
122°
128°
134°
140°
146°
152°
12°
18°
24°
30°
36°
a
b
c
d
e
f
g
h
i
g EASTERN BEACH
0 10 kms
Lake Wellington
L. Victoria
L. King
Ninety Mile Beach
Lakes Entrance
Eastern Beach
10m
20m
30m
40m
50m
N
i CABLE BEACH
Cape Latreille
Cable Beach
Field Site
BROOME
Roebuck Bay
Gantheaume Point
0 1 2 3 kms
5m
10m
15m
30m
50m
N
d SEVEN MILE BEACH
Shoalhaven R.
0 2 kms
10m
20m
N
a FENS BEACH
Myall River
Fens Beach
Pt. Stephens
0 1 2 3 kms
5m
10m
20m
N

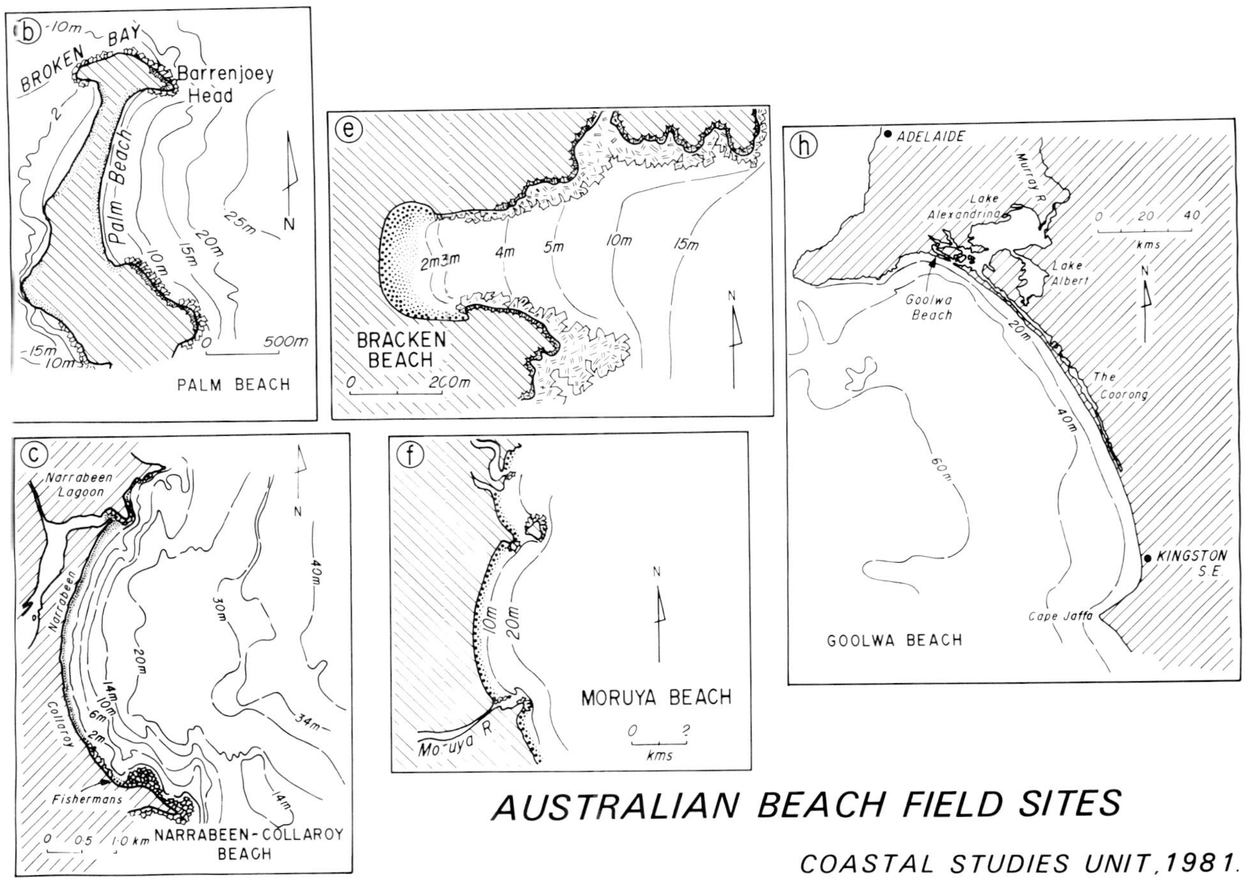

FIGURE 1. Locations and compartment shapes of major study sites.

The purpose of this paper is to illustrate some of the important roles that morphodynamic state play in modifying or controlling beach and surf-zone processes. Specifically, we will discuss the associations among morphodynamic state, surf-zone process signatures, modes of beach cut, and beach mobility based on field observations.

The Study Environments and the Data Set

This paper synthesizes field results obtained from contrasting environments around the coast of Australia. Although moderate to high wave energy is a general characteristic of most of the study sites, the study includes beaches with varying degrees of topographic sheltering and encompasses prolonged periods over which large temporal variations in energy occurred. Most of the sites are in the microtidal environment of southern and southeastern Australia; however, we include in our discussion some results from a macrotidal beach in northwestern Australia.

Figure 1 shows the locations of the main study areas. Several beaches on the New South Wales coast provide examples of the full range of dissipative, intermediate, and reflective states owing to varying sediment size and varying degrees of compartmentization and sheltering. These beaches include Seven Mile Beach (dissipative), Moruya (intermediate), Bracken (reflective), Narrabeen-Collaroy (intermediate), Fisherman's (reflective), Palm (intermediate), and Fens (intermediate). The New South Wales coast experiences an east coast swell regime characterized by a highly variable wind-wave climate superimposed on a persistent, moderate, long-period southeasterly swell.[16] Weakly embayed sand beaches alternate with rock headlands. Littoral drift tends to be weak or negligible in comparison to shore normal transport. Semidiurnal tides have a mean spring range of 1.6 m (microtidal).

The long, straight, relatively uninterrupted beach of East Gippsland (Ninety Mile Beach) provides an example of a storm-wave dominated beach with pronounced intermediate bar-trough topography, strong longshore currents, and littoral drift. Highly variable wave conditions result from the frequent passage, particularly in early winter, of southwesterly or southeasterly gales. Spring tide range averages 1.5 m.

The Goolwa site in the Coorong Region of South Australia is representative of high-energy, persistently highly dissipative beach conditions.[26] This beach is 210 km long and faces directly into the perennially high-energy west-coast swell of the Southern Ocean. Spring tide range is only 0.8 m.

A macrotidal environment (spring range 9.5 m) is the dominant feature of Cable Beach near Broome, in northwestern Australia. Low- to moderate-energy wind waves superimposed on low, long-period westerly swells characterize the summer wave conditions affecting Cable Beach. During winter, offshore winds prevail maintaining low-energy or calm conditions nearshore.

Data on morphologic change are most extensive for the Narrabeen-Collaroy-Fisherman's compartment,[12,13] Moruya Beach[20,24] and Seven Mile Beach. Monthly surveying transects of Narrabeen, Collaroy, and Fisherman's beaches have been replicated over a period of six years. Daily observations of beach state cover six years for Narrabeen Beach and nine months for Palm Beach. During intensive experiments on all beaches, levelling transects were normally repeated on a daily basis over the experiment period. At less frequent intervals, surveys were extended offshore to depths of 20 to 30 m by echo-sounding traverses.

Wave and current measurements from within and near the surf zone were made using the field monitoring system described by Bradshaw et al.[1] This system consists of up to 10 strain-gauge-type pressure transducers and 12 low-inertia bidirectional duct-impeller flow-meters. Sensors were connected by cable to chart recording and multiplexing digital tape recording facilities. Response characteristics of the sensors are described by Nielsen and Cowell.[11] In most experiments, data from all sensors in an array were logged simultaneously at intervals of 1 sec (greater in some cases) over run lengths of 30 min to 1 hr or more.

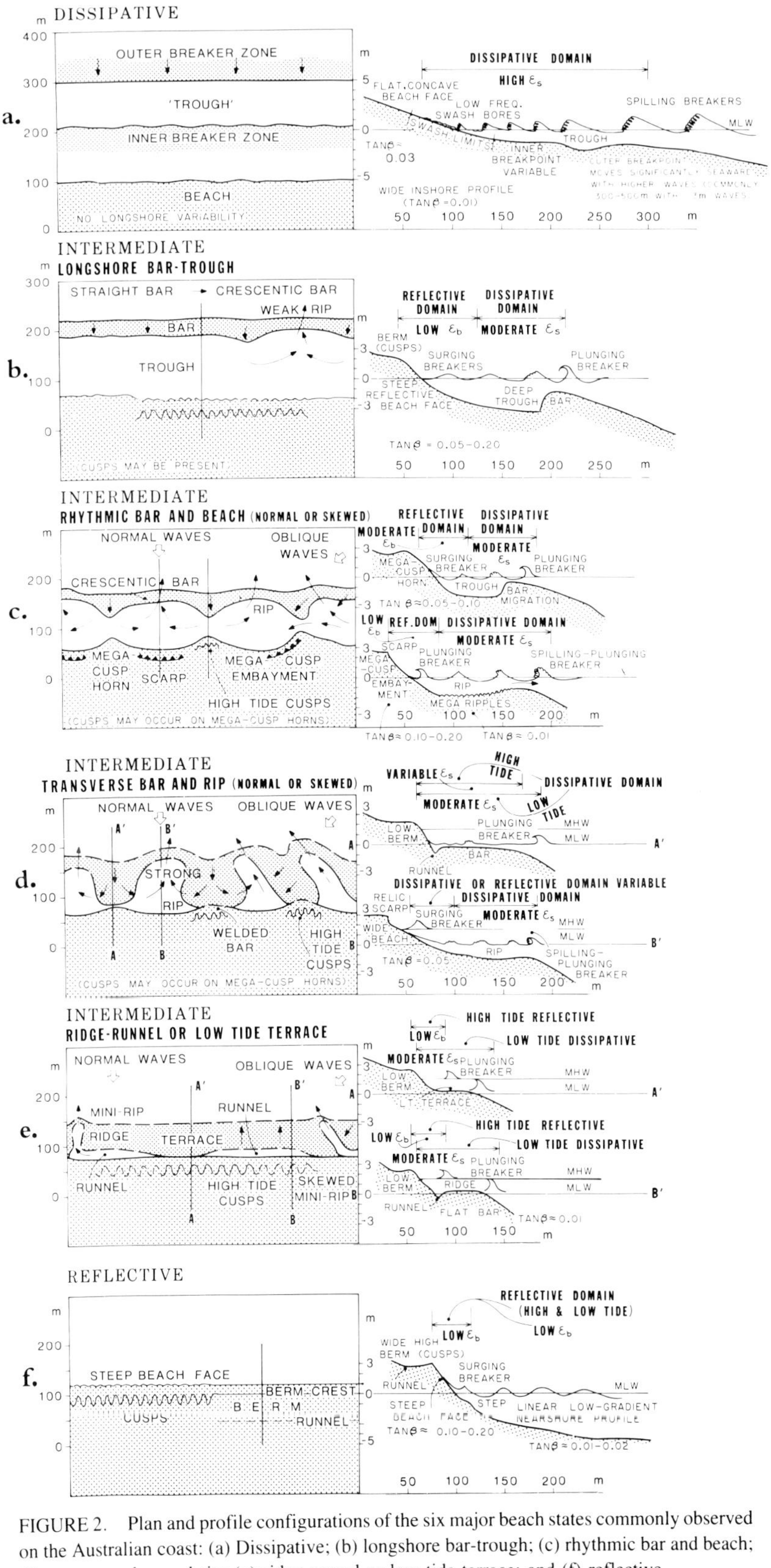

FIGURE 2. Plan and profile configurations of the six major beach states commonly observed on the Australian coast: (a) Dissipative; (b) longshore bar-trough; (c) rhythmic bar and beach; (d) transverse bar and rip; (e) ridge runnel or low tide terrace; and (f) reflective.

FIGURE 3. Photographic examples of the six major beach states: (a) Dissipative; (b) longshore bar-trough; (c) rhythmic bar and beach; (d) transverse bar and rip; (e) ridge runnel/low tide terrace; (f) reflective (with beach cusps).

FIGURES 3c and 3d.

FIGURES 3e and 3f.

Raw data were corrected for frequency response characteristics before spectral and statistical analyses were performed.

Morphodynamic States of Beaches and Surf Zones

The basic framework for this synthesis is inherited from the earlier morphodynamic beach model discussed by Wright et al.[24,25] and Short.[12,13] Wright et al.[24,25] recognized six commonly observed morphodynamic states, each of which is distinguished by a different association of morphology, circulation, surf behavior, and resonant frequencies. Short[12,13] showed how different morphological types relate to stages in erosional or accretionary temporal sequences. Both sets of studies were carried out over roughly the same period and in similar environments, and the morphologic patterns described by Wright et al.[24,25] and Short[12,13] were similar. However, because the earlier studies were carried out independently and with different aims the respective classification schemes exhibited same taxonomic differences which may have been a source of confusion to those wishing to compare results of the two studies. We have attempted to reconcile these discrepancies by means of the modified and updated beach state model shown in Figure 2. Photographs of each state are shown in Figure 3.

The two extreme states shown in Figure 2 and previously described by Wright et al.[24,25] and Wright[22] are (1) fully dissipative, and (2) highly reflective. Morphologically, these states correspond, respectively, to flat, shallow beaches with relatively large subaqueous sand storage and steep beaches with small subaqueous sand storage. Morphodynamically the two extremes are distinguished on the basis of the surf-scaling parameter[7]

$$\epsilon = a_b\, \omega^2/g\tan^2\beta \tag{1}$$

where a_b is breaker amplitude, ω is incident wave radian frequency ($2\pi/T$; T = period), g is acceleration of gravity, and β is beach/surf zone gradient. Complete reflection can be expected when $\epsilon < 1.0$; however Guza and Inman[7] and Guza and Bowen[5] note that so long as $\epsilon \leq 2.0$ to 2.5 strong reflection will continue to permit strong standing wave motion, surging breakers, and resonance, particularly at subharmonic frequencies. Such a situation defines the highly reflective extreme. When $\epsilon > 2.5$ then waves begin to plunge, dissipating energy, and when $\epsilon > 20$ then spilling breakers occur (e.g., Galvin[4]). The surf zone widens and turbulent dissipation of incident wave energy increases with increasing ϵ.

The dissipative extreme (Figures 2a and 3a) is distinguished by high ϵ values across the surf zone and on the beach face (ϵ ranges from 30 to over 100), spilling breakers and approximately saturated surf zones, across which bores decay progressively to become very small by the time they reach the subaerial beach.[27] Subtle longshore bars are commonly present but longshore rhythms such as "megacusps" are absent. The dissipative extreme is analogous to the "storm" or "winter" profile of seasonally varying beaches; it corresponds to the "Type 1" beach discussed by Wright et al.[24] and to Short's[12,13] "Stage 6". To avoid further confusion, we will simply refer to it as "dissipative".

The four intermediate states shown in Figures 2 and 3 all possess coexisting dissipative and reflective elements and ϵ varies significantly across the profile or alongshore. The longshore bar-trough state (Figures 2b and 3b) can develop from an antecedent dissipative profile in an accretionary sequence. It was originally classified by Wright et al.[24] as Type 2 and by Short[12,13] as Stage 5. Bar-trough relief is much higher and the beach face much steeper than on the dissipative profile. The bar is the locus of initial wave-breaking and is moderately dissipative. In contrast to the dissipative beach, the broken waves cease their depth-dependent decay after passing over the steep inner edge of the bar and reform within the deep (2 to 3 m) trough. With respect to the partially dissipated and subsequently reformed waves, the steepened beach face is reflective with local ϵ commonly around 2. Low-steepness waves surge up the beach face; steeper waves experience a violent collapsing or plunging shore break near the base of the beach face followed by an explosive surge up the subaerial beach. In both cases, runup is relatively high. Cusps are often present in the swash zone.

The "rhythmic bar and beach" state (Figures 2c and 3c, Type 3 of Wright et al.[24] and Stage 4 of Short[12,13]) is dynamically similar to the "longshore bar-trough" state inasmuch as a relatively deep trough separates a moderately dissipative and pronounced bar from a generally reflective beach. However, the distinguishing features of the rhythmic bar and beach state are the rhythmic longshore undulations of the crescentic bar and of the subaerial beach. Spacings between crescentic bar horns or shoreline protrusions are typically on the order of 200 to 300 m. Weak to moderate rip circulation, similar to that shown in Figure 2c, prevails and the rips are persistent in location. In addition to shore-normal variations in reflectivity similar to those which exist on the longshore bar-trough topography, there are also shore-parallel variations: embayments are more reflective and shore protrusions are more dissipative.

The most pronounced longshore segregation of dissipative and reflective regions occurs in association with the "transverse bar and rip" topography (Figures 2d and 3d, Type 4 of Wright et al.,[24] Stage 3 of Short[12,13] The topography and circulation of this state are generally identical to those of the rhythmic topography discussed by Sonu.[17,18] High dissipation and setup over the shallow, flat transverse bars alternate with reflective conditions, low setup, and higher runup in embayments which are occupied by relatively strong rips. In an accretional sequence of beach change, the spacings between rips or transverse bars are usually inherited from the antecedent "rhythmic bar and beach state". However, the transverse bar and rip state may persist for extended periods during which time it may undergo considerable evolution involving changes in rip frequency and spacing. Most commonly, rip intensity and spacing decrease with decreasing wave energy (e.g., McKenzie,[10] Short,[12] and Wright et al.[24,25]).

A relatively narrow, moderately dissipative and flat accumulation of sand at or just below low tide level, backed by a steeper beach face which is reflective at high tide characterizes the "ridge and runnel/low-tide terrace" state (Figures 2e and 3e, Type 5 of Wright et al.,[24] Stage 2 of Short[12,13]). Small, weak and irregularly spaced rips related to the atrophying of preexisting rips may be present. Shore parallel runnels formed by shoreward migration of swash bars over the flat low tide terrace are often present near the break in slope between the beach face and the low tide terrace. Variations in reflectivity vs. dissipativeness are largely related to tidal phase: the beach is typically dissipative at low tide and reflective at high tide.

The fully reflective beach state (Figures 2f and 3f) lacks any dissipative elements. Breakers are exclusively surging to collapsing and turbulence related to the "breaking" process is confined to the zone of high runup on the beach face. Immediately beneath the steep, usually linear, beach face is a pronounced step composed of coarser material. The step depth increases with increasing wave height. Seaward of the step the bed slope decreases appreciably. Pronounced and highly rhythmic beach cusps are often present in the swash zone (Figure 3f); however, under low-energy conditions the beach is often capped by a high straight-crested berm.

Surf-Zone Process Signatures and Beach State

Wind-generated waves are the main source of the energy which drives beach changes. However, the complex processes, which operate in natural surf zones and involve various combinations of dissipation and reflection, can lead to the transfer of incident wave energy to other modes of fluid motion, some of which may become dominant over the waves themselves. Since it is primarily the flows near the sediment-water interface which determine the entrainment and ultimate fate of beach and nearshore sediments, it is important to have some appreciation of the relative roles played by these different modes of fluid motion in contributing to the near-bottom flow field and bed shear stresses. For discussion purposes, we may group the important modes of motion into four broad categories:

1. Oscillatory flows corresponding directly to the incident waves
2. Oscillatory or quasi-oscillatory flows corresponding to standing waves and edge waves at frequencies lower than incident wave frequency
3. Net circulations generated by wave energy dissipation
4. Nonwave-generated currents

Storm surge is not included in the above list because it is not normally very consequential on the beaches of southeastern Australia; however in many other environments it is important.

The first category includes the sediment-agitating oscillations, in the frequency band of the deepwater incident waves, due to both broken and unbroken waves. Oscillations in the second category can exist over a wide range of frequencies; they may be either "leaky" mode standing waves or trapped edge waves. Included are the subharmonic edge waves (at twice the incident wave period) which are readily excited on reflective beaches,[6] long-period infragravity "surf beat" at periods on the order of 1 to 3 min,[9,27] and higher-frequency infragravity motions at periods of 30 to 50 sec.[23] The important wave-generated currents are longshore currents, rip currents, and rip "feeder" currents. The fourth category includes tidal currents and currents generated by local wind shear. Our experiments on the different beach states show that the relative velocities of these different modes of fluid motion are strongly dependent on the state that the beach is in at the time and on local environmental conditions.

The Dissipative Extreme

Two beaches — Seven Mile Beach on the Australian east coast south of Sydney and Goolwa Beach in South Australia (Figure 1) provide examples of long, fully dissipative beaches. Both beaches are sinks for fine sands and exhibit very low gradients ($\tan \beta$ = 0.01 to 0.02) and wide, multibarred surf zones (Figure 2a). No longshore rhythms or significant irregularities are present. During the experiments, significant breaker heights were about 2 and 3 m on Seven Mile Beach and Goolwa Beach, respectively; the respective wave periods were 10 sec and 12 to 15 sec. The surf scaling parameter, ϵ (Equation 1) was on the order of 30 on Seven Mile Beach and 250 to 400 on Goolwa Beach. On both beaches, waves broke by spilling and dissipated progressively as they crossed the wide surf zone to become very small at the beach face. The more comprehensive data and analyses were from Goolwa Beach; these results indicated a constant ratio between bore heights H and local depth, h, across the surf zone. The ratio $\gamma = H_b/h_b$ had an average value of 0.42, which is significantly less than is normally assumed.[26]

Power spectra of surface elevation η, and horizontal near-bottom currents u and v from Seven Mile Beach and Goolwa Beach have been described in detail by Wright[23] and Wright et al.,[26] respectively. Figure 4 shows typical u spectra from Goolwa. The important feature of these spectra and of all η, u, and v spectra from high-energy, fully dissipative surf zones is the dominance of wide-banded infragravity energy in the periodic region of 100 to 200 sec. Phase angles of $\pi/2$ between η and u for the infragravity frequencies on Seven Mile Beach and Goolwa Beach indicate that oscillations at those frequencies are consistently standing.[23,26] At Goolwa, the velocity amplitudes of u corresponding to the infragravity "surf beat" were on the order of 1.2 to 1.5 m/sec, accompanying surf beat heights of roughly 1 m on the beach.[26] Standing oscillations at subharmonic frequencies are consistently absent from dissipative surf zones. The likelihood that infragravity oscillations are edge waves has been discussed by Holman in Chapter 2 and for the specific case of Goolwa by Wright et al.[26]

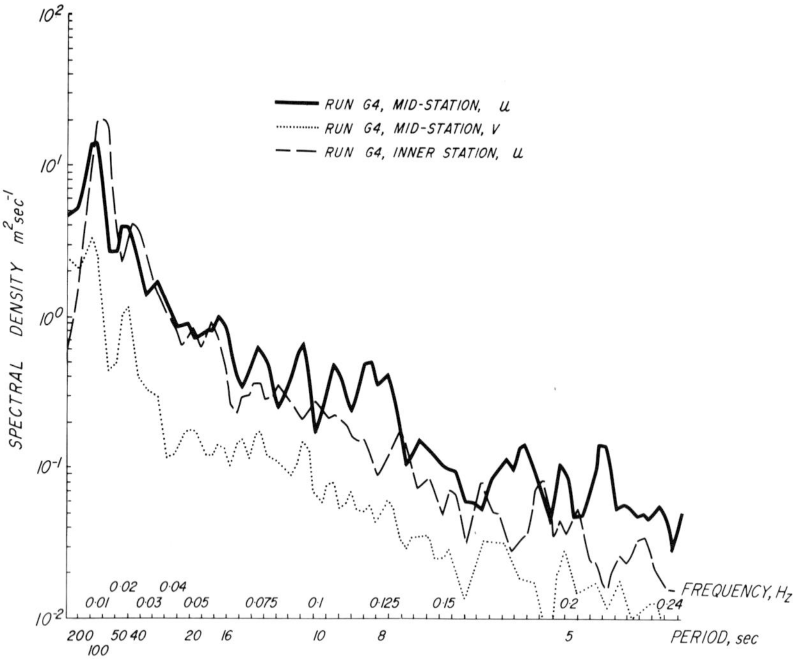

FIGURE 4. Power spectra of shore normal currents, u, and shore parallel currents, v, from mid and inner surf zone regions of a high-energy dissipative surf zone, Goolwa, South Australia, showing dominance of low-frequency infragravity energy. (From Wright, L. D., Guza, R. T., and Short, A. D., *Mar. Geol.*, 45, 41, 1982. With permission.)

The net time-averaged flows at Goolwa were weak relative to the surf-beat motions. There were no spatially persistent rips; vertical segregation of shoreward (upper layer) vs. seaward (bottom layer) transports generally characterized the shore-normal components of circulation.[26] It should be pointed out, however, that on strongly embayed and headland-bounded beaches such as those near Sydney, very strong, large scale rips frequently accompany the dissipative conditions that occur during storms.[12,24]

Figure 5 describes the "process signature" observed in the Goolwa surf zone. The bar graphs indicate the *relative* amplitudes of near-bottom flows associated with the incident waves, the surf beat, and the net transports. The bars are expressed in dimensionless form relative to the observed incident-wave related velocities, which are assigned a value of 1. The pronounced shoreward growth of flows due to infragravity oscillations and the strong dominance of those flows in the inner surf zone are the most important features of the graph, confirming the schematic Figure 4 presented in Chapter 2. This is typical of dissipative beaches in general (e.g., Seven Mile Beach). Recent studies on Torrey Pines Beach (California) by Guza and Thornton[8] also show that surf beat dominates runup and that it increases in amplitude linearly with increasing breaker amplitude.

The Reflective Extreme

Most of our experiments on reflective beach processes were carried out on Bracken Beach south of Sydney (Figure 1). This beach is similar in morphology and behavior to other reflective beaches in the region. The persistently low ϵ values (ϵ = 1 to 3) which characterize these beaches and define their reflective state are consequences of a combination of low-refraction coefficients related to the embayed nature of the compartments and steep beach faces ($\tan \beta$ = 0.1 to 0.15), commonly but not consistently associated with coarse material. Waves reach the beach face without breaking and surge strongly up the beach or collapse over the step.

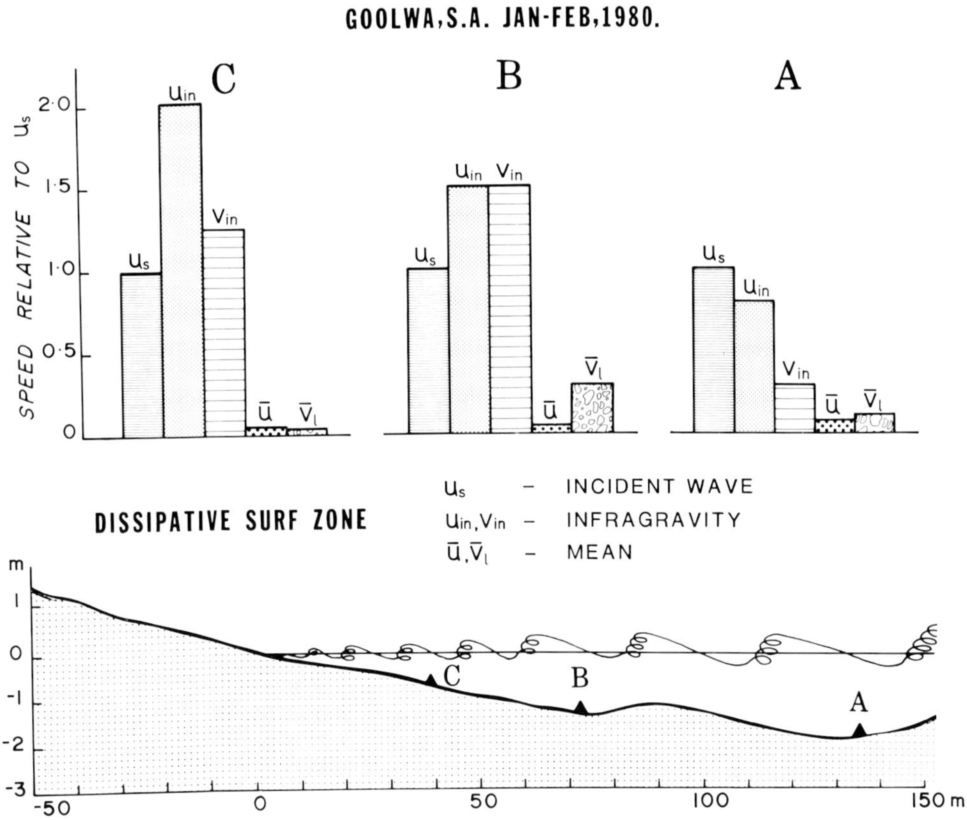

FIGURE 5. Process signature across a high-energy dissipative surf zone, Goolwa, South Australia. The bar graphs indicate the *relative* magnitudes (scaled relative to the significant velocity amplitude u_s of the incident waves) of u_s velocity amplitudes of shore normal and shore parallel flows at infragravity frequencies; and $\bar{u}$, $\bar{v}_e$ net time averaged shore normal and alongshore currents.

Figure 6 shows typical spectra of η and u from Bracken Beach under moderate swell conditions. In sharp contrast to the spectra from the dissipative extreme, most of the energy near the reflective beach is at incident wave frequency and at the first subharmonic of the incident waves (twice the incident wave period). Infragravity oscillations are very weak or negligible in most of our data from reflective beaches and are always subordinate to incident waves. The subharmonic oscillations are invariably standing as indicated by the 90° (π/2) phase angle between η and u. An interpretation of the subharmonic oscillations as edge waves standing alongshore is consistent with the conclusions of Guza and Davis[6] that subharmonic edge waves of mode zero are the most easily excited and can exist under highly reflective conditions. More detailed descriptions of the reflective beach spectra are given by Wright.[23] Our data indicate that the well-developed beach cusps that are normally present on reflective beaches are spaced at one half the predicted length of the zero-mode subharmonic edge waves. Under low-energy conditions subharmonic oscillations tend to be of low amplitude relative to incident wave amplitude. However, under conditions of a moderate-energy long-period swell, the subharmonic oscillations attain amplitudes greater than those of the incident waves at the beach, producing accentuated runup which may overtop the berm crest and initiate subaerial beach cutting.

Intermediate States

The most complex process signatures are associated with the intermediate beach states

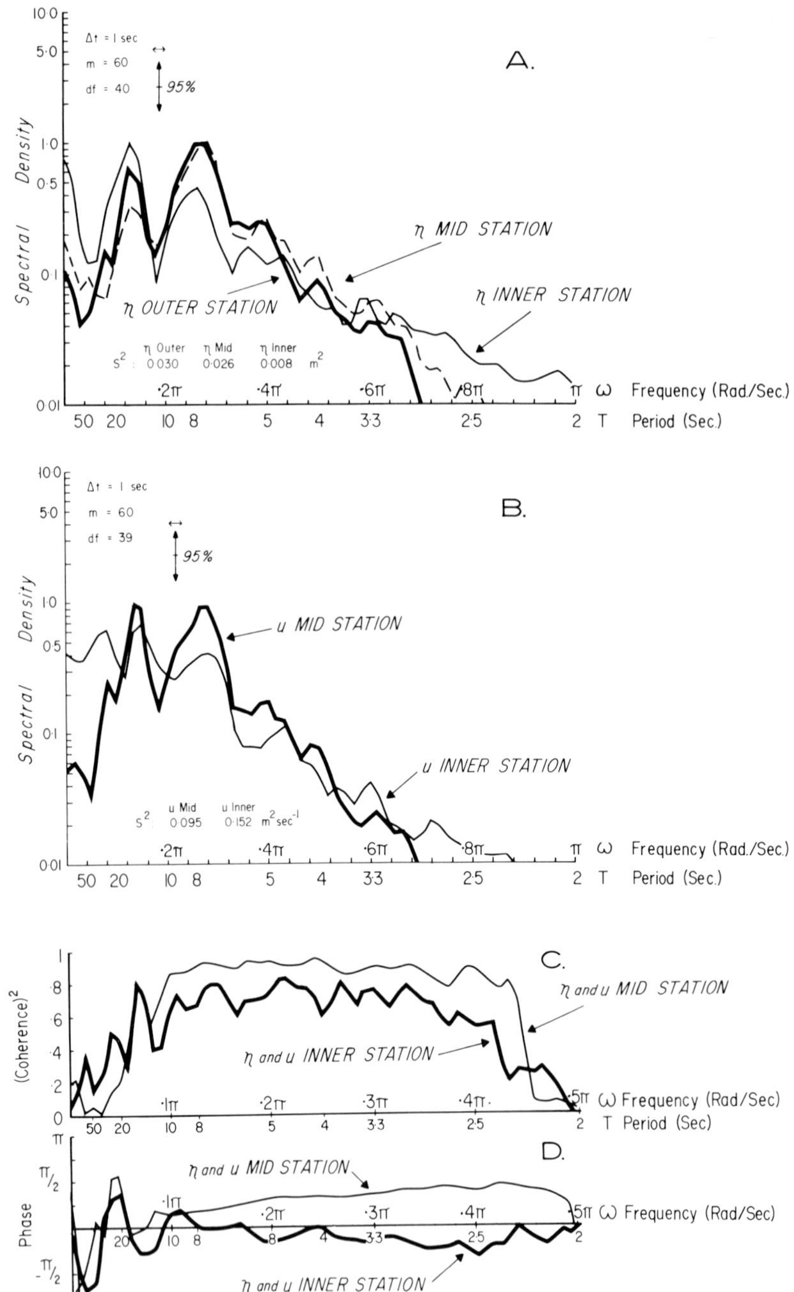

FIGURE 6. Spectra and cross spectra of water surface elvation, η and shore normal currents, u from a highly reflective beach (Bracken Beach, December 10, 1977). (A) Power spectra of η; (B) power spectra of u; (C) coherence squared between η and u; (D) phase angles between η and u. Prominent features are the dominant peaks at incident wave frequency (7.5 sec.) and at its first subharmonic (15 sec). The phase relationships of π/2 (90°) for the subharmonic peaks show these oscillations to be standing. (From Wright, L. D., *J. Mar. Freshwater, Res.*, 1982. With permission.)

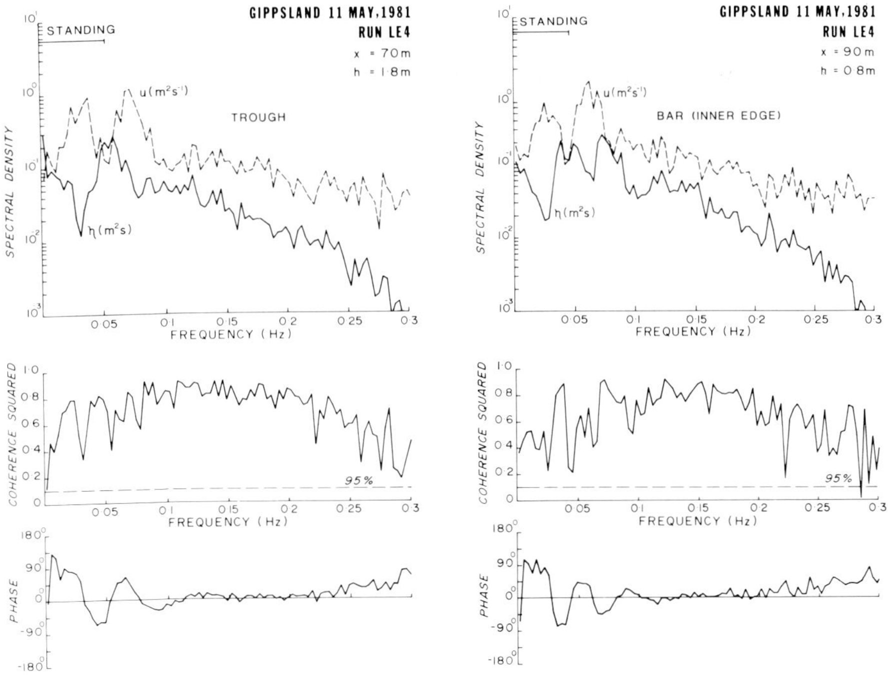

FIGURE 7. Spectra and cross spectra of η and u from the bar and trough regions associated with a beach state transitional between the longshore bar-trough and rhythmic bar and beach states (Eastern Beach, Gippsland, May 11, 1981). The broad peak at incident wave frequency is dominant; however there is also considerable standing wave energy at relatively short infragravity periods (e.g., around 0.025 Hz or 40 sec.).

owing to the fact that those states possess both dissipative and reflective elements. Processes associated with the longshore bar-trough and rhythmic bar and beach states are illustrated by results from Moruya Beach, some of which have been discussed by Wright et al.[24,25] and Wright,[23] and by more recent and more conclusive results from Gippsland (Figure 1). Figure 7 shows spectra of η and u from the trough and bar regions of pronounced rhythmic bar and beach topography on Eastern Beach, Gippsland, Victoria. The spectra in Figure 7, like all spectra from bar-trough and crescentic bar and beach topography on Eastern Beach and on Moruya Beach, show dominance of energy at swell frequency but with appreciable standing wave energy over a band that includes both subharmonic and infragravity frequencies. The infragravity standing waves occur at much higher frequencies than in fully dissipative surf zones: typical periods of the dominant infragravity oscillations are on the order of 35 to 50 sec. Chappell and Wright[2] and Wright[23] have discussed these intermediate frequency standing waves and believe that they may be edge waves which initiate the crescentic bar and beach rhythms and that these frequencies may be constrained or selected by the dimensions of the pronounced trough.

The net mean currents in the presence of bar-trough and crescentic bar and beach topographies are characterized by moderate strength, widely spaced rips, and longshore currents within the troughs. On straight bar-trough topographies the rips are relatively weak and usually do not remain fixed in position, but migrate alongshore. Their positions become topographically arrested by the crescentic bar and beach topography, however, resulting in intensification of rip currents. When waves break obliquely to the shore it is the bar trough, and to a lesser degree the crescentic bar and beach topographies, which exhibit the strongest longshore currents, with longshore velocity maxima concentrated in the trough. A typical vertical velocity profile of a longshore current, measured in the deep trough of Eastern Beach, Gippsland, Victoria is shown in Figure 8.

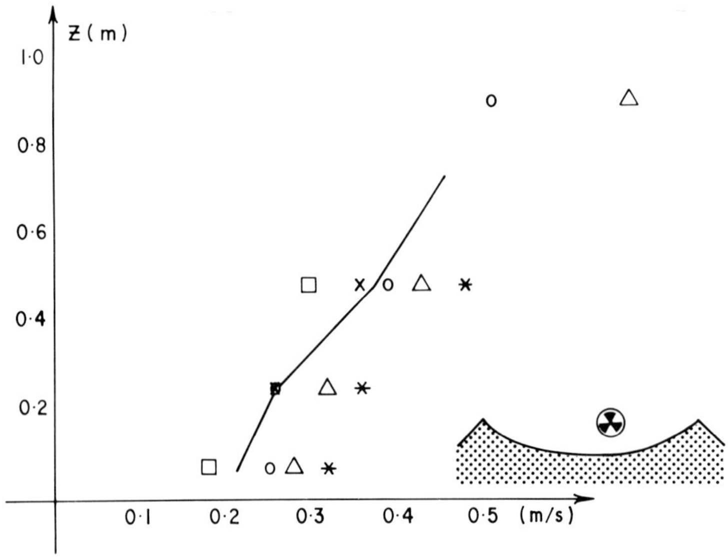

FIGURE 8. Vertical velocity profile of the longshore current as measured within the trough of Eastern Beach, Gippsland, May 13, 1981. Note the high current speeds on the order of 25 cm/sec which are present at an elevation of less than 10 cm above the bed. (Key: Δ — 10 sec filter, sampled every 6 sec; * — 100 sec filter, sample interval 1 min; ; □ — 100 sec filter, sample interval 1 min; x — fast, sample interval 6 sec; ○ — fast, sample interval 6 sec.)

Figure 9 illustrates the net circulation and the process signature associated with crescentic bar and beach topography with a pronounced trough based on our experimental results from Eastern Beach. The histograms indicate the magnitude, relative to the velocity amplitude u_s associated with the local incident waves, of near-bed velocities related to subharmonic oscillations (u_{sub}), infragravity oscillations (u_{in}, v_{in}), and net time-averaged longshore and rip currents (V) in different regions of the surf zone. Incident waves tend to dominate everywhere but are overwhelmingly dominant over the bar where they initially break. Subharmonic oscillations become nearly as strong as incident waves over the step near the beach face; subharmonic edge waves are probably responsible for well-developed cusps on the locally reflective beach face in the lee of the offshore bar. Longshore currents and offshore rip flows pulse strongly at infragravity frequencies in the rip embayment.

Once the horns of the crescentic bars have welded to the beach, usually in the vicinity of the beach "megacusp" horns, to produce alternating transverse bars and rips (Figures 2d and 3d) the most pronounced rhythmic longshore variations in dissipation vs. reflectivity, setup, and pressure result. It is under these conditions that the strongest rip circulations occur for any given breaker height. Figure 10 shows the process signature associated with transverse bar-rip topography as exemplified by experiments conducted on the moderately embayed Palm Beach (Figure 1). The important features are the extremely strong rip currents and rip "feeder" currents. Although these currents remain persistently strong and unidirec-

FIGURE 9. Net circulation pattern and process signature observed in the presence of rhythmic bar and beach topography, Eastern Beach, Gippsland. The broad arrows idicate general circulation based on dye released. Fine arrows indicate current speeds and directions measured at specific points. As in Figure 5, the bar graphs indicate the magnitudes relative to incident wave orbital velocities of different modes of horizontal current motion: u_{sub} shore normal currents at subharmonic frequency; u_{in}, v_{in} = shore normal and shore parallel currents at infragravity frequencies, V = net time averaged velocities.

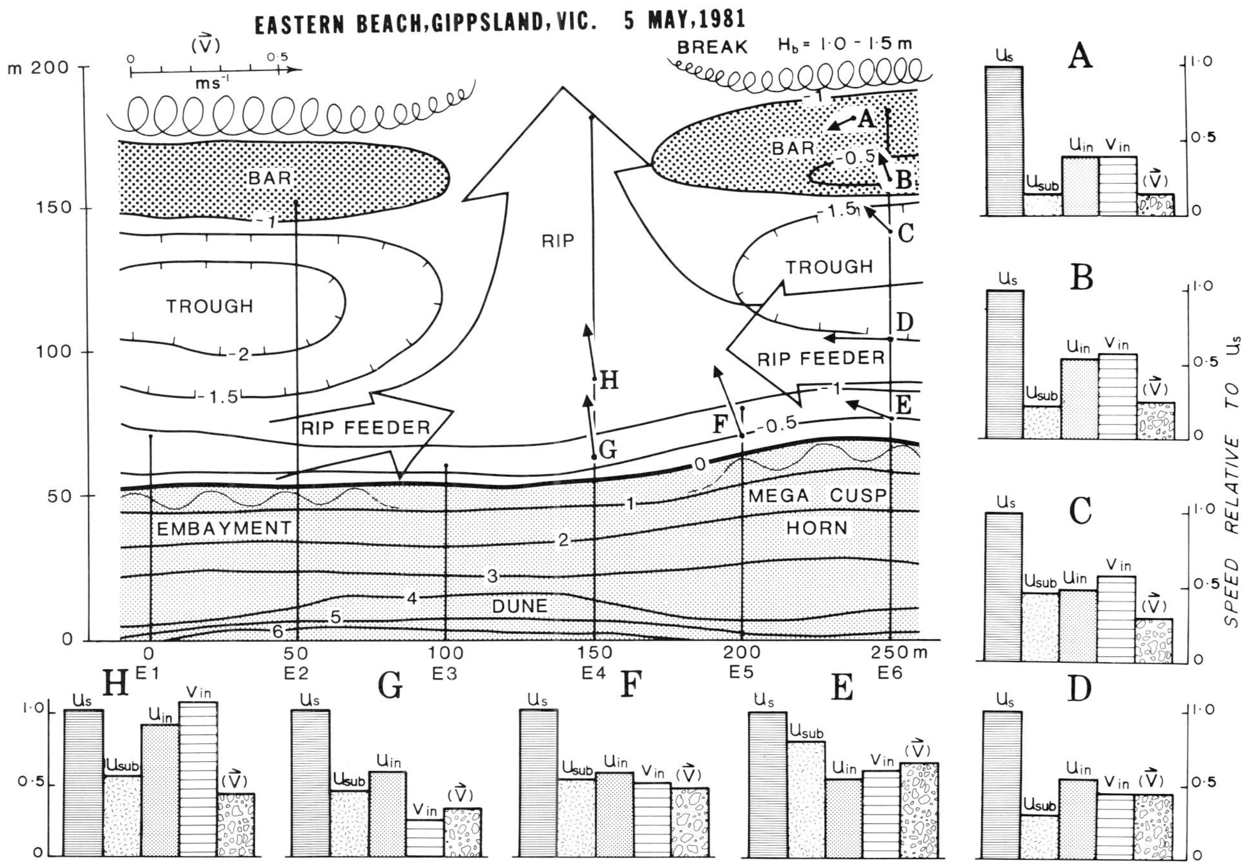

FIGURE 9. See caption p. 50.

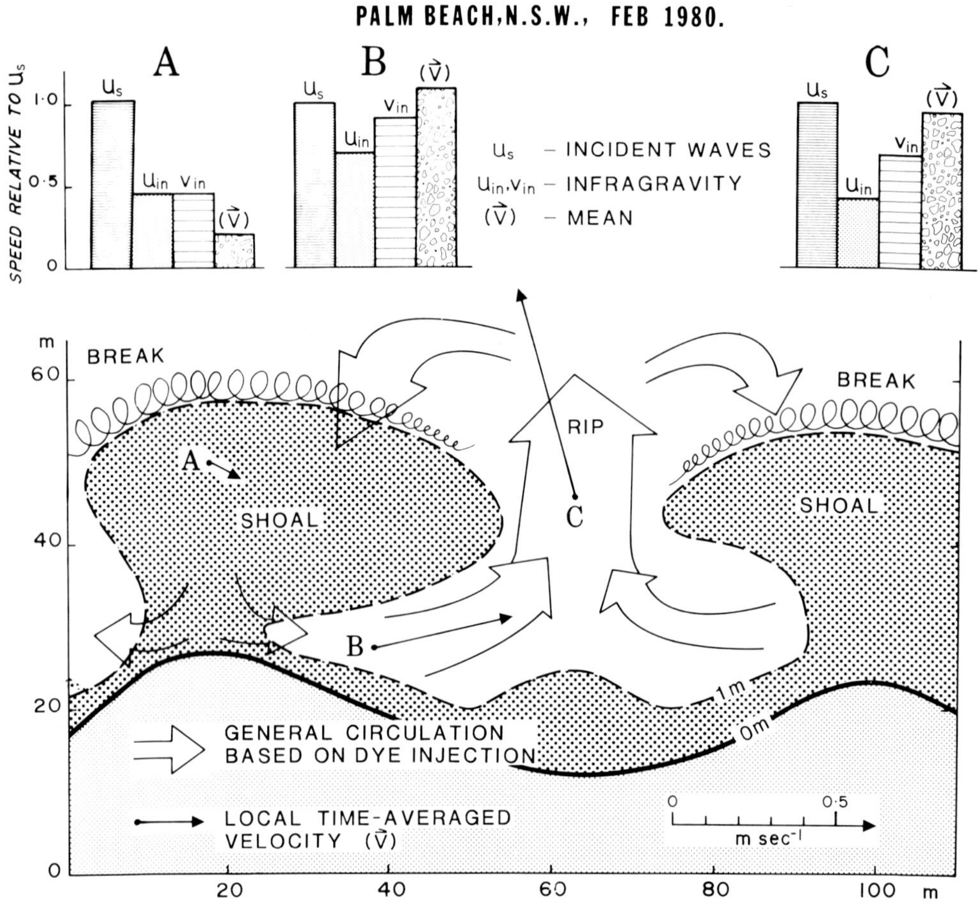

FIGURE 10. Process signature associate with transverse bar and rip topography (Palm Beach N.S.W., February 1980). Note the strong rips and rip feeder currents.

tional, they alternately strengthen and slacken at infragravity frequencies. On this type of topography, we observe rip current speeds to be typically on the same order as the breaking incident wave orbital velocities. The strong rips, localized by the rhythmic topography related largely to antecedent processes can cause significant scour of rip embayments, even when the adjacent transverse bars are accreting.[12,16,22]

The ridge-runnel/low-tide terrace state normally prevails only under moderate or low wave energy. Breaking incident waves and their dissipative bores dominate at low tide, with low amplitude and relatively high frequency infragravity oscillations making comparatively minor secondary contributions.[23] Rips are generally weak and irregularly spaced or absent altogether. At high tide, the process signature is similar to that of reflective beaches with incident waves and subharmonic edge waves dominating.

The Macrotidal Beach

The macrotidal beach (Cable Beach, Broome) had a concave upward profile consisting of a subtidal region below the extreme low-tide level, a low-tidal region between the levels of spring and neap low tides, a mid-tidal zone between neap low and neap high-tide levels, and a high-tide zone between neap high tide and spring high tide. The subtidal and low-tidal zones were highly dissipative and the high-tidal zone was reflective.[26] The process signature across the Cable Beach profile is shown in Figure 11 together with the relative frequency at which each point on the profile is active. Details of the morphodynamic processes on Cable Beach are discussed by Wright et al. in Reference 26.

As shown in Figure 11 the effects of incident waves dominate everywhere across the profile. However, time-averaged estimates of wave work over the lunar half-cycle[26] showed that most of the work over the subtidal, low-tidal, and mid-tidal zones was performed by unbroken shoaling waves rather than by surf-zone processes. Surf-zone processes only dominated the reflective high tidal zone which is active for less than 25% of the time (Figure 11). Standing wave energy, particularly at higher infragravity frequencies (40 to 50 sec) played important secondary roles on the high-tidal zone at spring high tide and over the mid-tidal zone at neap high tide, but was of minor importance over most of the profile. A feature which distinguishes the process signature of the macrotidal beach from those of microtidal beaches is the subdominant contribution of strong shore-parallel tidal currents over the subtidal and low-tidal zones. Over the subtidal zone asymmetrical tidal currents, skewed toward the north, attained maximum speeds of 0.5 m/sec just after high tide. A time series of the currents is shown in Figure 12. These currents were probably responsible for the net longshore transport of sand initially suspended by waves.[26] No significant rip cells or wave-induced longshore currents were observed on Cable Beach.

Modes of Beach Cut

In much of the older literature it was considered that whether a beach eroded or accreted depended solely on the height and steepness of the waves. However, it is now clear that the state of the beach immediately prior to and during a high-energy or potentially destructive event is equally important. For example, it is common for the same set of wave conditions to erode reflective beaches while producing accretion of neighboring dissipative or intermediate beaches.[22,25] One probable reason for this is the dependence of process signatures on beach state as just discussed in the preceding section. Just as the relative magnitudes of different modes of sediment-disturbing and sediment-transporting motion vary with beach state, so also do modes of beach cut. In a recent study, Wright[22] discriminated between three quasi-discrete but occasionally overlapping modes of state-dependent beach cut. These modes of cut are summarized in Table 1. The temporal sequence of beach erosion from a reflective state, through the intermediate to the dissipative state, was found to progress through five states,[12] each possessing elements of the three erosion modes.

Although the fully reflective beach states represent the accreted extreme of a "beach building" sequence, these states are also the most susceptible to being cut during rising wave energy owing, in part, to the lack of a subaqueous store of sand (all of the active sand is stored subaerially), and in part to the fact that reflective conditions are conducive to the growth of resonant subharmonic edge waves. Growth of subharmonic edge waves on reflective beaches to amplitudes greater than those of the incident waves under rising long-period swell on steep beaches causes accentuated runup. This can cause either beach face scarping or overtopping and initial cutting of the berm.[12] Large, widely spaced erosional cusps are often formed in the early stages of erosion.[22] Because subharmonic edge waves are easily excited by long-period, low-steepness swells,[6] this mode of erosion requires the least energy to induce.

Subharmonic resonance is suppressed in turbulent, highly dissipative surf zones. As shown earlier, long-period surf beat dominates runup and inner surf-zone flows on highly dissipative beaches. However, appreciable wave energy is required to cause surf beat to attain destructive amplitudes, so that beaches which are fully dissipative to begin with are the least sensitive to erosion. Nevertheless, growth of setup and infragravity oscillations under high energy conditions can allow the bores of partially dissipated waves to penetrate to the backshore or foredune base where scarping occurs (Figure 13). This mechanism can cause dune and backshore recession without necessarily producing a net offshore loss of beach material.[15]

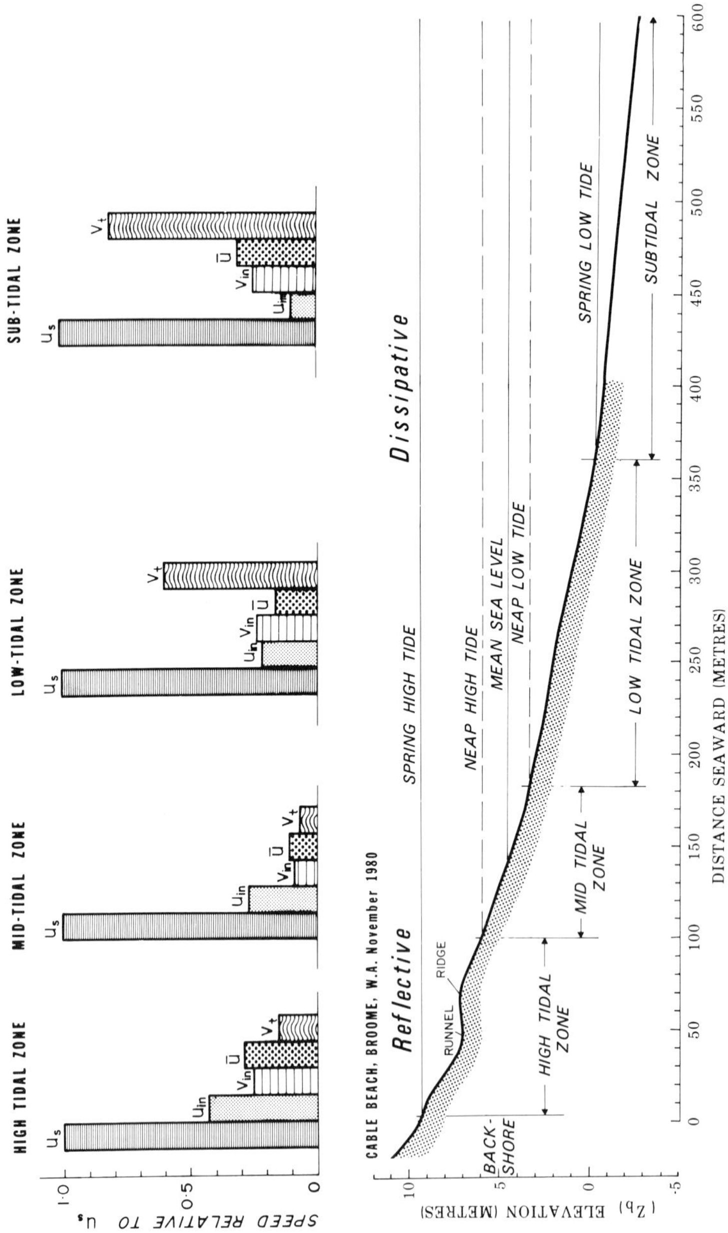
HIGH TIDAL ZONE
MID-TIDAL ZONE
LOW-TIDAL ZONE
SUB-TIDAL ZONE
SPEED RELATIVE TO U_s
CABLE BEACH, BROOME, W.A. November 1980
Reflective
Dissipative
SPRING HIGH TIDE
NEAP HIGH TIDE
MEAN SEA LEVEL
NEAP LOW TIDE
SPRING LOW TIDE
RUNNEL
RIDGE
BACK-SHORE
HIGH TIDAL ZONE
MID TIDAL ZONE
LOW TIDAL ZONE
SUBTIDAL ZONE
DISTANCE SEAWARD (METRES)
ELEVATION (METRES) (Z_b)

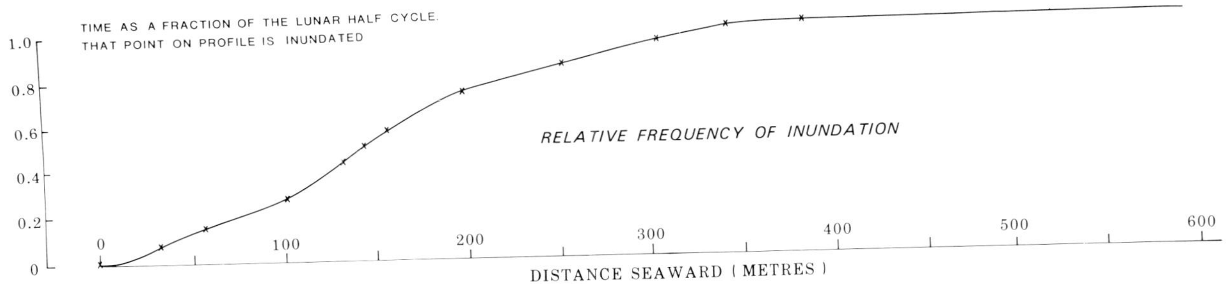

FIGURE 11. Process signature associated with a macrotidal beach (Cable Beach, Broome, Western Australia, November 1980). The lower curve indicates the fraction of time that corresponding points on the profile are inundated (or active) during a fortnightly neap to spring cycle. The bar graphs have the same meaning as previously except that V+ refers to shore parallel tidal currents.

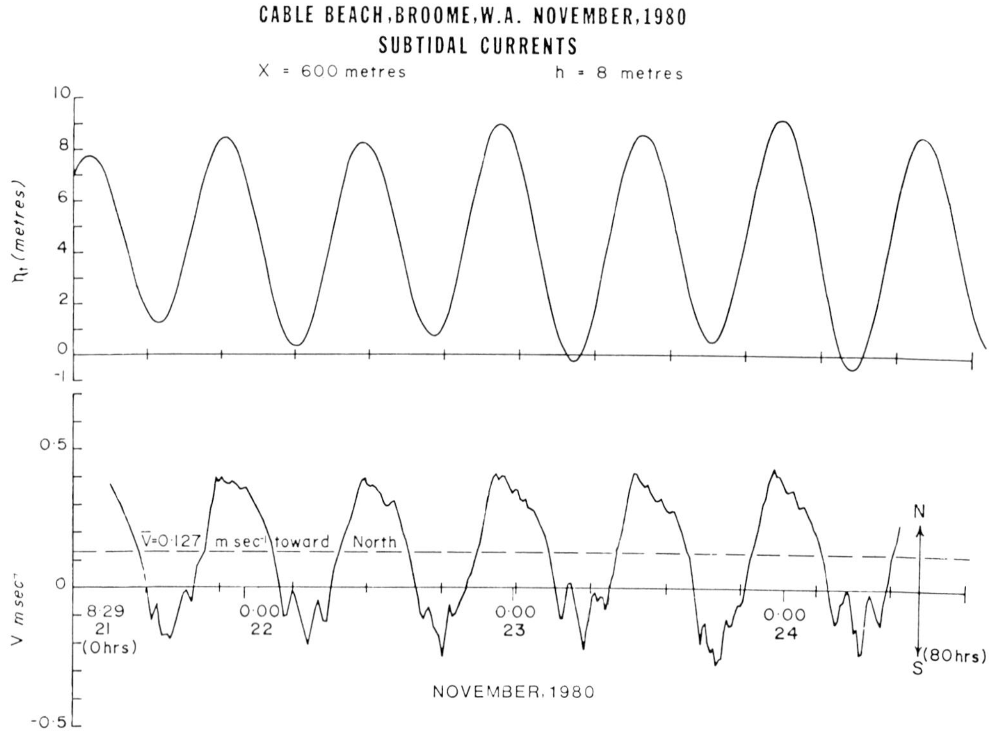

FIGURE 12. Shore parallel current velocities, V +, as recorded over the subdivided zone fronting Cable Beach from a depth of −8 meters below mean water level. The upper curve shows corresponding tidal elevations, η +. (From Wright, L. D., Nielson, P. N., Short, A. D., and Green, M. O., *Mar. Geol.*, 50, 97, 1982. With permission.)

Table 1
MODES OF BEACH CUT IN RELATION TO BEACH STATE

	Mode of beach cut	Cause	Associated beach surf-zone conditions	Relative energy required to induce beach cut
1	Accentuated runup and berm overtopping, formation of erosional cusps	Strong subharmonic resonance	Reflective	Lowest
2	Backshore scarping by bores superimposed on long period setup oscillations	Strong infragravity oscillations	Dissipative	Highest
3	Beach scarping in embayments of arrested rip cells	Radiation stress gradients induced by irregular surf zone topography	Intermediate, rhythmic surf-zone topography, regions adjacent to shoreline protrusions, inlets or river mouths	Intermediate

Both of the mechanisms of beach cutting just described can operate on intermediate states. In addition, the intermediate states are subject to a third mode of cutting: scour in the embayments of topographically arrested rips (Figure 14). This mode is most acute in the presence of transverse bar-rip topographies[12] particularly where the transverse bars remain fixed in position for prolonged periods. Rips can become quite large and highly persistent

FIGURE 13. Backshore scarping associated with mode 2 erosion on a dissipative beach.

causing localized chronic erosion problems in regions downdrift from the ebb tidal deltas of tidal inlets or other quasi-permanent or man-made shoreline protrusions.[22] Because the intermediate states experienced all three cutting modes, these states tend to have the highest erosion sensitivity.

On some of the beaches we have studied, notably the longer straighter ones (e.g., Gippsland), littoral drift plays varying roles in causing net input or output of sand to local beach segments. However, this process is generally not very important on beaches like Goolwa Beach which faces directly into the dominant swell, or most of the New South Wales beaches which occupy embayed compartments bounded by protruding headlands. On most of these beaches the most significant removal of sand, when it occurs, is associated with transport processes acting normal rather than parallel to the shore. During high-energy events, these processes may move sand offshore to depths of 10 to 20 m from which it requires 2 years or more to return.

Chronic, long-term recession of the macrotidal Cable Beach is related in large part to northerly longshore transport of sediment out of the region. However, this longshore transport is not attributable to oblique wave approach but rather to asymmetrical shore-parallel tidal currents. The longshore transport process in this case probably simply involves suspension of sediment by shore-normal waves and longshore advection of the sediment by tidal currents.[26]

Modal Beach State and Beach Mobility

The morphology of a beach at any particular time is a function of its sediment characteristics, immediate and antecedent wave, tide and wind conditions, and the antecedent beach state. However, over the long term a given beach will tend to exhibit a *modal* or most frequently recurrent state which depends on environment. The forcing wave climate can, in turn, be described in terms of the modal or most frequently occurring wave condition around which a spread of higher and lower waves may prevail. The beach in response to the wave climate consists of a modal beach type determined by the modal wave conditions, and a

FIGURE 14. Chronic scour and beach recession in the embayments of topographically arrested rips. This rip caused the destruction of two beach-front homes (Wamberal Beach, N.S.W.).

range of beach morphologies dependent on the range of the wave conditions. Associated with the temporal variation of beach state around the modal state is a corresponding variation in shoreline position and in profile shape. The range of this variability defines the mobility of a beach. Because the same environmental factors which determine the modal state of a beach also determine its variability, there is a close association between modal beach state and beach mobility (Figure 15).

For the medium-to-fine sand beaches we studied in microtidal southeast Australia the

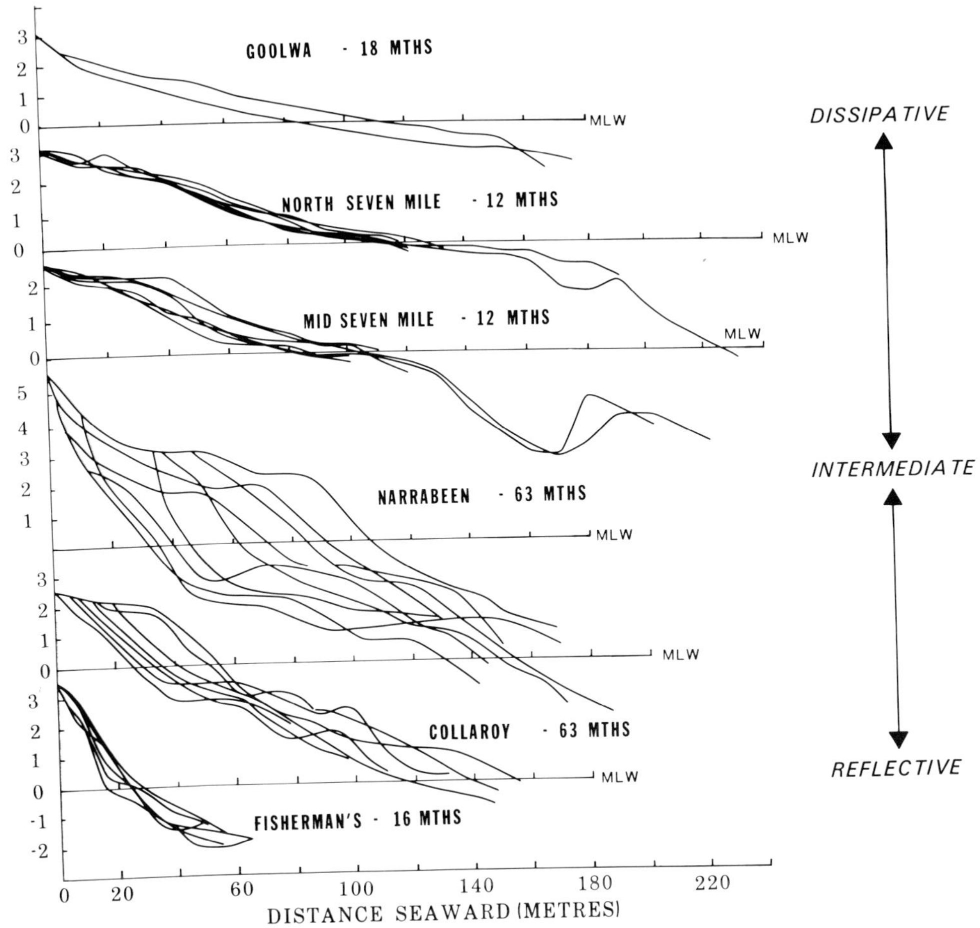

FIGURE 15. Successive beach and surf zone profiles of different dissipative, intermediate, and reflective beaches. The mobility of these beaches can be considered proportional to the sweep zones occupied by the profiles over prolonged periods.

modal wave height required to favor each beach type ranged from >2.5 m for dissipative beaches to <1 m for reflective beaches, with intermediate beaches favored by waves of 1 to 2.5 m in height.[14] The high Southern Hemisphere west coast swell and storm wave environments maintain dissipative conditions year round (e.g., Goolwa). The more moderate and highly variable east coast swell environments of the New South Wales coast favor intermediate modal beach types in the more exposed localities. Reflective beaches can prevail in sheltered embayments in virtually any wave climate.

The notion of a universal summer or winter beach morphology is only applicable to specific environments (e.g., California). It is inapplicable in Australia. Each wave climate has its own level of modal wave energy and signature of wave conditions. Breaker heights are functions of the deepwater wave climate and the degree of attenuation and refraction. Consequently, local changes in nearshore gradients may result in longshore gradients in breaker wave height and resultant beach morphology for identical deepwater wave conditions. On the New South Wales coast, it is very common for the modal or instantaneous beach state to vary spatially along the length of the same beach.[12,13,22,24]

Modally dissipative beaches are associated with persistent high waves and abundant fine sand to produce the wide, low-gradient beach and surf zone. They are stable beaches

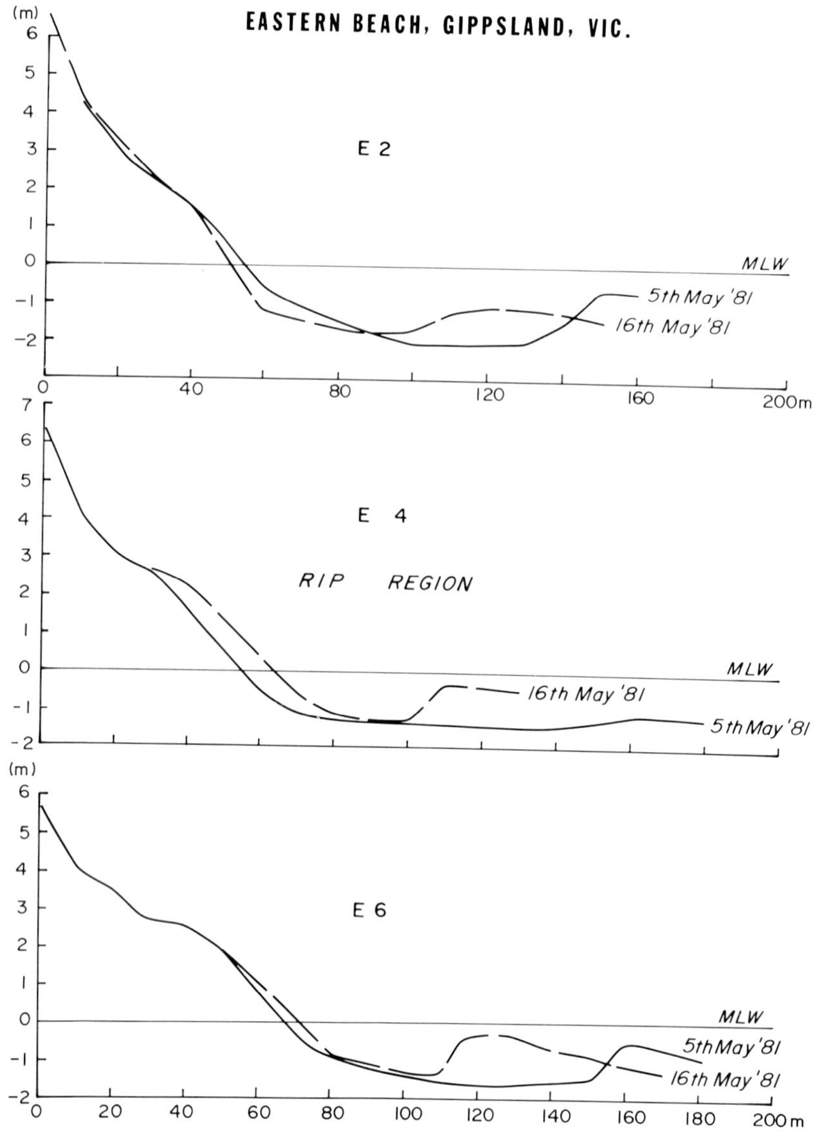

FIGURE 16. Successive beach and surf zone profiles from Eastern Beach, Gippsland, Victoria, illustrating the rapid bar migration which can occur in association with longshore bar-trough and rhythmic bar and beach states.

possessing low spatial and temporal variability. Because of the persistent high-energy wave conditions, fine sediments, and low gradients most of the active sediment is stored in and seaward of the surf zone. Increases in wave height widen the surf zone but cause only slight planing down of the beach and surf zone. The low mobility of modally dissipative beaches is evident from the narrow sweep zones of profile variation shown in Figure 15 for Goolwa and Seven Mile Beaches.

Reflective beaches represent low-energy beaches with low-modal wave heights (<1 m) and the end point in the accretionary sequence for periodically higher-energy beaches.[12,14] During the low summer waves many of the California beaches become reflective. In eastern Australia this may occur following a period of low late-winter swell. However, on exposed coasts, reflective conditions are usually short-lived since reflective beaches possess high

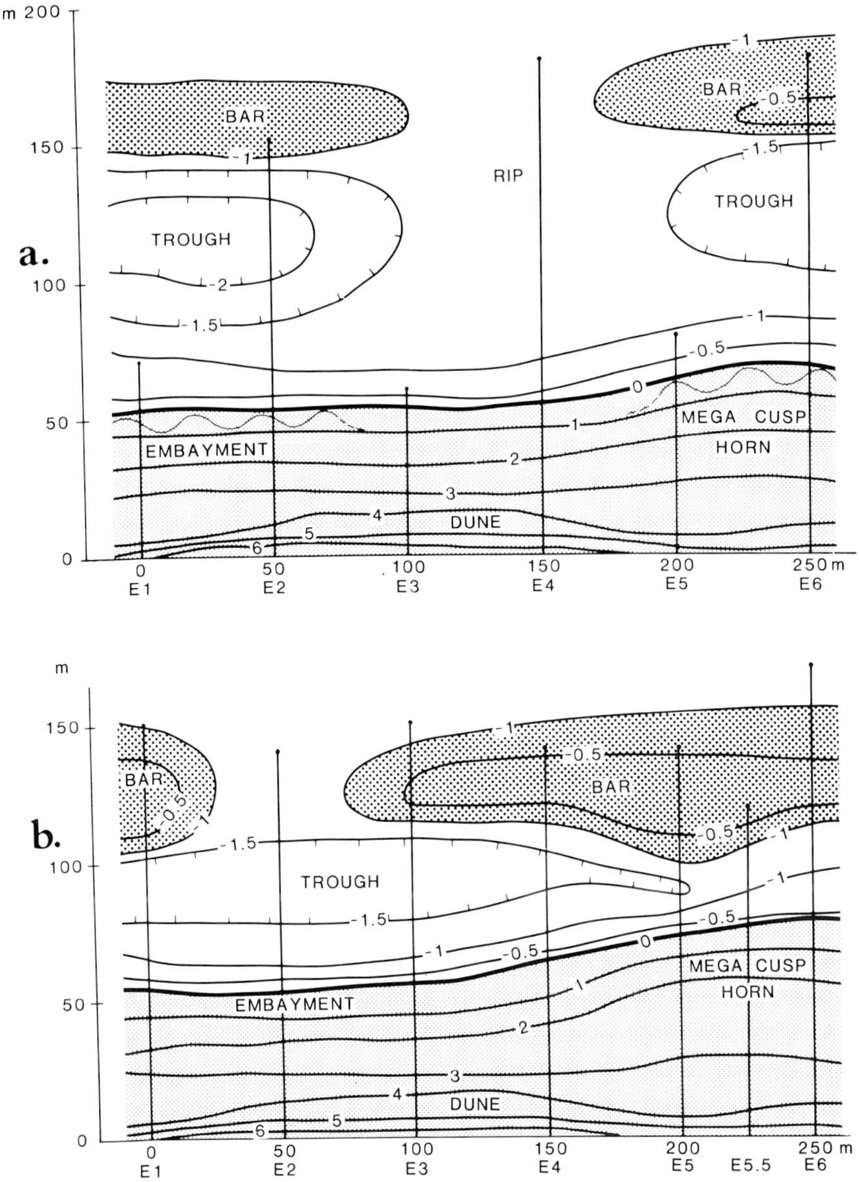

FIGURE 17. Temporal plan-view variability associated with rhythmic bar and beach topography, Eastern Beach, Gippsland; (a) May 8, 1981; (b) May 18, 1981.

erosion sensitivity and are quickly eroded by the return of higher waves. They can exist as the modal state only in sheltered compartments where sufficient coarse material is available. Under these conditions, the filtering out of high-energy events by surrounding headlands and by embayment geometry results in *modally* reflective beaches which also exhibit low temporal variability as indicated by the narrow sweep zone of Fisherman's Beach (Figure 15).

Intermediate beaches are spatially and temporally the most dynamic beach forms. With their characteristic rip circulation, dynamic bar forms, abundant sediments, and moderate waves they can undergo rapid changes as wave height fluctuates, causing rapid reversals in onshore-offshore and alongshore sediment transport. As one might expect considering the complex process signatures of these states, the mobility of modally intermediate states is higher than that of either of the two extremes. Longshore bar-trough and crescentic bar and beach morphologies tend to persist as the modal states in areas of variable moderate to high waves (e.g., Gippsland). They are also well developed in the presence of oblique waves which produce strong longshore currents in troughs and tend to be particularly favored by bimodal sediment populations with coarse fractions accumulating in the form of the steep subaerial beach face and finer fractions occupying the surf zone and composing the bars. Periodically high waves produce the straight longshore bar-troughs morphology, while subsequent prolonged periods (several days or longer) of more moderate waves encourage the development of crescentic bars and, eventually, megacusp shoreline protrusions and embayments.

Intermediate beaches are spatially and temporally the most dynamic beach forms. With their characteristic rip circulation, dynamic bar forms, abundant surf zone and beach sediment, and moderate waves they can undergo rapid changes as wave height fluctuates, causing rapid reversals in onshore-offshore and alongshore sediment transport. As one might expect considering the complex process signatures of these states, the mobility of modally intermediate states is higher than that of either of the two extremes. Longshore bar-trough and crescentic bar and beach morphologies tend to persist as the modal states in areas of variable moderate to high waves (e.g., Gippsland). They are also well developed in the presence of oblique waves which produce strong longshore currents in troughs and tend to be particularly favored by bimodal sediment populations with coarse fractions accumulating in the form of the steep subaerial beach face and finer fractions occupying the surf zone and composing the bars. Periodically high waves produce the straight longshore bar-troughs morphology, while subsequent prolonged periods (several days or longer) of more moderate waves encourage the development of crescentic bars and, eventually, megacusp shoreline protrusions and embayments.

Bars and surf-zone features are highly mobile on both longshore bar-trough and rhythmic bar and beach topographies. However, the straight reflective subaerial beaches of the longshore bar-trough state, partially sheltered by the offshore bar, normally exhibit only low to moderate variability. In contrast, the inherent spatial (longshore) variability of the rhythmic bar and beach states combines with the high temporal variability of the bar, producing high mobility of both the bar and the beach face. Figure 16 shows the extensive profile changes which occurred over a ten-day period in association with rhythmic bar and beach morphology on Eastern Beach, Gippsland. The corresponding three-dimensional changes over the same period are shown in Figure 17.

Transverse bar and rip morphology is a product of persistent moderate wave energy (modal wave height 1 to 1.5 m) following a period of decreasing wave height from a rhythmic or longshore bar form. Like the rhythmic bar-beach, the alternating bars and rips produce high spatial variability, and rearrangement of the bars and rips causes moderate to high temporal variability. On the summer sea-breeze-dominated Sydney coast, 1 to 1.5 m waves produce several weeks of transverse bars and rips, with bar-rip locations shifting on the order of once a week.

Ridge-runnel or low-tide terrace morphology occurs in areas of moderate waves following a period of lower waves (<1 m). The rip channels infill as the bar welds to the beach producing mini-rip, then a ridge and runnel or low-tide terrace, the latter having no runnel. The beach face steepens during low waves as cusps are deposited. The close proximity of the bar and beach face permits rapid exchange of sand leading to moderate to high temporal variability. The Collaroy Beach profiles (Figure 15) with a modal wave height of 1 m, illustrate this range of mobility.

The four intermediate beach types exhibit a wide range in form from longshore bar-trough to the low-tide terrace. However each is characterized by some degree of rip circulation and rhythmic longshore form. The longshore bar-trough is transitional between the fully dissipative type and initial rip circulations and crescentic bar formation. The rhythmic bar and beach and transverse bar or rip represent the classic rip-dominated rhythmic topography, which make them the most dynamic both spatially and temporally. On the ridge and runnel or low-tide terrace state the rips fill as the bar welds to the beach, ready to move up onto the beach to form a reflective beach.

ACKNOWLEDGMENTS

The study on which this paper is based has been supported by the Office of Naval Research, Coastal Sciences Program, Task NR 388-157, Grant N-00014-80-G-0001 and by the Australian Research Grants Committee (ARGC) and the Australian Marine Sciences and Technologies Advisory Committee (AMSTAC). Valuable and competent assistance in the field and in data analysis was provided by P. Nielsen, M. P. Bradshaw, M. O. Green, F. C. Coffey, G. Lloyd, and other members of the Coastal Studies Unit (past and present). Figures were drawn by J. Roberts and J. de Roder.

REFERENCES

1. **Bradshaw, M. P., Chappell, J., Hales, R. S., and Wright, L. D.,** Field monitoring and analysis of beach and inshore hydrodynamics, *Proc. 4th Aust. Coastal Ocean Eng. Conf.*, American Society of Civil Engineers, New York, 1978, 171.
2. **Chappell, J. and Wright, L. D.,** Surf zone resonance and coupled morphology, *Proc. 16th Int. Conf. Coastal Eng.*, American Society of Civil Engineers, New York, 1978, 1359.
3. **Dean, R. G.,** Heuristic models of sand transport in the Surf Zone, *Proc. Conf. Eng. Dyn. Surf Zone*, American Society of Civil Engineers, New York, 1973, 208.
4. **Galvin, C.J.,** Wave breaking in shallow water, in, *Waves on Beaches*, Meyer, R W., Ed., Academic Press, New York, 1972, 413.
5. **Guza, R. T. and Bowen, A. J.,** Resonant interactions from waves breaking on a beach, *Proc. 15th Int. Conf. Coastal Eng.* American Society of Civil Engineers, New York, 1977, 560.
6. **Guza, R.T. and Davis, R. E.,** Excitation of edge waves by waves incident on beach, *J. Geophys. Res.* 79(9), 1285, 1974.
7. **Guza, R. T. and Inman, D. L.,** Edge waves and beach cusps, *J. Geophys. Res.*, 80(21), 2997, 1975.
8. **Guza, R. T. and Thornton, E. B.,** Swash oscillations on a natural beach, *J. Geophys. Res.*, 87, 483, 1982.
9. **Huntley, D. A., Guza, R. T., and Thornton, E. B.,** Field observations of surf beat. I. Progressive edge waves, *J. Geophys. Res.*, 86, 6451, 1981.
10. **McKenzie, P.,** Rip current systems, *J. Geol.*, 66(2), 103, 1958.
11. **Nielsen, P. and Cowell, P. J.,** Calibration and data correction procedures for flow meters and pressure transducers used by the coastal studies unit, *Coastal Stud. Unit, Tech. Rep. 81/1*, University of Sydney, 1981.
12. **Short, A. D.,** Wave power and beach stages: a global model, *Proc. 16th Int. Conf. Coastal Eng.*, American Society of Civil Engineers, New York, 1979, 1145.
13. **Short, A. D.,** Three dimensional beach stage model, *J. Geol.*, 87, 553, 1979.
14. **Short, A. D.,** Beach response to variations in breaker height, *Proc. 17th Int. Conf. Coastal Eng.*, American Society of Civil Engineers, New York, 1980, 1016.
15. **Short, A. D. and Hesp, P. A.,** Wave, beach and dune interactions in Southeastern Australia, *Mar. Geol.*, 48, 259, 1982.
16. **Short, A. D. and Wright, L. D.,** Beach systems of the Sydney Region, *Aust. Geol.*, 15, 8, 1981.
17. **Sonu, C. J.,** Field observations of nearshore circulation and meandering currents, *J. Geophys. Res.*, 77, 3232, 1972.
18. **Sonu, C. J.,** Three dimensional beach changes, *J. Geol.*, 81(1), 42, 1973.
19. **Sonu, C. J. and James, W. R.,** A Markov model for beach profile changes, *J. Geophys. Res.*, 78, 1462, 1973.

20. **Thom, B. G. and Bowman, G. M.,** Beach erosion at two time scales, *Proc. 17th Int. Conf. Coastal Eng.*, American Society of Civil Engineers, New York, 1980, 934.
21. **Winant, C. D., Inman, D. L., and Nordstrom, C. E.,** Description of seasonal beach changes using empirical eigenfunctions, *J. Geophys. Res.*, 80, 1797, 1975.
22. **Wright, L. D.,** Beach cut in relation to surf zone morphodynamics, *Proc. 17th Int. Conf. Coastal Eng.*, American Society of Civil Engineers, New York, 1981, 978.
23. **Wright, L. D.,** Field observations of long-period surf-zone standing waves in relation to contrasting beach morphologies, *Aust. J. Mar. Freshwater Res.*, 33, 181, 1982.
24. **Wright, L. D., Chappell, J., Thom, B. G., Bradshaw, M. P., and Cowell, P.,** Morphodynamics of reflective and dissipative beach and inshore systems: Southeastern Australia, *Mar. Geol.*, 32, 105, 1979.
25. **Wright, L. D., Thom, B. G., and Chappell, J.,** Morphodynamic variability of high energy beaches, *Proc. 16th Int. Conf. Coastal Eng.*, Hamburg, 1978, 1180.
26. **Wright, L. D., Nielson, P. N., Short, A. D., and Green, M. O.,** Morphodynamics of a macrotidal beach, *Mar. Geol.*, 50, 97, 1982.
27. **Wright, L. D., Guza, R. T., and Short, A. D.,** Dynamics of a high energy dissipative surf zone, *Mar. Geol.*, 45, 41, 1982.

Chapter 4

THE EROSION OF SILETZ SPIT, OREGON

Paul D. Komar

The severest case of recent erosion on the Oregon coast is that which has been occurring on Siletz Spit, 5 km south of Lincoln City and 160 km south of the Columbia River (Figure 1). This erosion has been documented and its causes investigated as a part of two theses,[15,18] a series of technical reports,[7,11,19] and as published papers.[8,12] The present chapter summarizes the findings of these studies together with new material on subsequent erosion events.

The erosion of Siletz Split is of particular interest in that it provides a good example of several natural factors combining to cause beach and property losses; rip currents hollowing out embayments, unusual storm waves, and tide levels. There was also an unnatural factor — beach sand mining.

DEVELOPMENT AND EROSION HISTORY

Prior to 1960, Siletz Spit appeared much as it had for at least the last century. The earliest aerial photographs of the area date back to 1939, but detailed maps go back to the late 1800s. These show a high degree of stability to the overall spit shape, with the bay mouth at the far north end identical to its present position (Figure 1). The only noteworthy long-term change is the occurrence of erosion on the bay side of the spit where water from the Siletz River impinges on the spit after it crosses the bay and just prior to exiting through the bay mouth. This bay-side erosion has significantly decreased the spit width at its already narrowest location.[18,19] It is probable that the relief of the spit has increased somewhat since the early 1930s when dune grass was introduced into Oregon,[2] resulting in foredune development and growth along much of the coast. This would inhibit spit washovers, but there is little evidence for washovers having occurred prior to the 1930s. Sea level is 'stable' along the Oregon coast (Chapter 1, Figure 8), not providing optimal conditions for the occurrence of washovers and breaching.

Although Siletz Spit extends in a northward direction with the possible implication of a net littoral drift in that direction, several lines of evidence indicate that there is in fact no net drift. Other nearby spits point toward the south, and the Oregon coast is divided into a series of embayments by large rocky headlands that would block any net drift of sand (Figure 1). A study of shoreline changes due to jetty construction most clearly demonstrates that although there is a seasonally reversing littoral drift, the net drift is zero.[9]

Development of Siletz Spit began in the early 1960s. A road was built along its length and artificial lagoons were carved into the backside of the spit. Houses began to appear in the late 1960s, and today most of the spit length is developed with houses on 28-m-wide lots. The majority of seafront houses were built in the zone of foredunes.

The first postdevelopment erosion occurred early in 1971 when a series of storms cut away at the dunes. The erosion was not severe, however, and riprap placement protected the homes before they were seriously threatened.

Major erosion occurred during the winter of 1972-73, principally associated with a storm covering much of the north Pacific from the 23rd to the 25th of December. This erosion was centered in a 350-m-long portion of the spit, threatening a series of three houses in a group as well as a fourth house under construction (Figure 2). The house under construction was left unprotected and so was undermined by the retreating dune bluff and collapsed onto the beach. Riprap was hastily placed to protect the completed homes. However, they were

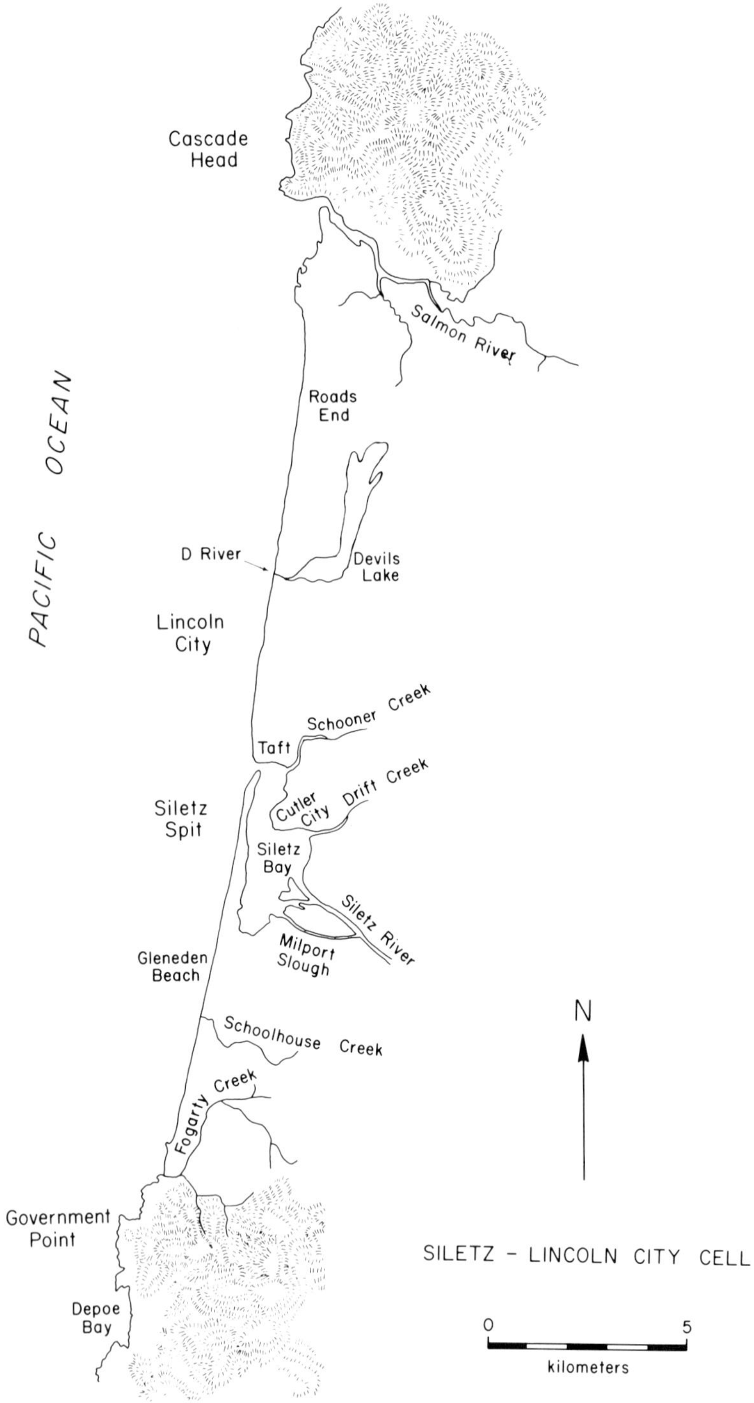

FIGURE 1. Location of Siletz Spit on the mid-Oregon coast.

initially protected only on their seaward sides, empty lots to either side being left unprotected. Dune erosion and retreat of course continued in these empty lots, flanking the riprap fronting the homes, thus necessitating riprap placement along their sides as well. Three homes thereby ended up on promontories out onto the beach, supported by riprap (Figure 2). Dune bluff

FIGURE 2. Property losses during the December 1972 storm. The house that had been under construction is lying in ruins on the beach and another house on a promentory of riprap. The dunes retreated several more meters subsequent to this photograph, entirely eliminating the lots and threatening the access road.

recession amounted to a maximum of 30 m in a 3-week period. More than a dozen lots were lost and the erosion was halted only when riprap was placed to protect the access road and utility lines.

The winters of 1973-74 and 1974-75 were mild with few storms, and the storms that did occur were less intense. Erosion returned in February 1976 and again in February 1978; both occurrences resulting from unusual storm conditions. The increased use of riprap along the spit length offered protection to the houses and none was lost. A minor washover occurred during the 1976 storm, but resulted mainly from bulldozing down the foredunes to provide an access onto the beach for equipment installing the riprap. Although some erosion occurred in the same area that eroded during 1972-73, the main erosion was centered at other stretches of the spit. As a rule, erosion does not generally occur along the entire spit length but instead focuses in one or two 300- to 600-m-long stretches, different areas being eroded by separate storms in different years. This shifting pattern of erosion results from the role of rip current embayments in the erosion process.

NATURAL PROCESSES OF EROSION

Several natural processes interact to produce the most severe episodes of erosion on Siletz

FIGURE 3. Embayments cut into the beach and foredunes on Siletz Spit leading to property losses during the December 1972 storm. As shown in the opposite photograph, each embayment is the location of a rip current which transports sand offshore.

FIGURE 3.

Spit, and at other locations on the Oregon coast. As already indicated, a long-term sea-level change is not one of them, unlike many other coastal regions and in particular the barrier islands of the U.S. east and Gulf coasts (Chapter 5). Instead, the erosion is in response to the combined actions of nearshore currents, waves, and tides.

The stage is set for coastal property erosion by the general retreat of the beach as the winter months are entered with their increased wave-energy levels and by the greater erosion occurring at rip current locations. It has been shown[1] that the Oregon beaches have a seasonal cycle, as is typically found on other coasts (Chapter 1, Figure 6). There is a general increase in the wave-energy levels from summer to winter and a notable increase in the number and strength of storms.[10] The beaches respond in the usual way with sand shifting offshore and a decrease in the berm width so that the wave swash zone becomes closer to the coastal property. The seasonal shift in beach profiles at Siltez Spit is somewhat larger than normal due to the coarseness of the beach sand (median diameter = 0.4 mm), and when a particular storm hits the coast the beach profile responses are large and rapid.

In the case of Siletz Spit, more important than the general beach retreat during the early winter is the much larger retreat at the rip current positions. As described in Chapter 1, the rip currents transport sand offshore and in the process hollow out embayments, giving rise to a rhythmic shoreline of alternating bays and cusps. Typically, the bays are not all developed to the same degree so the shoreline may become quite irregular (Figure 3). In some years the largest of the embayments may cut back entirely through the beach and begin to impinge on the coastal property. This is the case for the central embayment shown in Figure 3, photographed at the time of major erosion in December 1972.

The exceptionally large rip-current embayments thereby become the focal points for property erosion, explaining why the severest erosion is generally limited to 300 to 600 m of spit length. Since the rip currents have different positions from one year to the next, their occurrence also explains the shifts in maximum erosion.

The rip-current embayments reach their greatest development when the rips are stable in position. That occurs when the waves are predominantly parallel to the coastline (Chapter 1). An oblique wave approach tends to produce a longshore migration of the rip currents which inhibits embayment development. However, once the embayments are reasonably well developed the rips become locked in position by the topography, even under an oblique wave approach.

The general retreat of the beach and development of rip-current embayments as the winter months are entered controls, in large part, the time of year when severe episodes of property erosion can occur. Although major storms commonly hit the coast in October and November, these seldom result in property losses as there is still enough beach berm to act as a buffer. But by December the rip embayments may have cut through the beach and have reached the property, so that storms during December through about March can produce property losses.

In themselves, the rip-current embayments cause very little direct property erosion. But their presence sets the stage for erosion by wave attack. In addition to hollowing out the embayments, the rip currents deepen the adjacent offshore water by cutting through offshore bars. As a result, the ocean waves are able to travel into the embayments and break very close to shore, producing more intense swash and greater runup within the embayments. The rip embayments thereby become the focal points for greater wave attack.

Severest erosion at Siletz Spit occurs when unusual storm waves are generated at a time of maximum rip embayments. Waves are measured on a routine basis four times a day at the OSU Marine Science Center in Newport, 32 km south of Siletz Spit. The measurements are obtained with a microseismometer system which has been in operation since 1971. Based upon the work of Longuet-Higgins[14] on the generation of microseisms by reflected ocean waves, the system was calibrated by empirically correlating the microseism records with direct wave measurements obtained with a pressure transducer, so that now the microseism record alone yields values of the significant wave height, H_s, and period.[5,17] Such a ''remote-sensing'' mechanism is required on the high-energy Oregon coast as it is difficult to maintain a pressure transducer in operation for very long. Data from the system are summarized in Komar et al.[10] and in Creech.[3] There is an annual cycle in the wave-energy level, wave breakers averaging about $H_s = 2$ m during the summer months and 3 to 4 m during the winter. But important to coastal erosion are the exceptional storms that generate breakers with significant wave heights in excess of 7 m. These have occurred only five times since the microseismometer system went into operation (1971 to 1981). Severe erosion resulted in four of these five occurrences, the fifth storm being too early in the season to cause property losses.

Komar and McKinney[11,12] analyzed the storm systems of the 23rd to 25th of December 1972 and the 17th of February 1976 that generated high wave conditions and caused erosion on Siletz Spit and elsewhere along the Oregon coast. These two storms were contrasting in character, demonstrating that different circumstances can lead to large waves. The storm system of December 1972 covered most of the north Pacific, the low-pressure center being located just south of Alaska, so that eastward-directed winds occurred across thousands of kilometers of ocean and were aimed directly at the Pacific Northwest coastline. Thus, the storm provided a tremendous fetch and duration and transferred considerably energy into the waves even though winds of only 30 to 40 kn were recorded by ships. In contrast, the storm of February 1976 was characterized by high winds (50 to 60 kn) and a much smaller fetch area, but still a long duration in that the associated storm front moved rapidly across the Pacific, keeping pace with the generated waves and so was able to continue to pump energy into them. Waves hindcasted with the computer-prediction analysis developed by Enfield[5] showed reasonable comparisons with the waves actually measured at Newport.

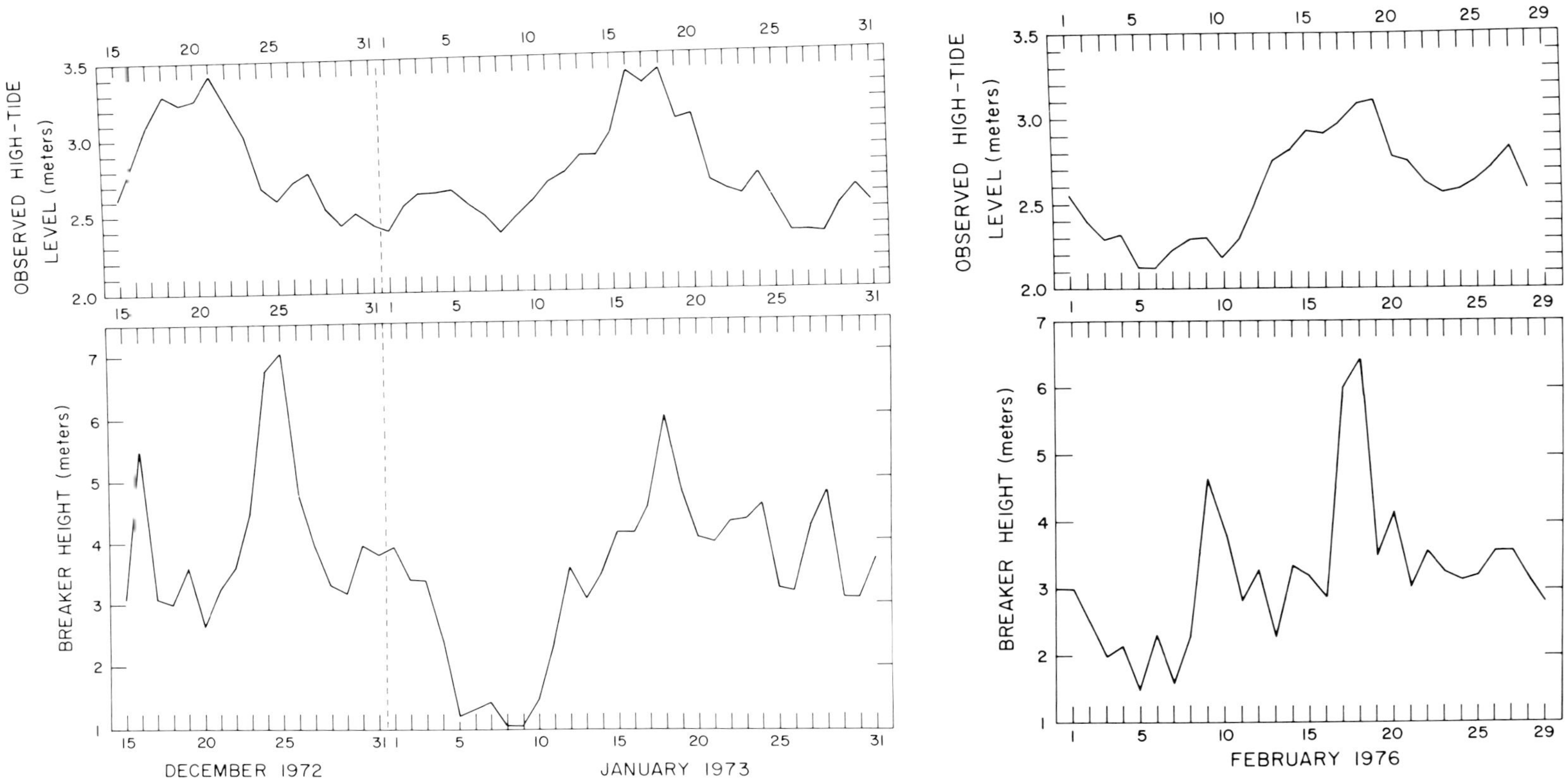

FIGURE 4. Measured high tide levels and breaker heights during the December 1972 and February 1976 storms which caused erosion on Siletz Spit. Although having greater wave heights, the 1972 storm occurred at a time of neap tides whereas the 1976 storm took place under near spring tide conditions.

A third factor affecting the amount of resulting erosion is the state of the tide levels. This factor is difficult to establish quantitatively, but intuitively the higher waters associated with spring tides raise the water levels closer to or higher up on the coastal properties, thereby enhancing the erosion (Chapter 1). This seems to have been the case in the erosion that has occurred at Siletz Spit. Neap tides prevailed during the December 1972 storm whereas tides were nearly at spring levels during the February 1976 storm (Figure 4). Due to the higher water levels of the 1976 storm, the one small washover of Siletz occurred, and many drift logs were thrown atop the dunes, neither of which occurred during the 1972 storm. In spite of the waves being somewhat smaller in the 1976 storm than in 1972 the dune retreat was comparable, the higher tide levels of 1976 perhaps making up for the lower wave heights. Large +7-m storm waves combined with high spring tides in February 1978 to produce their strongest combined effects. By 1978 much of Siletz Spit had been riprapped so erosion was confined to only the northerly portion of the spit. But this is also the narrowest portion, and breaching nearly occurred. Under the same storm, Nestucca Spit, 32 km to the north, did breach although it has subsequently closed.[6] This was the first historic natural spit breaching on the Oregon coast (not counting that at Bayocean Spit, which was caused by jetty construction). The unusual combination of large rip embayments, high storm waves and spring tides seems to have been responsible for this rare breaching event.[6]

One other natural process that could potentially play a role is a storm surge, raising the water to still higher levels than produced by the tides alone (Chapter 1). However, the analysis of tide measurements at Newport for the 1972 and 1976 storms revealed no significant storm surges.[11] This was because the erosion waves were generated by distant storm fetches, not by local onshore coastal winds that are necessary to produce a storm surge. However, it is possible that a storm surge could become a significant factor in future erosion events if the circumstances are right. Storm surges are certainly an important factor on other coasts, especially those affected by hurricanes (Chapter 5).

CYCLES OF EROSION

During the erosion episodes it was noted that numerous drift logs became exposed in the retreating dune bluff, and, of particular interest, many had been cut with a saw, clearly indicating that the foredune areas upon which houses were constructed must have accreted recently.

This was confirmed by an investigation utilizing available aerial photographs, 1939 to the present.[18,19] These revealed many past erosion eposides, much as had occurred in 1972, 1976, and 1978 described above. Except in about 1960, when erosion occurred simultaneously over much of the spit length, the erosion was concentrated in 100- to 600-m-long segments indicating a role played by rip-current embayments. The aerial photographs also revealed a cycle of deposition following a particular erosion occurrence. After erosion removed the foredunes, drift logs accumulate in the eroded areas (Figure 5). These crisscrossing masses of logs are effective in trapping sand washed there by waves or blown into the zones by the predominately onshore winds. Within less than ten years the foredunes were thereby naturally restored. Therefore, a natural cycle existed with the erosion followed by dune reformation, only to have the erosion reoccur. Most of the homes built on the spit are in this area of foredunes that has been repeatedly eroded, in many cases less than ten years prior to development.

The aerial photographs also indicate that throughout the cycles of foredune erosion and reformation there was not a measurable net landward migration of the spit as indicated by the dune edge. This contrasts with the East and Gulf coasts where there are significant net erosion rates (Chapter 14). It would appear that the differences in local ''sea level'' changes are responsible.

FIGURE 5. Drift logs washed into an embayment cut by a rip current into the foredunes on Siletz Spit, now actively trapping wind-blown sands and beginning to reform the dunes.

This natural erosion-deposition cycle on Siletz Spit has now been eliminated by the placement of riprap along much of the spit length. Although the riprap has been successful in protecting the homes, its installation has been expensive and it has to be restored frequently due to rock losses under wave attack. Also, its placement has detracted from the natural beauty of the area and has inhibited access to the beach. Since the erosion episodes involved only the seawardmost 30 m of foredunes, a preferable solution would have been to establish set-back lines, keeping homes out of this ephemeral zone. No costly riprap would have been needed and the natural beach-foredune environment could have been maintained.

COMPARISONS WITH OTHER AREAS

The processes of erosion on Siletz Spit occur elsewhere on the Oregon coast and in other coastal areas. However, there are some differences which result principally from contrasting beach sediment grain sizes and thus beach morphology. As pointed out above, the beach at Siletz Spit is unusually coarse and is therefore steeper than most Oregon beaches. There are also differences in the morphologies of the rip-current embayments. Although narrow and deep embayments can be cut by rip currents into fine-sand low-sloping beaches, more commonly the embayments are broader and shallower than at Siletz Spit. When they play a role in focusing wave attack the resulting erosion tends to extend over a longer stretch of coast. A good example is the erosion that occurred on Nestucca Spit during the February 1978 storm which also eroded Siletz Spit.[6] Other than the rip-current embayments being wider, the processes of erosion were much the same. In this case, however, the erosion led to the breaching of the spit as already described. The eroding foredunes similarly exposed sawed drift logs indicating a comparable cycle of foredune erosion and reformation.

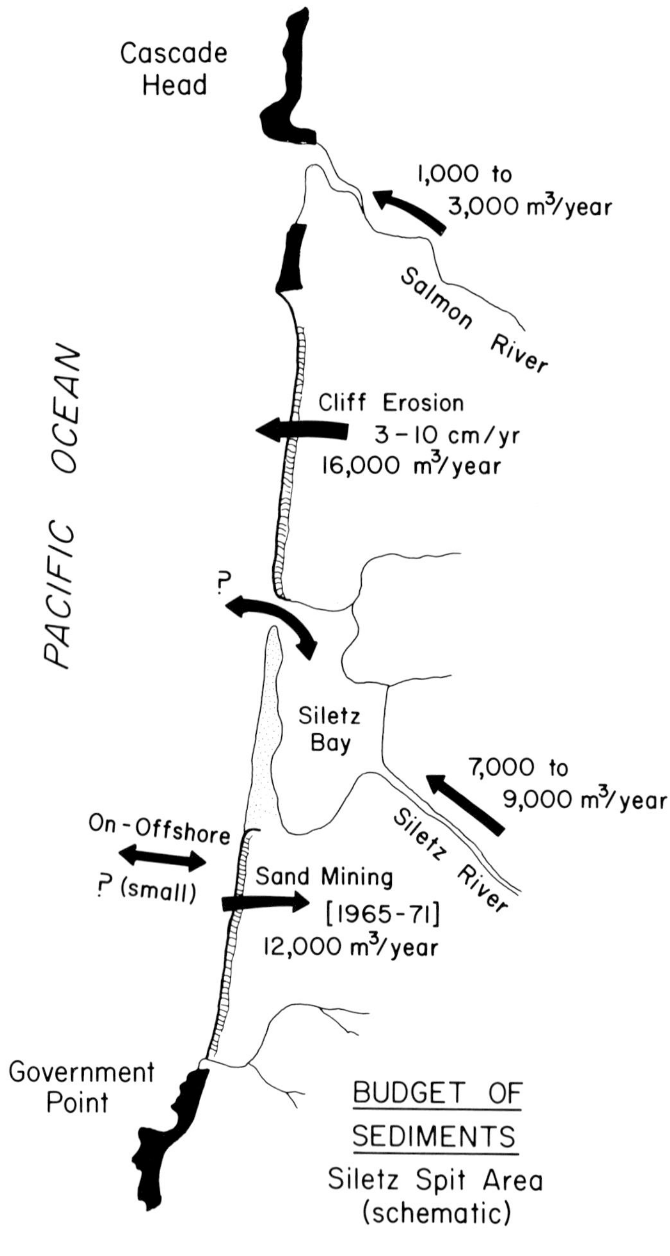

FIGURE 6. An approximate budget of beach sands for the stretch of coast including Siletz Spit. Shown are estimates of the various sources and sinks of beach sand, including that due to sand mining.

Similar observations have been made on other coasts. Dolan,[4] in particular, documented the role of shoreline embayments in the erosion of Cape Hatteras, N.C., embayments which were probably formed by rip currents. The investigations of Wright et al.[21] and Short[20] in Australia, summarized in Chapter 3, similarly document the importance of rip currents in beach erosion, formulating a cycle for the overall beach configuration which involves the presence or absence of rip currents and their embayments.

BEACH-SAND MINING AND THE BUDGET OF SEDIMENTS

One additional factor in the erosion of Siletz Spit has been a sand mining operation, removing sand from the beach at Schoolhouse Creek, 3 km south of the spit (Figure 1). In operation from 1965 through 1971, the mining records indicate that on average 12,000 m^3 of beach sand were removed each year. Beach-sand mining can of course decrease the total volume of sand on the beach, thereby reducing its effectiveness as a buffer between the ocean waves and coastal properties.

One way to evaluate the significance of this 12,000 m^3/year sand loss is to develop a budget of littoral sediments for the area, an approach discussed in general terms in Chapter 1. Siletz Spit is part of a 24-km-long stretch of beach extending from Cascade Head on the north to Government Point on the south (Figures 1 and 6). It is unlikely that beach sand can bypass these headlands, so we have an ideal cell in which to confine our sand budget estimations.

It would appear that the Siletz River is the principal contributor of sand to the beach. However, a series of investigations have shown that very little riverine sand is able to bypass the Oregon coast estuaries. Instead, the sand appears to become permanently deposited within the estuaries. Utilizing heavy minerals and grain roundness as tracers of sand sources and final deposition areas, Kulm and Byrne[13] demonstrated that this is the case for Yaquina Bay at Newport, and Peterson et al.[16] show that it is also true for Alsea Bay. Our unpublished results similarly indicate that little if any sand from the Siletz River is reaching the adjacent ocean beaches. Rather than being a sand source, Siletz Bay is in fact a sink for beach sand in that tidal currents carry beach sand up into the estuary where it appears to be permanently deposited. Evaluation of this sink is not presently possible, but it is probably small when considered on a yearly basis.

The main contribution of sand to the beach comes from sea-cliff erosion. Excepting the headlands which are composed of resistant basalt, most of the beach is backed by cliffs cut into Pleistocene marine terraces composed of shallow marine, beach and dune sands. They are retreating on average by about 10 cm/year supplying some 16,000 m^3/year of sand that is sufficiently coarse to remain on the beach.

Prior to sand mining the beach appears to have neither increased nor decreased in overall width over the years, indicating that the natural losses of sand approximately balance the gains from the sources; these losses must be to the offshore, perhaps in the past to dune growth on Siletz Spit, and some beach sand movement into Siletz Bay.

The budget of sediments for the Siletz Spit area is summarized in Figure 6, and although we have been unable to evaluate many of the sources and sinks, the results reveal that the sand-mining operation became a major component in the budget, being on the order of the volume of the principal source. In that there appears to have been a natural balance prior to mining, the sand removal by mining most likely tipped the balance of the budget into the red, resulting in a long-term cumulative loss of beach sand volume.

Beach-sand mining, however, could not have been the principal factor in the erosion of Siletz Spit. As discussed above, the old aerial photographs reveal that the spit suffered many episodes of erosion prior to development and prior to beach-sand mining. Although not the primary factor, the beach-sand mining certainly was an aggravating factor in producing increased amounts of erosion on Siletz Spit. For this reason, the state of Oregon did not renew the operater's permit, and sand removal ceased in 1972.

The budget of sediments for the Siletz Spit area, Figure 6, points out another possible long-term problem. The sea-cliff erosion is threatening many homes and commercial buildings in Lincoln City, Taft, and Gleneden Beach. As a result, riprap and other defensive structures are being placed to stop this erosion. Although this effort does help to protect the property, it is apparent that it will also cut off the main source of sand to the beach, the sand supplied by the cliff erosion. It is probable that we may see the effects of this loss in future years, assuming that the natural sinks continue to operate. The changes will be gradual and cumulative over the years, and decades may elapse before there is a noticeable decrease in the beach width.

REFERENCES

1. **Aguilar-Tunon, N. A. and Komar, P. D.,** The annual cycle of profile changes of two Oregon beaches, *Ore Bin,* 40(2), 25, 1978.
2. **Cooper, W. S.,** *Coastal Sand Dunes of Oregon and Washington,* GSA Mem. 72, Waverly Press, Baltimore, 1958.
3. **Creech, C.,** *Nearshore Wave Climatory, Yaquina Bay, Oregon (1971-1981),* Sea Grant Coll. Prog., ORESU-T-81-002, Oregon State University, Corvallis, 1981.
4. **Dolan, R.,** Coastal landforms: crescentic and rhythmic, *Geol. Soc. Am. Bull.,* 82, 177, 1971.
5. **Enfield, D. B.,** Prediction of Hazardous Columbia River Conditions, Ph.D. thesis, School of Oceanography, Oregon State University, Corvallis, 1974.
6. **Komar, P. D.,** Wave conditions on the Oregon coast during the winter of 1977-78 and the resulting erosion of Nestucca Spit, *Shore Beach,* 46(4), 3, 1978.
7. **Komar, P. D. and Rea, C.C.,** *The Causes of Erosion to Siletz Spit, Oregon,* Sea Grant Col. Prog., ORESU-T-75-001, Oregon State University, Corvallis, 1975.
8. **Komar, P. D. and Rea, C. C.,** Erosion of Siletz Spit, Oregon, *Shore Beach,* 44(1) 9, 1976.
9. **Komar, P. D., Lizarraga, J. R., and Terich, T. A.,** Oregon coast shoreline changes due to jetties, pap. 11933, *J. Waterways Harbors Coastal Engr.,* 102(WW1), 13, 1976.
10. **Komar, P. D., Quinn, W., Creech, H. C., Rea, C. C., and Lizarraga-Arciniega, J. R.,** Wave conditions and beach erosion on the Oregon coast, *Ore Bin,* 38(7), 103, 1976.
11. **Komar, P. D. and McKinney, B. A.,** *The Spring 1976 Erosion of Siletz Spit, Oregon,* Sea Grant Coll. Prog., ORESU-T-77-004, Oregon State University, Corvallis, 1977.
12. **Komar, P. D. and McKinney, B. A.,** The Spring 1976 erosion of Siletz Spit, Oregon, with an analysis of the causative storm conditions, *Shore Beach,* 45(3), 23, 1977.
13. **Kulm, L. D. and Byrne, J. V.,** Sedimentary response to hydrology in an Oregon estuary, *Mar. Geol.,* 4, 85, 1966.
14. **Longuet-Higgins, M. S.,** A theory of the origin of microseisms, *Philos. Trans. R. Soc. London Ser. A,* 243, 1, 1950.
15. **McKinney, B. A.,** The Spring 1976 Erosion of Siletz Spit, Oregon, with an Analysis of the Causative Wave and Tide Conditions, M.S. thesis, School of Oceanography, Oregon State University, Corvallis, 1977.
16. **Peterson, C., Scheidegger, K. F., and Komar, P. D.,** Sand dispersal patterns in an active margin estuary of the northwestern United States as indicated by sand composition, texture and bedforms, *Mar. Geol.,* in press.
17. **Quinn, W. H., Creech, H. C., and Zopf, D. E.,** Coastal wave observations via seismometer, *Mar. Weather Log,* 18, 367, 1974.
18. **Rea, C. C.,** The Erosion of Siletz Spit, Oregon, M.S. thesis, School of Oceanography, Oregon State University, Corvallis, 1975.
19. **Rea, C. C. and Komar, P. D.,** *The Erosion of Siletz Spit, Oregon,* Ref. 75-4, School of Oceanography, Oregon State University, Corvallis, 1975.
20. **Short, A. D.,** Three dimensional beach-stage model, *J. Geol.,* 87, 553, 1979.
21. **Wright, L. D., Chappell, J., Thom, B. G., Bradshaw, M. P., and Cowell, P.,** Morphodynamics of reflective and dissipative beach and inshore systems: Southeastern Australia, *Mar. Geol.,* 32, 105, 1979.

Chapter 5

BARRIER ISLANDS

Dag Nummedal

Barrier islands defend low-lying coastal lands against storms, they enclose and protect the rich natural resources of estuaries and marshes, and their sandy beaches and jungle-like interior provide scenic qualities unparalleled elsewhere in the coastal zone. In recent decades, barrier islands have increasingly become the focus of intense real estate development. The desirability of a beach home is obvious, yet the fragile nature of the barriers and their associated environments does not permit high-density development without seriously degrading the natural ecosystem.

The resolution of conflicts regarding barrier island utilization must be based on a full understanding of the dynamic nature of these islands. This chapter attempts to provide such an understanding by reviewing the state of knowledge regarding the physical processes that have led to the development of our present barriers, and the natural forces of sediment dispersal which continue to modify them. Although barrier islands are of bewildering complexity at first glance, it is the thesis of this paper that their stratigraphy and morphology are controlled by relatively few factors, the most important of which are (1) Holocene sea level history, (2) the longshore distribution of sediment sources and sinks, (3) the volume of water being exchanged through flanking tidal passes (the tidal prism), and (4) the nearshore wave energy.

Most of the North American barrier islands appear to have had fluvial deltaic (or related) sand bodies as their ultimate source. Through erosion of such deposits and associated local shoreline retreat, sediments have been provided for the littoral transport system to build wide beach-ridge barriers in more sheltered locations. A transgressive barrier stratigraphy typically has developed near the headlands, a regressive stratigraphy may prevail some distance away. The Holocene coastal stratigraphy is replete with examples of alternating periods of transgressive and regressive development. Rapid fluctuations both in sea level and sediment supply rates have contributed to this state of affairs. The present situation of widespread shoreline erosion is therefore not geologically unique. The current rapid rise in local sea level along many segments of the American coastline is probably a major cause. Man can do little (in the short run) to change the trends of global sea level rise. We can, however, take steps to reduce the rate of land subsidence, a factor that contributes greatly to local changes in the relative elevation of sea and land.

With an understanding of the dynamics of coastwise sand transport we can also take proper measures to avoid inadvertent accelerations in shoreline retreat. Barrier islands are part of a physical continuum. Changes in the system at one place have effects down the entire chain. This linkage between barriers is most readily observed near tidal passes where sand from an updrift beach is transferred to the downdrift island via a series of complex wave-current interaction mechanisms. A man-made perturbation of an inlet invariably has effects on the stability of the adjacent island beach.

The longshore profile of an island (the overall shape of its ocean front) is in fact controlled by sediment dispersal within the tidal passes. With an increase in either tidal prism or wave energy one can recognize systematic changes in barrier island characteristics. The existence of such correlations call for a search for physical causes and effects. This research is complex and much remains to be done, yet present understanding permits us to predict, with some degree of confidence, where the stable and unstable barrier beaches are located. Such stability assessments should be made the basis for land-use planning on all coastal barriers.

Barrier islands protect the coastal wetlands during storms, yet are themselves considerably modified in the process. This modification may be transient, as in the case of high-profile barriers where the sand is moved from the front dune-scarp to the nearshore bars during the storm just to be subsequently returned to the beach; or it may be permanent, as on low-profile barriers where the sand is washed across the barrier flats by the storm surge and dumped in the back-barrier environments. Such islands suffer net landward movement during each major storm. The damage potential to man-made structures is clearly different on these two barrier types.

The intent of this chapter is to synthesize the results of recent years' research pertinent to the problem areas just discussed. The author hopes to demonstrate that our understanding of barrier island evolution and continuing modification, although incomplete, is adequate for the increasingly pressing task before us: to provide the scientific basis for an optimum utilization of America's barriers, recognizing that their role in our economy derives from their recreational, scenic, ecologic, and strategic values.

GEOLOGIC SETTING OF NORTH AMERICAN BARRIER ISLANDS

A barrier island is an unconsolidated elongate body of commonly sandy (or gravelly) sediments lying above high tide level and separated from the mainland by a lagoon or a marsh. (See Shepard,[91] Kumar,[51] or Coates[11] for more detailed definitions.) A glance at the map of North America will convince one that such features are not distributed at random along our coasts. Barrier islands prevail along the east and Gulf coasts of the U.S., along the Arctic coast of Alaska, and locally in eastern Canada and the Gulf of Alaska.

This distribution owes its origin to the regional tectonic evolution of the continent. During Mesozoic and Cenozoic time (the last 225 million years) the North American continent has developed a distinct asymmetric relief: the west is high and mountainous, the east is composed of large interior plains and coastal lowlands. Alaska is similarly asymmetric with the coastal lowlands in the Arctic. This terrain distribution is a direct consequence of the movement and geometry of the North American plate. The west was adjacent to an active convergent margin throughout much of this time (the Oregon and Washington coasts constitute a convergent margin even today); the east has been within the stable center of the plate ever since the breakup of Pangea in the Triassic.[19] In terms of this tectonic picture it has become common to refer to the American west coast as a "leading edge coastline" whereas, the east and Gulf coasts form "trailing edge coasts".[46]

Because of the topographic asymmetry and the stability of the intraplate eastern margin, the trailing edge coastlines have received the bulk of the sediment eroded off the continent. Most of this sediment has been deposited in shallow coastal and nearshore settings along the gradually subsiding continental margin geosynclines, forming low-relief coastal plains and wide continental shelves. As the sea level rose and fell in response to tectonically induced changes in the volume of the ocean basins[41] and waxing and waning of continental ice sheets[58], the shorelines migrated back and forth across the continental margin. Barrier islands were common features along these past shorelines, a fact demonstrated by the existence of numerous barrier island deposits throughout the coastal plain stratigraphic sequence. Some of these ancient barriers form distinct sandy topographic highs on the present coastal plain,[5,106] others form some of the best hydrocarbon reservoirs in the subsurface of the Gulf Coast basin.[30]

At present, barrier islands dominate the coast along the slowly subsiding, passive margins of the North American continent. The barrier sediments were initially deposited in fluvial deltas, then redistributed by marine currents and waves into barrier spits, islands, and bars. The sand was blown from the beach by onshore winds and trapped by grasses and maritime forests. Many of our present barriers came into existence on what is today the continental

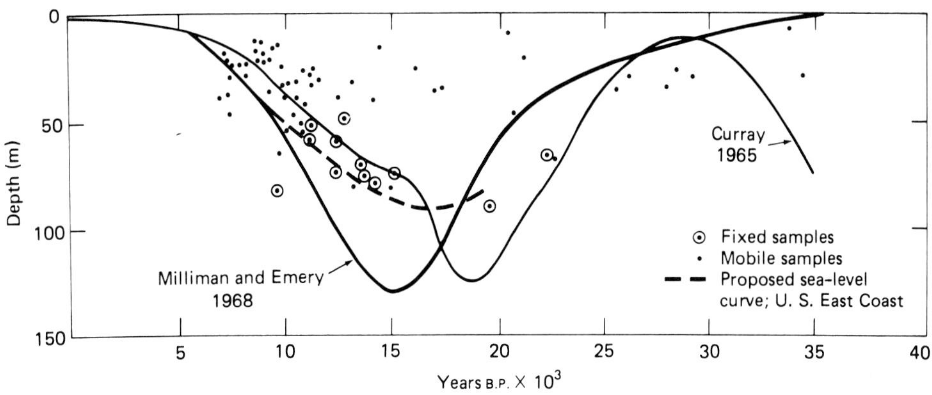

FIGURE 1. Sea level curves for the late Quaternary inferred from radiocarbon-dated samples along the coast of the U.S. The most recent curve (dashed in figure) suggests a late Wisconsin low stand of less than 100 m below present sea level. (From Dillon, W. D. and Oldale, R. N., *Geology*, 6, 56, 1978. With permission.)

shelf. The "primary barriers" may have formed thousands of years ago when the sea level was lower than at present. As the sea level rose during the Holocene many barriers migrated landward to their present positions.[24,53,70,97]

SEA LEVEL CHANGES

The sea level controls the location of the shoreline on the continental margin. Furthermore, the rate with which sea level has been changing relative to the local sedimentation or erosion rate determines the stratigraphy of the coastal deposits as well as the topographic profile of the barrier island surface. Sea level fluctuations and sediment supply constitute the primary factors controlling barrier development. Second-order variability in the coastal system is caused by longshore variations in tidal and wave energy, factors which will be discussed later in this chapter.

Sea level, that universal elevation datum, is neither level nor constant. Spatial and temporal fluctuations in sea level occur at all scales and frequencies. Global sea-level variations on a time scale of tens of millions of years occur in response to changes in the volumes of the ocean basins and subsidence of the continental margins. On passive margins (such as the east and Gulf coasts of the U.S.) the rates of tectonically controlled changes in the relative elevations of sea and land are quite slow, typically a few millimeters per 1000 years.[84] Superimposed on these essentially tectonic sea-level changes are higher-frequency fluctuations of a multitude of origins. Periodic formation of continental ice sheets and attendant deglaciations have, at least since the Pliocene and possibly throughout the Cenozoic,[58] been responsible for major sea level changes of a relatively high frequency. Recent seismic stratigraphic investigations by Beard et al.[1] have provided evidence that there may have been as many as three glaciations with attendant major lowerings of global sea level during the last 100,000 years.

Present sea level appears to be at an elevation comparable to that reached during earlier major interglacials. The latest low stand of sea level occurred at the end of the late Pleistocene Wisconsin glaciation. Early sea-level curves[14,61] indicated that this low stand was as much as 130 m below present sea level. Recent work by Dillon and Oldale[20] and Blackwelder,[6] however, strongly suggests that sea level may have risen much less than 100 m since the late Wisconsin low (Figure 1).

Regardless of the absolute magnitude of the sea level rise over the last 18,000 years, the

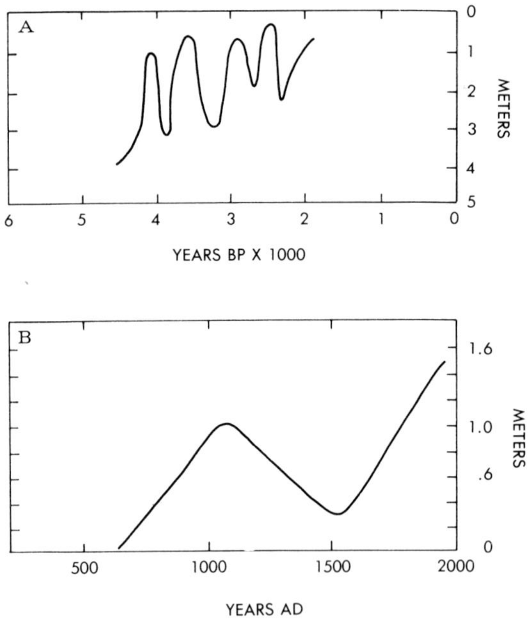

FIGURE 2. A. Sea level fluctuations on the central South Carolina coast over the last 4000 years. Curve is based on radiocarbon-dated archaeological samples and basal peats. (Modified from Brooks, M. N., Colquhoun, D. J., Pardi, R. R., Newman, W., and Abbott, W. J., *Fla. Anthropologist,* 32, 85, 1979. With permission.) B. Sea level fluctuations on the North Sea coast of Germany since 650 A.D. The curve is based on historical data. (Modified from Rhode, H., *Die Küste,* 32, 7, 1978. With permission.)

''Holocene transgression'' is responsible for the existence of a multitude of coastal sedimentary sequences (deltas, fluvial channel fills, marsh deposits, tidal channel fills, and transgressed barrier islands) on the present shelf floor.[13,23,83,98] At the time of maximum ice retreat, the sea level rose at a rate of about 1 m/century, a rate which is about 4 orders of magnitude faster than the longer-term tectonic sea level changes.

The sea level has been subjected to frequent fluctuations of great amplitude throughout the Cenozoic. Recent data also increasingly support the view that the latest rise of sea level, the Holocene transgression, was not a smooth, monotonous rise of the sea. Rather, the rise appears to have been characterized by a series of oscillations with an amplitude of a few meters on a typical time scale of hundreds of years (Figure 2A). Data supporting this view are both of archaeological and geological nature.[9,21,62] Historical data in Europe suggest that there has been a 1-m sea level fluctuation within the last millenium.[87] The most recent sea level minimum corresponds to the peak of the ''little ice age'' at the end of medieval time (Figure 2B).

Over the last century an increasing number of tide gages have been installed in harbors around the world. Records from such gages yield information about the local relative change in level of the sea and the land upon which the gage is placed. All such records demonstrate a large fluctuation in mean annual sea level, and generally these fluctuations are superimposed on a secular trend. The annual fluctuations derive from long-term meteorological tides (atmospheric pressure variations), continental runoff, and winds.[22] The longer term trend is more controversial, yet of paramount significance to human efforts at developing the coastal zone.

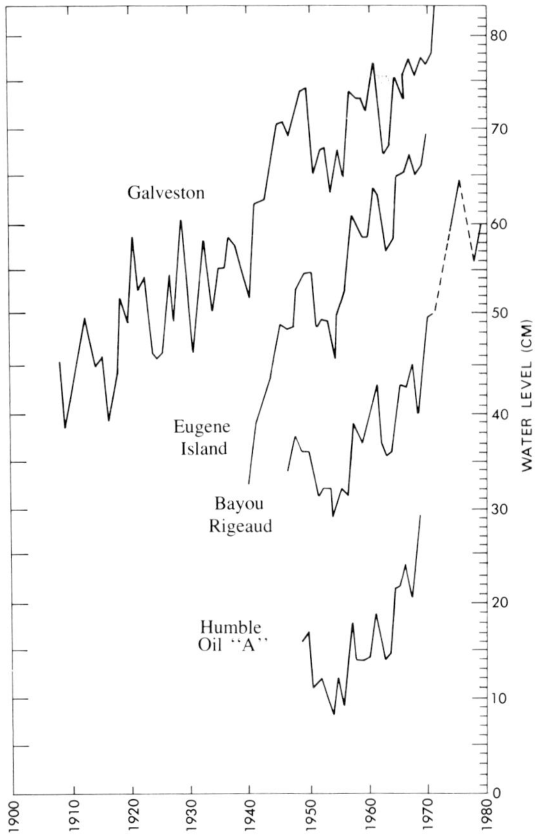

FIGURE 3. Yearly mean sea level series for four stations along the northcentral Gulf coast. Data from from tide gages at Galveston, Eugene Island (at the entrance to Atchafalaya Bay), Bayou Rigeaud (Grand Isle), and Humble Oil Platform "A" (13 km off Grand Isle). (From Nummedal, D., *Proc. Conf. Coastal Erosion Wetlands Modification in Louisiana*, Louisiana Universities Marine Consortium, Cocodrie, 1982, 164; U.S. Fish and Wildlife Service, FWS/OBS-82-59, Washington, D.C.)

Figure 3 presents three tide gage records from the central Louisiana coast as well as a longer time series from Galveston, Tex., all of which document a history of rapid local relative sea-level rise. The longer Galveston record documents well the temporal changes in observed rates of sea-level rise. For example, if the entire Galveston record is averaged, one finds a rate of rise of 5.5 mm/year. If one only considers the 20–year time span from 1950 to 1970, the rate then was 2.5 mm/year. The rapid local change in sea level at Galveston between 1940 and 1945 might be due to man's activities in the area, although sea-level curves from as far away as Pensacola show a rapid rise during the same time-period.[42] In view of these frequent changes the predicted subsidence rates in the following paragraphs should be considered very tentative.

From the Humble Oil "A" and the Bayou Rigeaud tide gage records (Figure 3) one finds a rate of local sea-level rise of between 10 and 11 mm/year for the period of duration of the two records. By subtracting a rate of 1.2 mm/year for eustatic rise,[33] one arrives at a subsidence rate of about 9 mm/year for the southcentral Louisiana coast. Farther west, at Eugene Island at the entrance to Atchafalaya Bay, one finds a subsidence rate of 7.3 mm/year.

A longer-term average subsidence rate can be derived from a ^{14}C-based local relative sea-level curve determined for the Caminada-Moreau beach ridge plain in southern Lafourche Parish by Gerdes.[31] Gerdes' data suggest that local relative sea level in that region has risen a total of 2.75 m during the last 1000 years. This corresponds to an average rate of 2.75 mm/year. If one assumes a eustatic rise of 0.5 m over the same time interval (Figure 2B) then one finds a local long-term subsidence rate of 2.25 mm/year. This is much less than the rate derived from local tide gages, an observation which has two alternative interpretations: (1) natural processes of subsidence in coastal Louisiana are highly time dependent, or (2) the rapid subsidence over the last few decades is largely man-induced. Whichever cause is the dominant, neither is likely to alter the current subsidence rate dramatically over the next century. One should therefore expect continued rapid rise in local sea level along the Louisiana coast.

Recent modeling of the climatic impact of the increased CO_2 content in the terrestrial atmosphere suggests a global warming of perhaps 2 to 3° C over the next century.[34,57] This warming is predicted to have a significant effect on eustatic sea level both because of further melting of polar ice-sheets and because of the steric expansion of the upper layer of the ocean predicted by Gornitz et al.[33] In fact, this prediction suggests that the eustatic rise in the sea level over the next few decades may be on the order of 1 cm/year.

These estimates of future sea level changes do not represent a worst-case scenario, they represent our best current predictions. The studies suggest that the rapidly subsiding Louisiana coast may experience a local sea-level rise (eustatic plus subsidence) of about 2 cm/year over the next few decades.[73] The more stable U.S. east coast and Gulf coast regions flanking the Mississippi delta may experience a rate of rise of about 1 cm/year. With such rates of sea-level rise it is imperative that plans for development and stabilization of the American barrier islands take sea-level changes into account.

EVOLUTION OF HOLOCENE BARRIERS

Stratigraphy

The distinction between transgressive and regressive stratigraphic sequences is fundamental to the understanding of the evolution of a barrier island shoreline. A transgressive sequence is one where the shoreline has moved in such a way that sediments deposited in seaward environments stratigraphically end up on top of sediments from more landward environments. A regressive sequence, on the contrary, is one where landward sediments are deposited on top of more seaward ones. During a transgression the shoreline moves landward; during a regression it moves seaward. Curray[13] formalized the connection between sea-level changes, the rate of sedimentation or erosion, and the consequent formation of transgressive and regressive coastal deposits (Figure 4). Sea-level change refers to local relative sea level. It makes no difference to the pattern of coastal sedimentation whether sea level change is eustatic (global) or due to local factors of subsidence.

Curray[13] envisioned that the local shoreline is going to transgress if the rate of sedimentation is less than the local rate of sea-level rise. Conversely, the shoreline will regress if the vertical rate of sedimentation exceeds the rate of sea-level rise. It is essential to realize that coastal evolution is not controlled directly by the absolute rates of sea-level rise or fall, but by the relative rates of sedimentation and sea-level change. In general, of course, a large-scale rise of the sea level will cause widespread transgression and sea-level fall will cause regression. However, during the general Holocene transgression many local, thick, regressive sedimentary deposits have been formed. The extensive growth of the south Louisiana coast due to Mississippi delta sedimentation over the last 6000 years is a good example. In fact, during the Holocene evolution of the Louisiana coastal plain regressions have occurred within active delta lobes concurrent with transgressions of the adjacent shorelines within

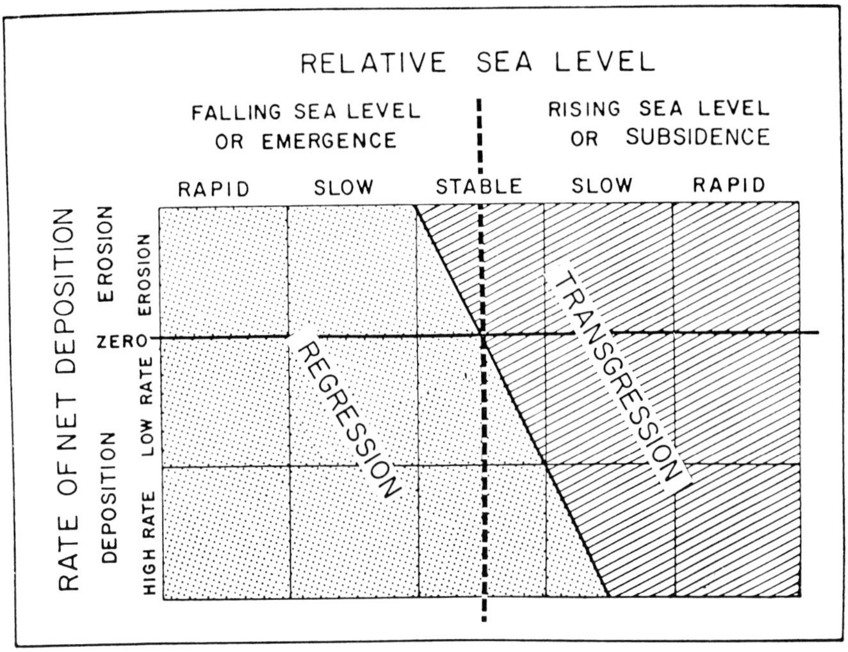

FIGURE 4. Coastal transgressions and regressions are defined as functions of local relative sea level rise and the rate of sediment deposition or erosion. (From Curray, J. R., in *Papers in Marine Geology — Shepard Commemorative Volume*, Miller, R. L., Ed., Macmillan, New York, 1964, 175. Copyright© 1964 MacMillan Publishing Co., Inc. With permission.)

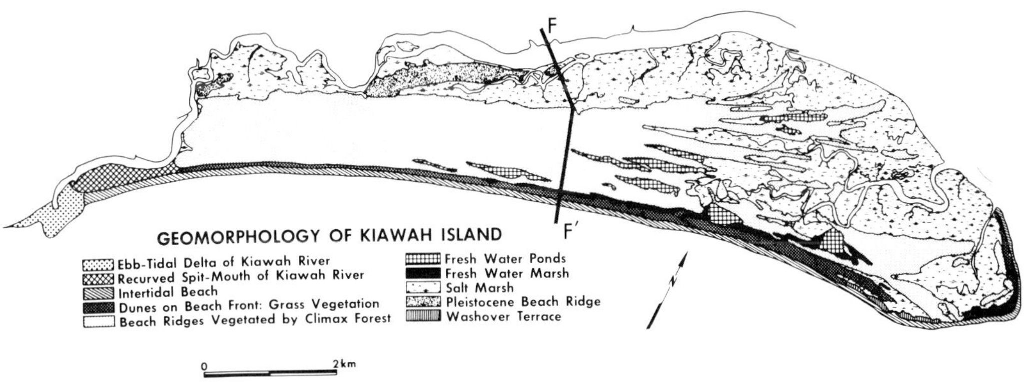

FIGURE 5. Geomorphic map of a tide-dominated, regressive beach-ridge barrier island on the South Carolina coast. Line F-F′ is the location of the cross section shown in Figure 6. (Map modified after Hayes, M. O., FitzGerald, D. M., Hulmes, L. J., and Wilson, S. J., *Terrigenous Clastic Depositional Environments*, Hayes, M. O. and Kana, T. W., Eds., University of South Carolina, Columbia, 1976, II-80. With permission.)

abandoned delta lobes. This dichotomy of local land gain and land loss also characterizes the Louisiana coast today.[75]

Detailed stratigraphic studies of numerous barrier island sequences have been performed over the last decade and a half. Most of this work is based on coring (recently vibracoring) through the Holocene coastal sequence and ^{14}C-dating of shells, wood, or peat samples within the cores. One of the earliest studies of this kind was done by Bernard et al.[4] who demonstrated that the east end of Galveston Island, Tex. was a regressive barrier island; the oldest deposits of which dated back some 3500 years b.p. Any one core through east

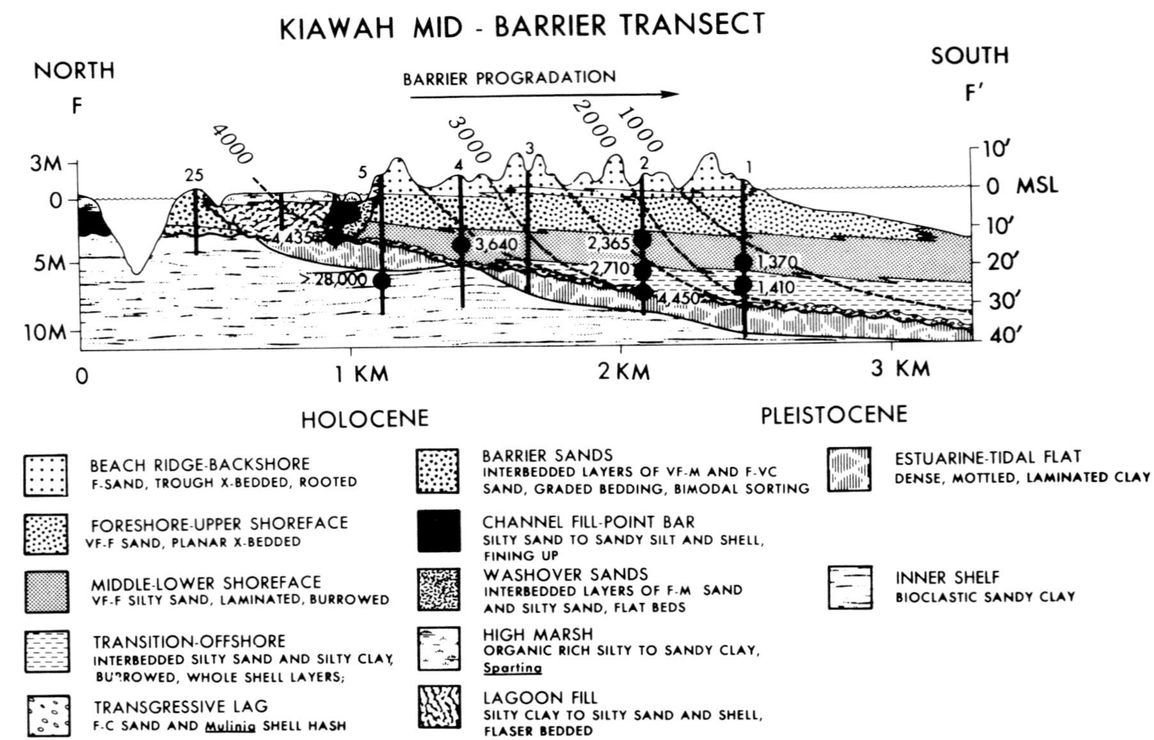

FIGURE 6. Midbarrier stratigraphic cross section of Kiawah Island. Note the basal lag indicating transgression prior to the emplacement of the thick regressive barrier sand body. (From Moslow, T. E., Stratigraphy of Mesotidal Barrier Islands, Ph.D. thesis, University of South Carolina, Columbia, 1980. With permission.)

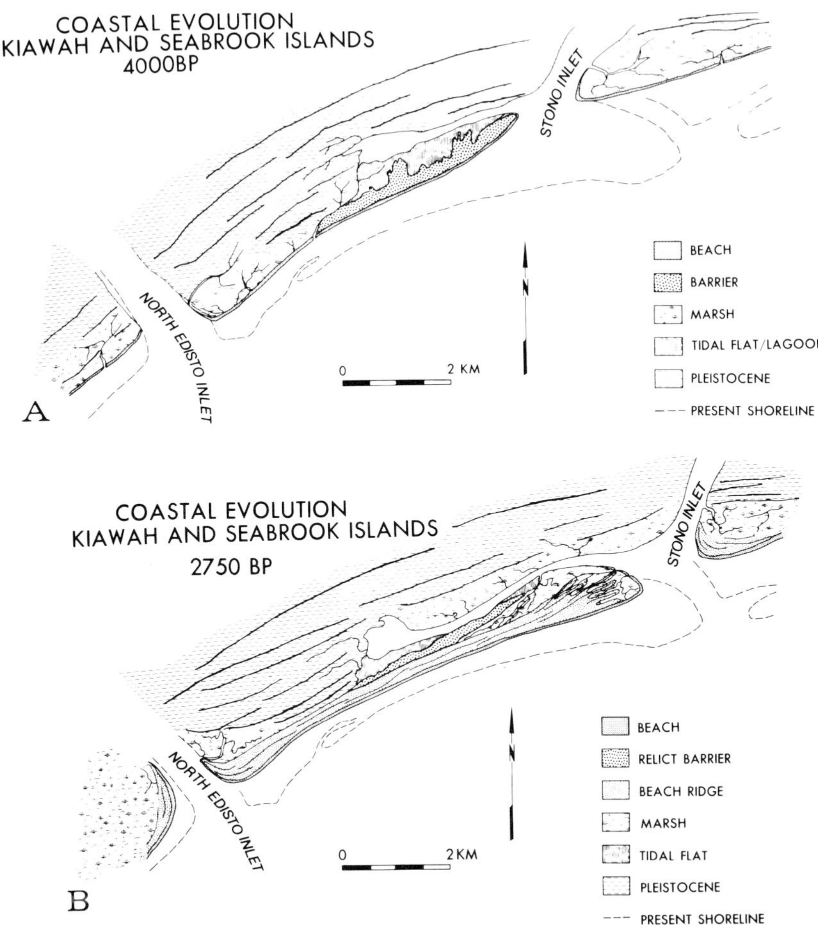

FIGURE 7. Inferred paleogeography of Kiawah/Seabrook Islands, S.C. A: 4000 years b.p. (end of coastal transgression); B: 2750 years b.p. (in the middle of the regressive phase). (Modified from Moslow, T. E., Stratigraphy of Mesotidal Barrier Islands, Ph.D. thesis, University of South Carolina, Columbia, 1980. With permission.)

Galveston Island shows beach sand on top of shoreface on top of inner shelf sediments, producing classical coarsening-upwards, regressive sequences. Subsequent studies of other regressive barriers including Matagorda Island in south Texas[104] and Kiawah Island in South Carolina[69] permit further refinement of this picture. The stratigraphy of Kiawah Island, a large regressive beach-ridge barrier island on the South Carolina coast (Figure 5), is shown in cross section in Figure 6. Stratigraphically the island is composed of two distinctly different sequences. The oldest part of the island is a transgressive lag overlying estuarine and tidal flat sediments. The oldest (most landward) subaerial ridge, Shulbred Ridge (at core location 25 in Figure 6), transgressively overlies back-barrier facies (not documented in the section shown in Figure 6). The bulk of the island farther seaward is a coarsening-upwards regressive sequence. Kiawah Island is about 4000 years old. The stratigraphy, together with the late Holocene sea-level curve of Figure 2A led Moslow[69] to propose the following scenario for the evolution of Kiawah Island.

The island began forming as a primary barrier some distance seaward of the present shoreline, perhaps about 6000 years b.p. Its origin may be related to the reduced rate of sea-level rise between 7000 and 5000 years ago. This primary barrier provided protected back-barrier settings in which tidal flats, marshes, and lagoonal conditions were established.

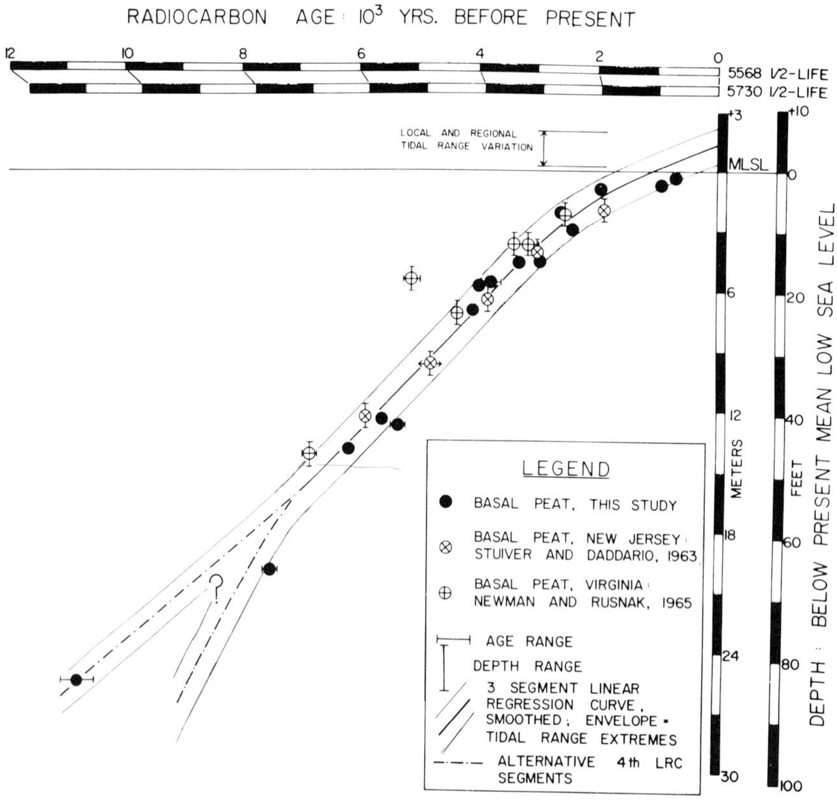

FIGURE 8. Local relative sea-level rise curve for coastal Delaware; based on carbon-14 dating of basal peats. (From Belknap, D. F. and Kraft, J. C., *J. Sediment. Petrol.*, 47, 610, 1977. With permission.)

With continued rise in sea level the barrier moved landward, eventually transgressing these tidal flat deposits forming Shulbred Ridge (Figure 7A). The ridge was probably stabilized about 4000 years b.p. following the very rapid rise in eustatic sea level which appears to have occurred between years 4500 and 4000 b.p.[9,21,62]

A standstill or drop in eustatic sea level after 4000 b.p. provided the conditions necessary for initiation of barrier shoreline progradation leading to development of the regressive part of Kiawah Island. Although intermittent periods of shoreline retreat undoubtedly occurred, Kiawah Island remained essentially accretionary until the present time. An intermediate stage in the evolution of this island is presented in Figure 7B.

As an example of an essentially transgressive barrier shoreline we will look at the coast of Delaware. Stratigraphically this is probably the best-known shoreline in the U.S. due to extensive work by J. C. Kraft and students over many years. The local relative sea-level rise curve for Delaware is presented in Figure 8. The data in Figure 8 support a relatively smooth, continuous rise in sea level till the present, with a decrease in the rate of rise over the last 8000 years. Because of fluctuating sedimentation rates many areas of the Delaware coastal plain contain both transgressive and regressive sequences developed during this interval of proposed continuously rising sea level.

The Atlantic shore of Delaware is composed of a transgressive barrier complex containing tidal delta deposits, linear barriers, and a major spit (Cape Henlopen). A stratigraphic cross section across Rehoboth Bay and its bay-mouth barrier is shown in Figure 9. The extant barrier in this location is quite young and overlies lagoonal and marsh muds previously

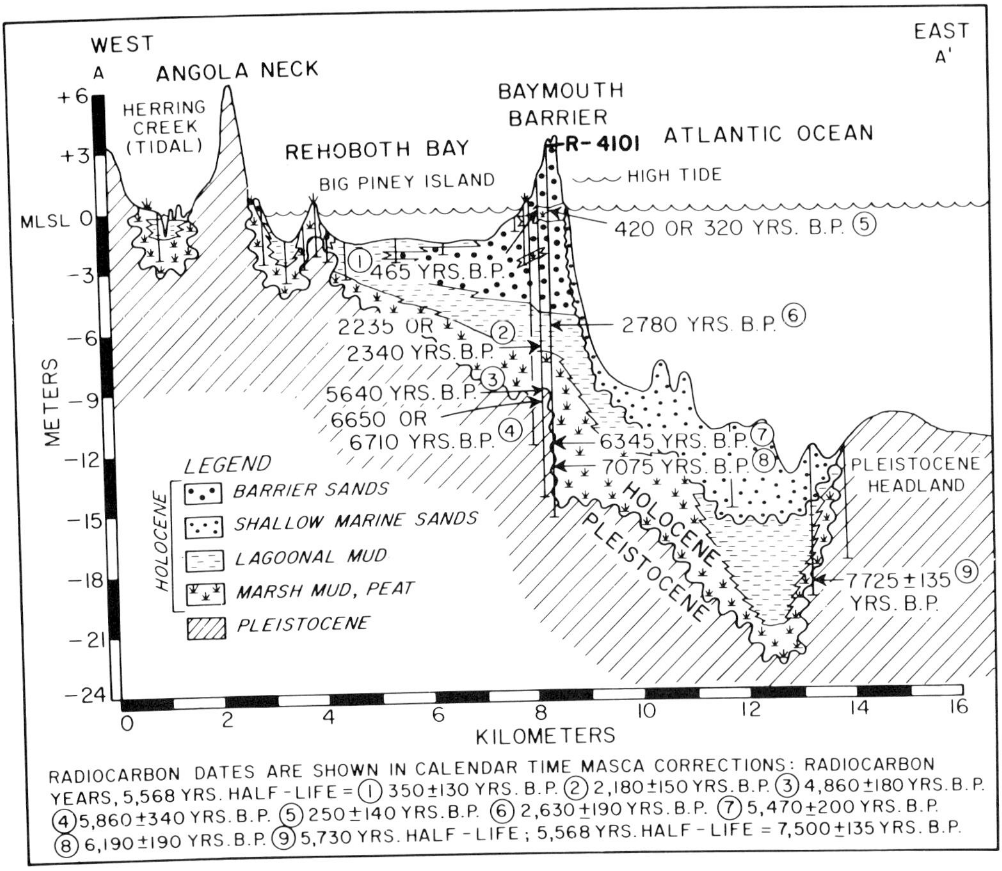

FIGURE 9. Stratigraphy of a transgressive coastal lagoon-barrier complex, Rehoboth, Del. (From Kraft, J. C. and John, C. J., *Am. Assoc. Petrol. Geol.*, 63, 2145, 1979. With permission.)

deposited in a sheltered back-barrier environment very similar to that of the present Rehoboth Bay. This implies that Delaware's Atlantic barriers must have migrated landward from a more seaward position, and that this migration pattern is continuing today.[54] Figure 10 shows the inferred late Holocene evolution of the Cape Henlopen segment of Delaware's Atlantic coast.

From the data presented above one may construct two generalized cross sections for transgressive and regressive barrier sequences (Figure 11). The sections provide an illustration of Walther's law of the succession of facies[7] which states that "only those facies can be superimposed, without a break, that can be observed beside each other at the present time." Regressive coastal evolution commonly leads to preservation of the whole barrier sequence. During transgressions, however, marine processes continue to erode the shoreface as the barrier retreats, thus destroying much of the upper part of the sequence. The depth of shoreface reworking is largely a function of the rate of sea-level rise relative to the rate of erosion. Rapid sea-level rise is more apt to drown, and thus preserve, the nearshore sedimentary sequence. Slow sea-level rise, on the contrary, may lead to complete erosional removal of the barrier and shoreface sequence if sediment supply to the shore is slow or nonexistent.[3,25,97]

The Sources of Sand

In the previous discussion we assumed that sediment was available in the marine envi-

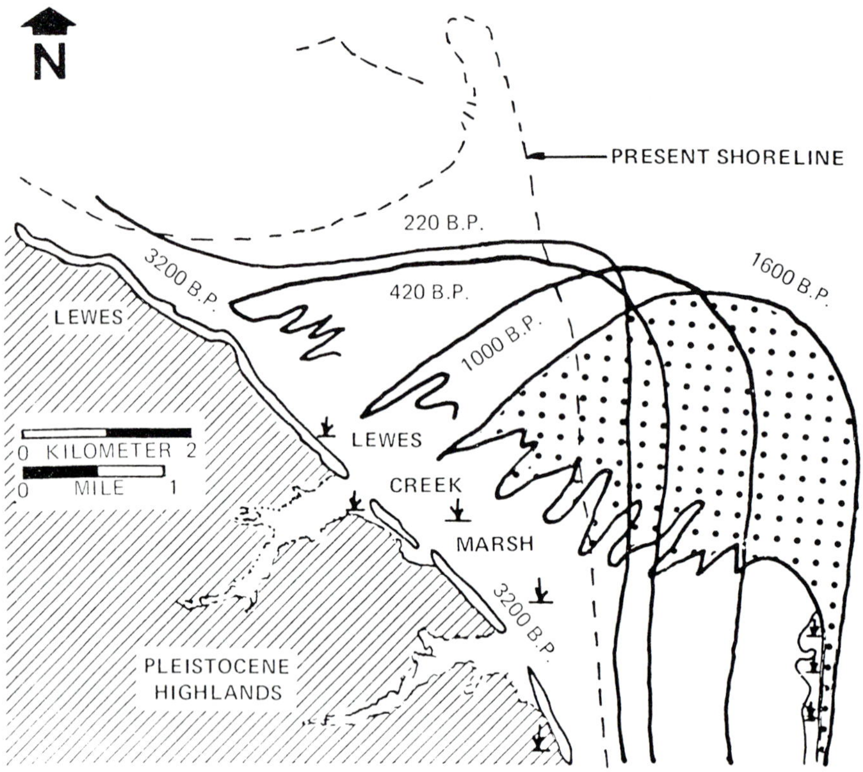

FIGURE 10. Migration of the Cape Henlopen spit and the adjacent coastal barrier of Delaware's Atlantic coast since year 1600 b.p. Reconstruction based on stratigraphic work. (From Kraft, J. C., Allen, E. A., Belknap, D. F., John, C. J., and Maurmeyer, E. M., Delaware's Changing Shoreline, Tech. Rep. No. 1, Delaware Coastal Zone Management Program, Lewes, Del., 1976. With permission.)

ronment to build the sandy (or gravelly) barrier islands. Where does this coarse-grained sediment come from? In the case of many barriers in the northeastern U.S., barrier spits and islands extend downdrift (in the direction of the prevailing longshore drift) away from Pleistocene erosional headlands. These headlands may be composed of glacial deposits, as on Cape Cod,[80] or they may be older barrier systems truncated by the eroding Holocene shoreline (e.g., Atlantic coast of Delaware[52]). Sandy shoals on the inner continental shelf, of a diversity of origins, may also effectively supply sediment to the adjacent modern shoreline.

River deltas, however, appear to be the single most important source of barrier island sand (Figure 12). The relationship between fluvial deltaic sediment sources and barrier island sinks is exceptionally well illustrated along the northwest coast of the Gulf of Mexico. The present coastline of Texas consists of three river deltas actively supplying sediment to the longshore transport system through the process of shoreface erosion. From the south to the north these deltas are the Rio Grande and Brazos-Colorado (both Holocene), and the Trinity (Pleistocene). Seismic surveys and bottom sediment sampling on the Texas shelf have demonstrated that all of these deltas were pronounced depositional headlands on the coast in earlier Holocene time (Figure 13). As the rate of sediment supply through coastal streams decreased through the late Holocene, these deltaic headlands were eroded back by the persistent wave action, a process which continues till the present. Shoreface retreat removed the sedimentary wedge of the deltaic plain down to the average Gulf fair-weather wave-base (about 10 m). The sediment removed was dispersed in different pathways depending on its

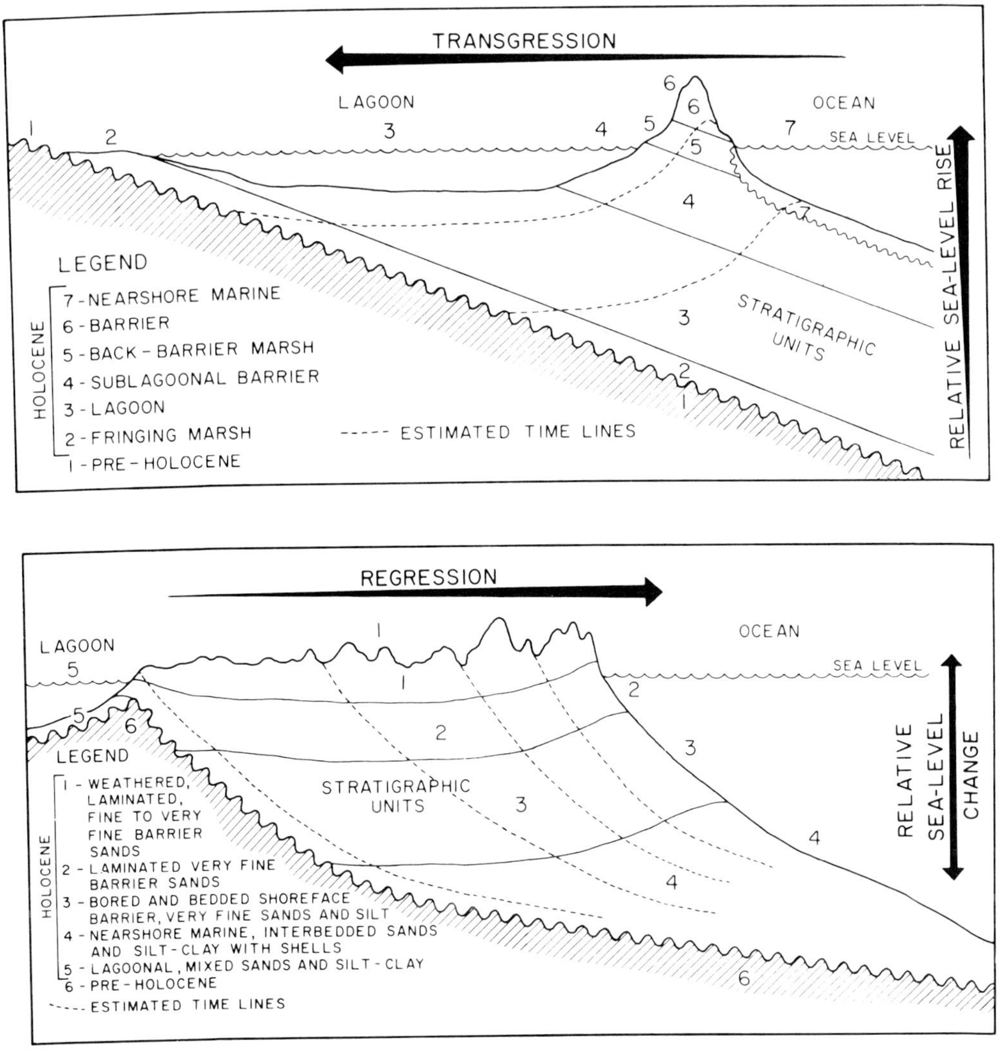

FIGURE 11. Generalized models of transgressive (top) and regressive (bottom) barrier complex stratigraphies. Note the cross-cutting relationships between lithostratigraphic boundaries and time lines. (From Kraft, J. C. and John, C. J., *Am. Assoc. Petrol. Geol.*, 63, 2145, 1979. With permission.)

grain size. Fine sediments were deposited on the shelf and to a lesser extent in marshes and lagoons, the sandy fraction formed the primary source for the barrier islands.[68]

A shoreline consisting of deltaic headlands and interdeltaic embayments will have a longshore sediment transport system characterized by divergence away from the deltas and convergence into the interdeltaic embayments. Geological evidence in favor of such a dispersal system along the Texas coast is found in the presence of a Rio Grande mineral suite northward from South Padre Island towards Mustang Island, whereas a Brazos-Colorado suite dominates further north. Along the south Texas coast this same transport system is in operation today. In response to the present wind regime in the Gulf of Mexico the prevailing longshore currents converge at central Padre Island near latitude 27°N.[55,59] A textural study along the Padre Island beaches[36] demonstrated how the northward-moving Rio Grande sand population became mixed with the southward-moving Brazos-Colorado population on central Padre Island. In earlier Holocene times both the Brazos-Colorado delta and the Trinity delta

FIGURE 12. Pitt Point on the Beaufort coast of Alaska. The delta of the Smith River illustrates nicely the flanking barriers built of sediment derived from distributary mouth bars. This westward-looking oblique airphoto was taken in July 1977. Note the drift ice.

acted as bidirectional sediment dispersal centers with a longshore transport pattern similar to that found at the Rio Grande delta today (Figure 13).

Although the history of Holocene sea-level rise has been essentially the same along the entire Texas coast, one encounters two distinctly different types of coastal stratigraphic sequences:

1. Overlying the deltaic headlands one encounters a transgressive barrier stratigraphy. The shoreline is invariably highly erosive,[64] consisting of a narrow and thin beach overlying back-barrier marsh sediments. Old marsh sediments commonly outcrop on the beach face (Figure 14). We refer to this type of barrier as low-profile (Figure 15). The most rapidly eroding segments of the Texas shoreline are found at South Padre Island (Rio Grande delta), the beach off Freeport (Brazos-Colorado delta), and between Galveston Bay and the Sabine River (Trinity delta). Flanking these highly erosive, transgressive, low-profile barrier beaches are barrier spits such as Matagorda Peninsula and Follets Island. These spits basically act as longshore sediment transfer systems where the sand moves along the beach from the updrift deltaic headlands to the downdrift barrier island. The spits are generally transgressive, moving landward together with the deltaic headlands. Matagorda Peninsula was eroded 250 m by *Hurricane Carla* in 1961.[60] These barrier spits commonly overlie back-barrier lagoonal or bay sediments.[67]
2. The central portion of each interdeltaic embayment is occupied by wide, high-profile, and generally regressive, barrier islands. In the case of Texas these are north Padre, Mustang, San Jose, and Matagorda Islands between the Rio Grande and Brazos-

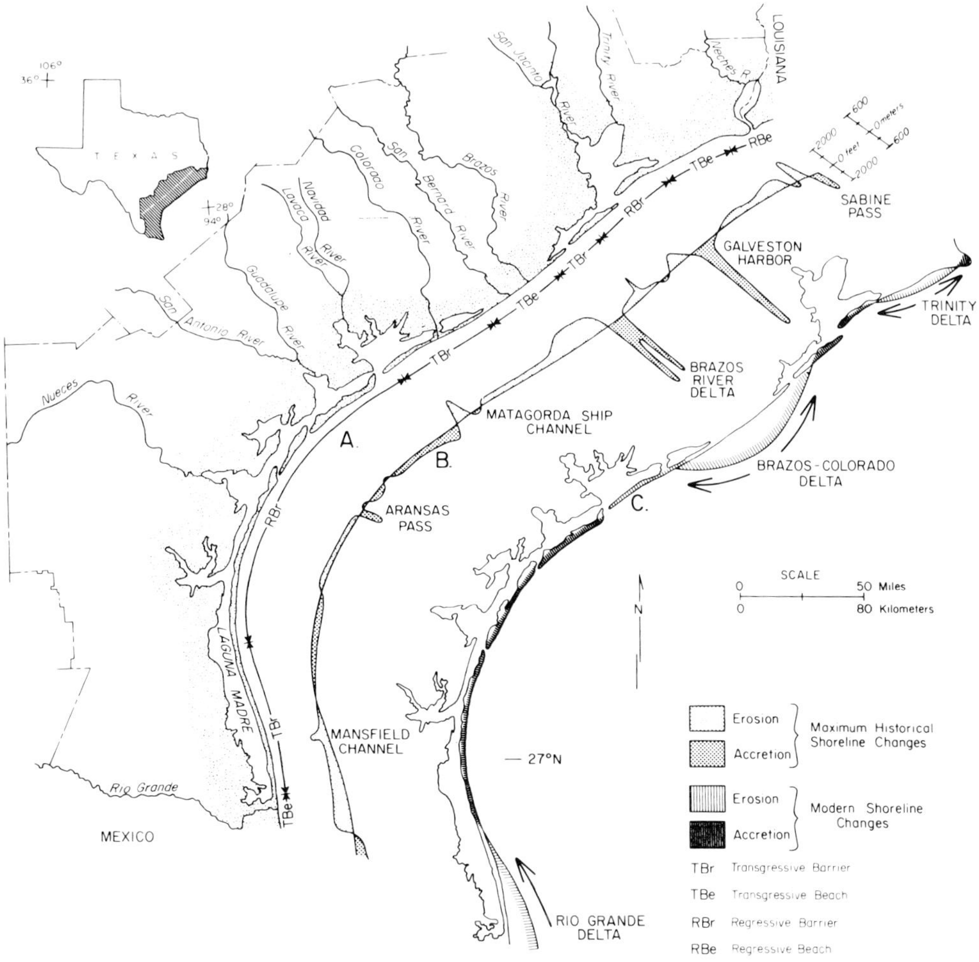

FIGURE 13. Map of Texas coast showing (A) distribution of transgressive and regressive shorelines, (B) maximum shoreline changes between the years 1850 to 1883 and 1973 to 1975, and (C) the hypothesized late Holocene shoreline with three deltaic headlands and littoral drift cells. (From Morton, R. A., *J. Sediment. Petrol.*, 49, 1101, 1979. With permission.)

Colorado deltas, and east Galveston Island to the east of the Brazos-Colorado delta. Stratigraphic studies of the Pleistocene Ingleside strandplain[105] and the Oligocene-Miocene Frio Formation[30] present detailed evidence that interdeltaic embayments have been the preferred sites for the development of broad regressive barrier strandplains throughout the evolution of the Texas coastal plain.

The concurrent development of regressive and transgressive barriers along the Texas coast over the last few thousand years illustrates the principle set forth in Curray's[13] diagram (Figure 4): the rate of sea level rise has been essentially the same along the whole coast, yet transgressive sequences have evolved along the deltaic headlands because of sediment removal by marine processes, whereas regressive barriers evolved in the adjacent embayments which have acted as sediment sinks. The regressive Texas islands are fairly stable. Significant shoreline erosion is restricted to areas where man has altered the natural shoreline (Figure 13B). Later in this chapter we will discuss the natural hazards associated with developments on barrier islands; the nature of such hazards can be directly related to the stratigraphy of the different barriers.

FIGURE 14. Marsh outcropping on the beach at Sargent Beach, Tex. (on the Brazos-Colorado deltaic headland). (Photo by R. A. Morton, December 1981.)

THE MISSISSIPPI DELTA BARRIERS

Most barrier islands are probably derived through the direct erosion of coarse sediments from abandoned fluvial deltas or other sedimentary headlands. The result of this evolutionary process is well illustrated on the Texas coast. The individual stages in this chain of barrier island evolution can be traced in the island systems flanking the Mississippi delta.

The Holocene history of the Mississippi River has been one of frequent channel switching and delta lobe abandonment. Stratigraphic studies and ^{14}C-dating has lead to the recognition of five major delta complexes (Figure 16) dating back some 6000 to 7000 years.[29,49] Within these larger delta complexes one can identify a total of 18 distinct episodes of distributary flow. After an average occupancy of about 1000 years, each delta complex appears to have become hydraulically disadvantaged compared to some new river outlet to the Gulf, and a major channel switch has occurred. Once a distributary became abandoned by the river, marine processes took control of the sediment dispersal. Sand, originally deposited as distributary mouth bars, now became subject to marine reworking through shoreface erosion. The incorporation of such multiple sand sources into an integrated longshore transport system ultimately produced extensive barriers along the periphery of the abandoned deltas. Because we know the age of each individual delta lobe on which these barriers have developed we can use the morphologic and stratigraphic data from the Louisiana coast to derive a model for the temporal evolution of barrier islands.

The modern Mississippi delta consists of massive, thick "bar-finger sands" along each of the three major distributaries.[26] Rapid progradation of the river mouths prevents the development of significant barriers during this phase of delta building. However, the source material for future barriers is being deposited. A small barrier spit is presently in existence at the jettied entrance to South Pass.

The late Lafourche delta lobe (Lobe 15, Figure 16) was abandoned by the Mississippi River a few hundred years ago. Today, the periphery of this lobe consists of a rapidly eroding section of transgressive mainland beach (the Caminada-Moreau headland), flanked to the east by Caminada spit and Grand Isle, and to the west by East Timbalier and Timbalier

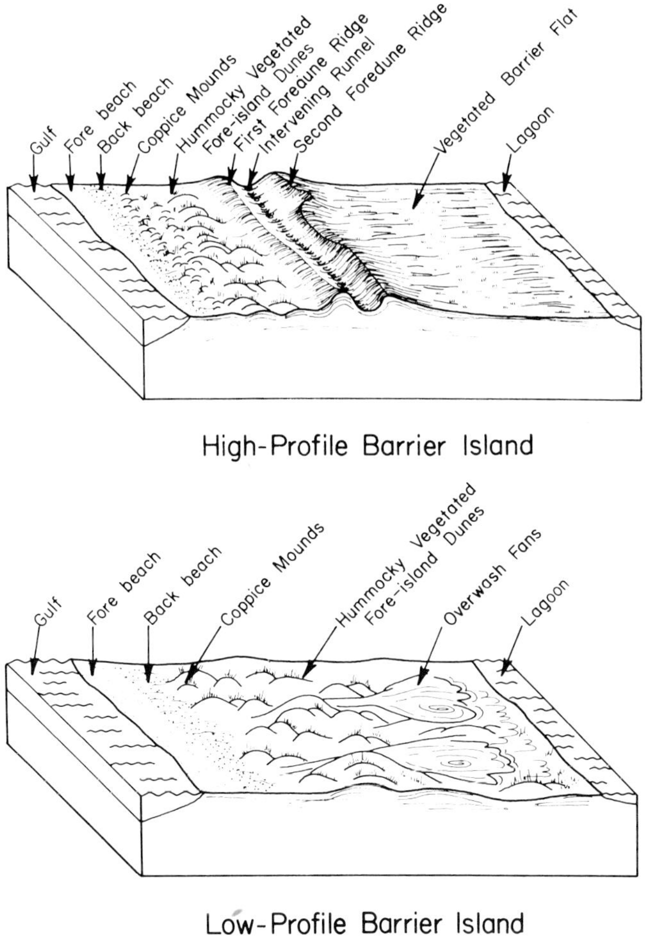

FIGURE 15. Comparative morphologic sketches of high- and low-profile barrier islands. (From White, W. A., Morton, R. A., Kerr, R. S., Kuenzi, W. D., and Brogden, W. B., *Land and Water Resources, Historical Changes, and Dune Criticality, Mustang and North Padre Islands Texas,* University of Texas, Austin, 1978. With permission.)

Islands (Figure 17). Erosion is extremely rapid at the center of the headland, locally in excess of 15 m/year (Figure 18). The beach there is a thin wedge of sand overlying the mainland marsh (Figure 19). The rates of erosion decrease in both directions away from the central Caminada headland. Net accretion characterizes the two termini of the barrier system (Figure 18). The west end of Timbalier Island is a naturally accreting shore; the east end of Grand Isle, in contrast, is heavily influenced by a jetty built to stabilize the entrance to Barataria Bay. Sand is presently supplied to this barrier system from erosion of an older beach ridge plain on the central Caminada headland[31] as well as from sand bodies of probable distributary mouth bar origin outcropping in the shoreface of Grand Isle. A generalized model for the development of barrier islands from the erosion of distributary mouth bar sands is shown in Figure 20. As the sediment abundance increases downdrift from the central headland, the barriers gradually attain a higher foredune profile. Because of their young age, however, they never attain the morphology characterizing the high-profile interdeltaic barrier islands of Texas.

The extremely rapid rate of shoreline change in this area becomes evident when we compare maps from 1853 and 1965. The central part of the headland retreated about 3 km during

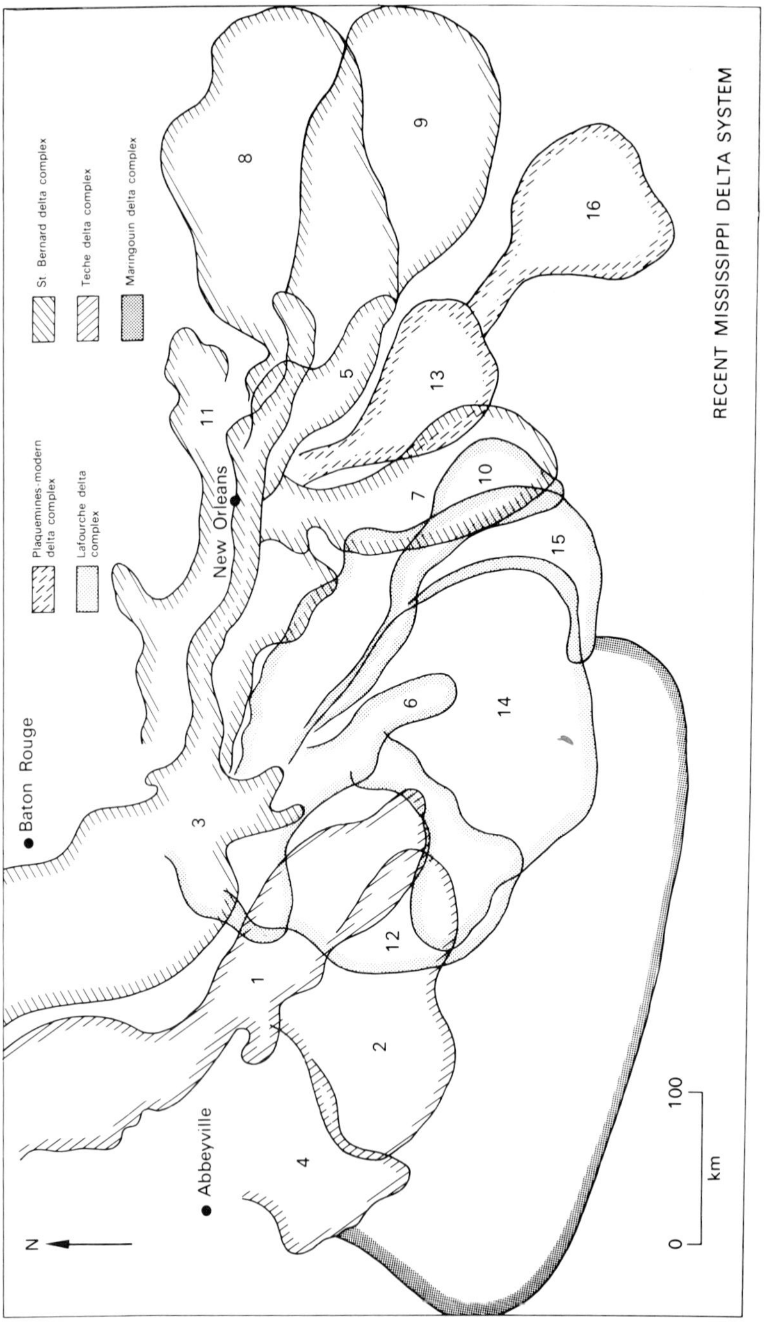

FIGURE 16. Distribution of delta complexes (five major ones) and individual delta lobes during the late Holocene evolution of the Louisiana coastal plain. (Modified after Frazier, D. E., *Gulf Coast Assoc. Geol. Soc. Trans.*, 17, 287, 1967. With permission.)

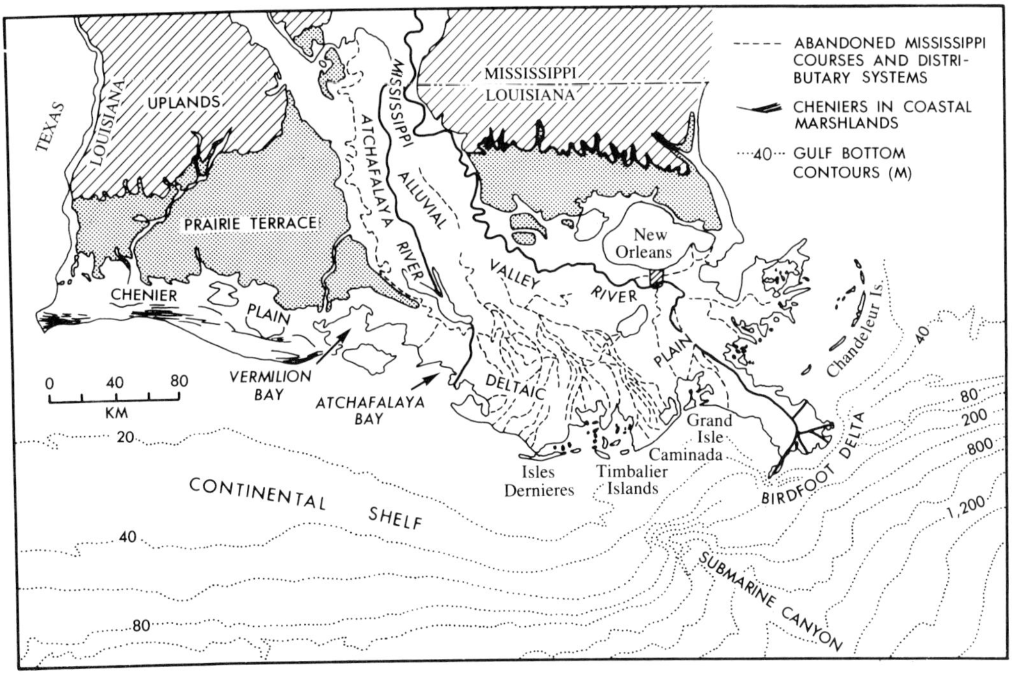

FIGURE 17. Map of the Louisiana coastal plain emphasizing the barrier islands.

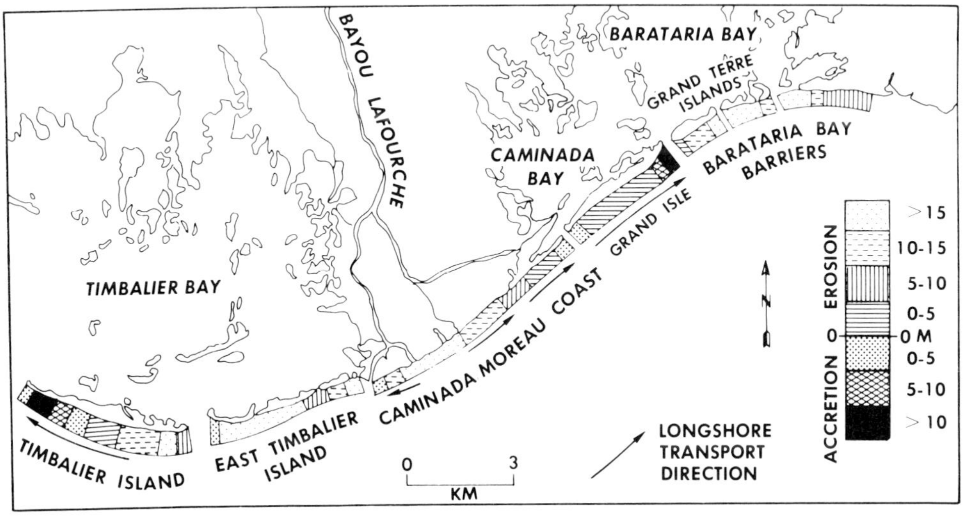

FIGURE 18. Rates of shoreline change between 1934 and 1974 along the coast from Timbalier Island to Barataria Bay. (From Penland, S. and Boyd, R., *Deltaic Sedimentation on the Louisiana Coast,* Nummedal, D., Ed., Society of Economic Paleontologists and Minerologists, Tulsa, Okla., 1982. With permission.)

this time. The west end of Timbalier Island accreted about 7 km westward since 1887, while its eastern end eroded.[82] The eastern half of Grand Isle has formed since 1842.[31] This whole island is probably only some 300 or 400 years old. We refer to the barrier evolution stage exemplified by the Caminada-Moreau system as stage 1: an erosional headland with flanking barriers (Figure 21).

The next stage in the evolution of Louisiana barriers is exemplified by the margins of the

FIGURE 19. (Top) Oblique aerial photo of the Caminada-Moreau coast. View towards the southwest. Note the transgressive "perched" beach and the multiple oil platforms on the inner continental shelf in the background. (Bottom) Close-up aerial view of a segment of the Caminada-Moreau headland. Note washovers infilling an old channel (Bayou Moreau) and the marsh outcrops on the beach.

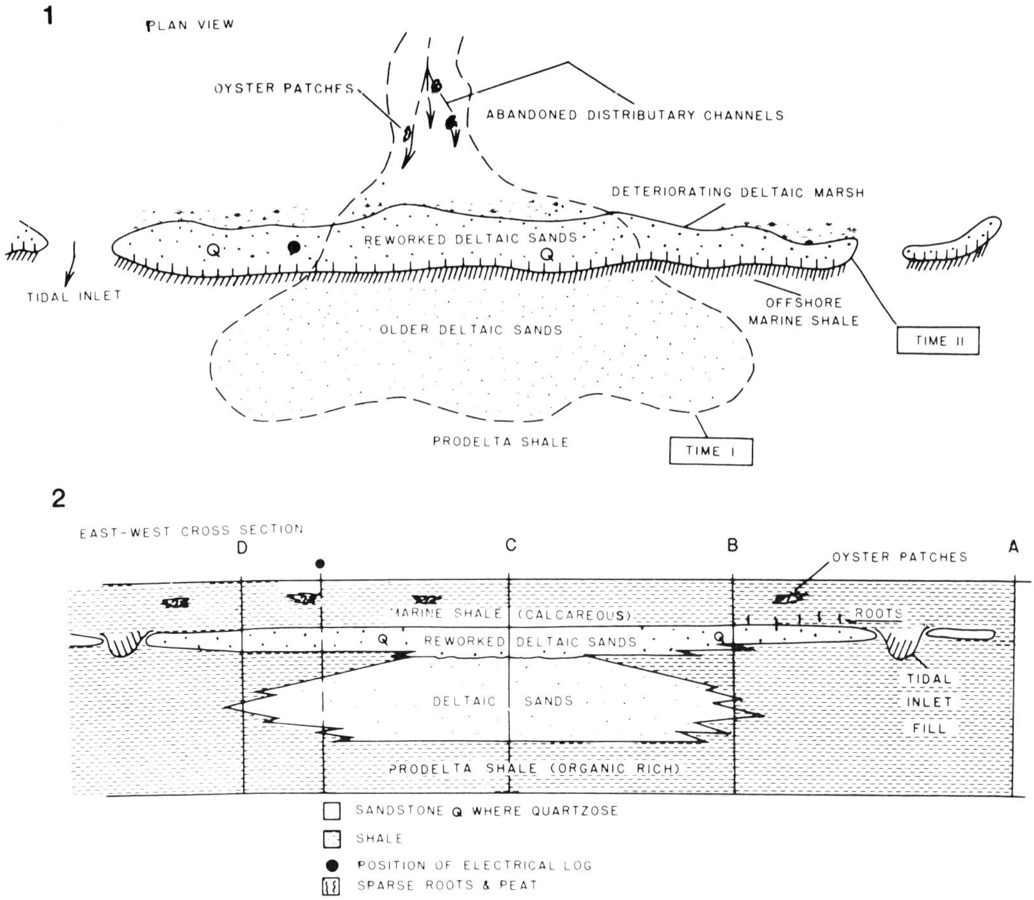

FIGURE 20. Schematic representation of the stratigraphic development of a barrier island from deltaic distributary mouth bar sands. (From Saxena, R. S., *Modern Mississippi Delta — Depositional Environments and Processes*, American Association of Petroleum Geologists, Tulsa, Okla., 1976. With permission.)

Early Lafourche delta lobe (abandoned about 1000 years ago) and the St. Bernard delta (abandoned 1800 years ago). In contrast to the erosional headland stage described above, these older barrier systems are entirely detached from land. We refer to them as transgressive barrier island arcs (Figure 21). The detachment of the Isle Dernieres has occurred fairly recently through the rapid enlargement of Pelto Bay into Lake Pelto (Figure 22). The wide Chandeleur Sound which separates the Chandeleur Islands at the margin of the St. Bernard delta lobe from the mainland (Figure 17) probably developed the same way but at an earlier date.

The rapid widening of these back-barrier lagoons is probably due to the high rate of subsidence of the Mississippi delta lobes. Parts of the Louisiana coastal plain may have been subsiding at a rate of 1 cm/year for the last few decades. The long-term average subsidence rate is probably somewhat less. The rapid subsidence of abandoned Mississippi delta lobes has two major consequences for barrier island evolution. First, the back-barrier environment is rapidly transformed from mainland marsh to an open marine sound. The barriers in this stage evolve through a detachment process as envisioned by Hoyt[45] and Swift.[97] Although the barrier continuously retreats through shoreface erosion at this stage (the Gulf beach of Isle Dernieres migrated 2 km landward between 1853 and 1974, Figure 22), the islands become progressively more separated from land since the mainland shore retreats faster

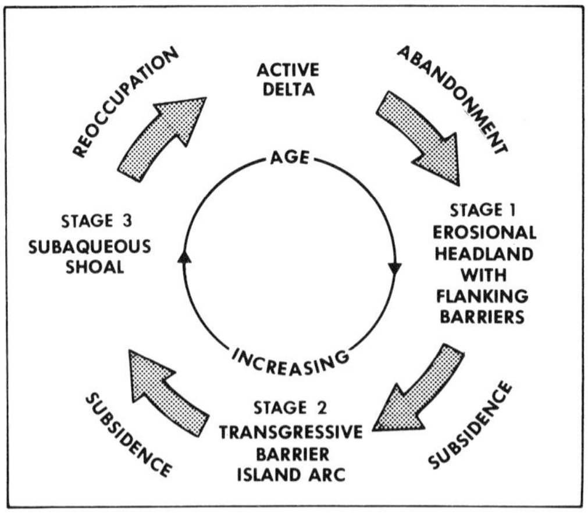

FIGURE 21. The cyclic evolution of Louisiana's barriers is controlled by abandoned delta lobe subsidence and ultimate reoccupation by a new active river distributary. (From Penland, P. S., Boyd, R., Nummedal, D., and Roberts, H., *Gulf Coast Assoc. Geol. Soc. Trans.*, 31, 471, 1981. With permission.)

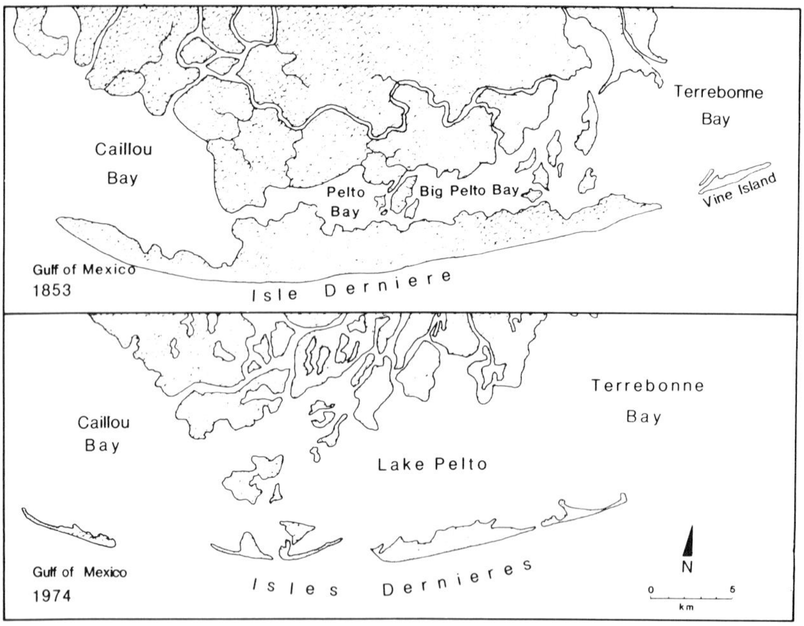

FIGURE 22. Historical shoreline changes in the Isles Dernieres. Note the rapid growth of Lake Pelto as well as the landward migration of the island system. (From Penland, P. S., Boyd, R., Nummedal, D., and Roberts, H., *Gulf Coast Assoc. Geol. Soc. Trans.*, 31, 471, 1981. With permission.)

under continued subsidence. Secondly, the sand bodies of the deltaic source sink progressively deeper. Ultimately, they subside below the level of the retreating shoreface, at which point they cease to be active sources of barrier sand. Although stratigraphic data are limited, suggestions are that shoreface erosion no longer reaches the distributary sand bodies of the St. Bernard delta beneath the Chandeleur Islands. The islands are transgressing landward into the gradually deepening Chandeleur Sound, and they lose sand to an offshore sand sheet.[90] Because there is no present sand source balancing this loss to the marine environ-

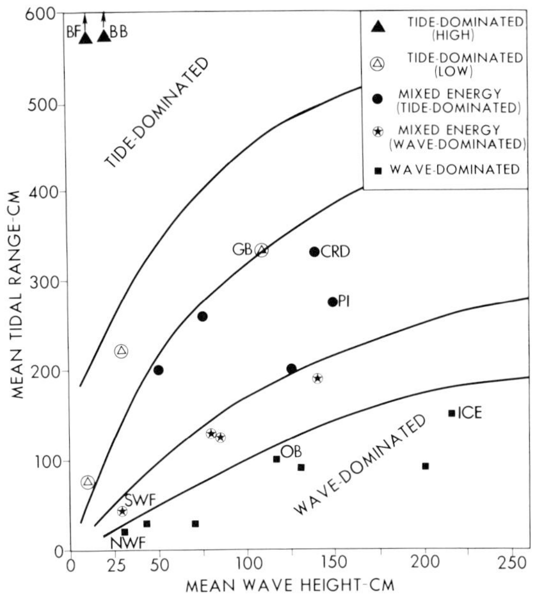

FIGURE 23. Barrier island morphology as a function of the tide range and mean wave height. 19 barrier island shores in North America and western Europe were used as the basis for this diagram. See text for wave-dominated and tide-dominated barrier characteristics. Modified from Nummedal and Fischer[76] and Hayes.[38]

ments, the subaerial portions of the islands must be rapidly decreasing in size. Historically, this decrease in island size is quite evident.[82]

The last stage identified within this evolutionary sequence for Louisiana's barrier islands is the inner shelf sand shoal (stage 3, Figure 21). Shoals along the margin of Teche and Maringouin deltas, abandoned 4000 and 6000 years ago, respectively, are probably transgressed barrier islands. The detailed stratigraphy of these shoals, however, is still incompletely known.

The evolutionary model derived for Louisiana's barrier islands sheds considerable light on the long-standing controversy regarding the origin of barrier islands. It is not appropriate to consider any barrier island as originating through either De Beaumont's[17] vertical accretion model, Gilbert's[32] spit growth model, or Hoyt's[45] detachment theory. At different stages in their evolution the Louisiana barriers have been subject to a combination of these modes of evolution. A logical discussion of barrier genesis should consider each system within the framework of its geological evolution and sediment source. At any point in its history, any single formative mode or combination of modes may dominate.[89]

MORPHOLOGY AND SEDIMENT TRANSPORT

Barrier Island Shapes

The discussion to this point has demonstrated how the barrier island topographic profile depends on its regressive or transgressive nature. Furthermore, we have demonstrated that regressive and transgressive barrier islands occur in a distinct spatial relationship relative to longshore sources and sinks of sediment. Barrier island plan form, however, depends less on the stratigraphic nature; it appears to be primarily controlled by the relative magnitude of tidal and wave energies. The nature of the tidal passes and their modes of sediment bypassing largely determines the stability of the adjacent barrier island beach.

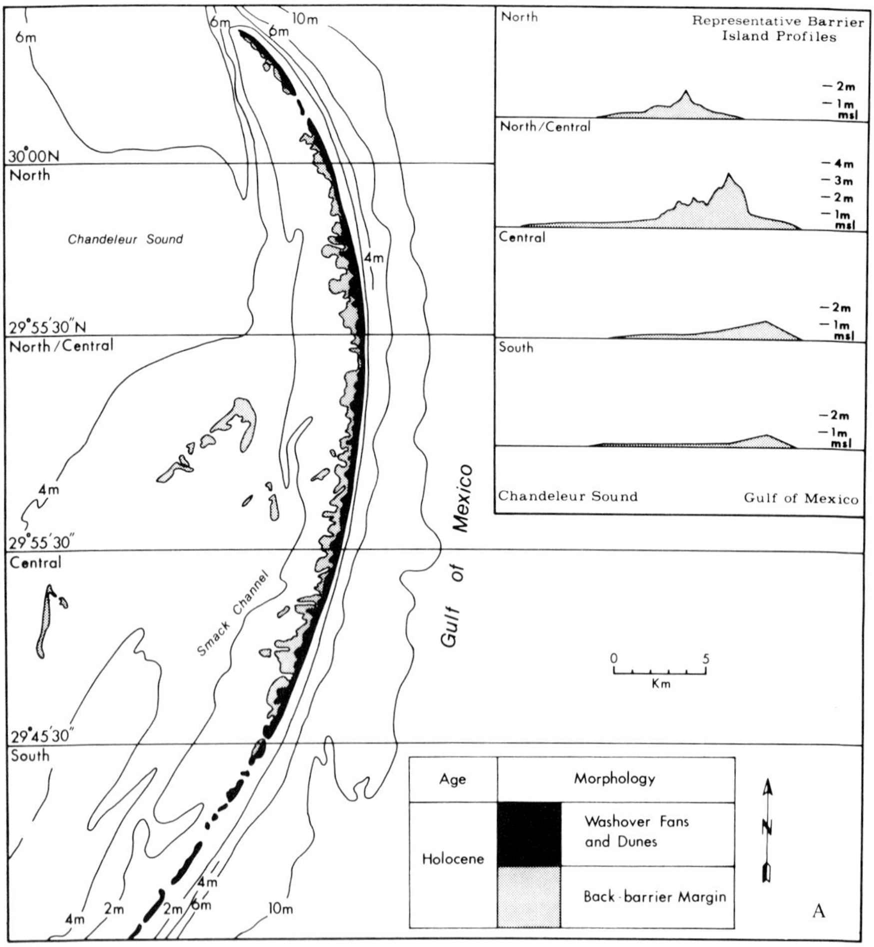

FIGURE 24. (A) Morphologic map of the Chandeleur Islands, a wave-dominated barrier island complex along the margin of the abandoned St. Bernard delta of the Mississippi River. (From Nummedal, D., Penland, S., Gerdes, R., Schramm, W., Kahn, J., and Roberts, H., *Gulf Coast Assoc. Geol. Soc. Trans.*, 30, 183, 1980. With permission.) (B) Oblique, westward-looking airphoto of Cross Island, a small wave-dominated gravel barrier off Prudhoe Bay on the Beaufort coast of Alaska. (Photo taken in July 1977.)

Initial attempts at barrier island morphologic classification were based strictly on tide range, following Davies'[15,16] concept of micro-, meso-, and macrotidal shorelines (see also Hayes and Kana[40]). Subsequent investigations, however, have made it clear that tide range and the mean annual nearshore wave energy play an equal role in the morphologic evolution of a barrier.[38,76,78] It is here recommended that depositional coastlines should be morphologically classified as wave-dominated, mixed, or tide-dominated (Figure 23).

Islands along the south Texas coast typify the wave-dominated coastline: they are long, generally narrow, and cut by widely separated tidal inlets with large sand accumulations in the back-barrier bays (flood-tidal deltas), and small or nonexistent ebb-tidal deltas (seaward shoals). The mean tide range along this coast is about 0.5 m, the mean annual breaker height is about 60 cm. Similar wave-dominated barrier islands are the rule also for parts of Louisiana (Chandeleur Islands) and the Florida panhandle (Santa Rosa Island), as well as southeast Florida, North Carolina's Outer Banks, the south shore of Long Island, and the Cape Cod segment of the Massachusetts coast.

Islands along the Georgia coast typify a tide-dominated coastline: they are relatively short and wide and separated by stable tidal inlets with an average spacing of 15 km. All inlets

FIGURE 24B.

are associated with large ebb-tidal deltas. The mean tide range along the Georgia coast is about 2 m, the nearshore mean annual breaker height is about 30 cm.[77] A tide-dominated coastline also is found adjacent to Georgia in northeast Florida, most of South Carolina, along the Delmarva, and parts of New Jersey peninsulas and in Massachusetts. Rather surprisingly, one also encounters typical tide-dominated morphology along the southwest coast of peninsular Florida, in spite of a tide range of only about 1 m.[95] The reason for this lies in the extremely low wave energy along that coast. Some Louisiana and upper Texas coast barriers also have a tide-dominated morphology (Galveston Island, Grand Isle) in spite of a tide range less than 50 cm. Prior to jetty construction at Bolivar Roads, this pass at the entrance to Galveston Bay had a large ebb-tidal delta.[63] The present ebb-tidal delta off Barataria Pass has a volume about equal to that of the adjacent Grand Isle barrier. The islands themselves are relatively short and stubby. The tide dominance in these areas is due to the large tidal prisms provided by such major embayments as Galveston and Barataria Bays.

The morphologic difference between tide-dominated and wave-dominated coastlines reflects the relative sediment transport capacity of the tidal currents vs. the wave-generated longshore currents. Along wave-dominated coasts the longshore currents dominate the sediment dispersal pattern. Consequently, islands become long and continuous while the ebb-tidal deltas remain small because of rapid dispersal of their sediment by waves. Sediment moved by tidal currents accumulates in large back-barrier flood-tidal deltas where it is sheltered from wave dispersal. The island shape will be controlled by the longshore profile of dynamic equilibrium. This profile is straight or gently seaward concave where sediment supply is large, inlets relatively small, and inlet sediment bypassing effective (Padre Island). The wave-dominated island profile is crescentic and seaward convex where sediment supply is limited, tidal passes wide, and bypassing ineffective (Chandeleur Islands, Figure 24A; most barriers of the Beaufort Sea coast of Alaska, Figure 24B; Figures 25A and B).

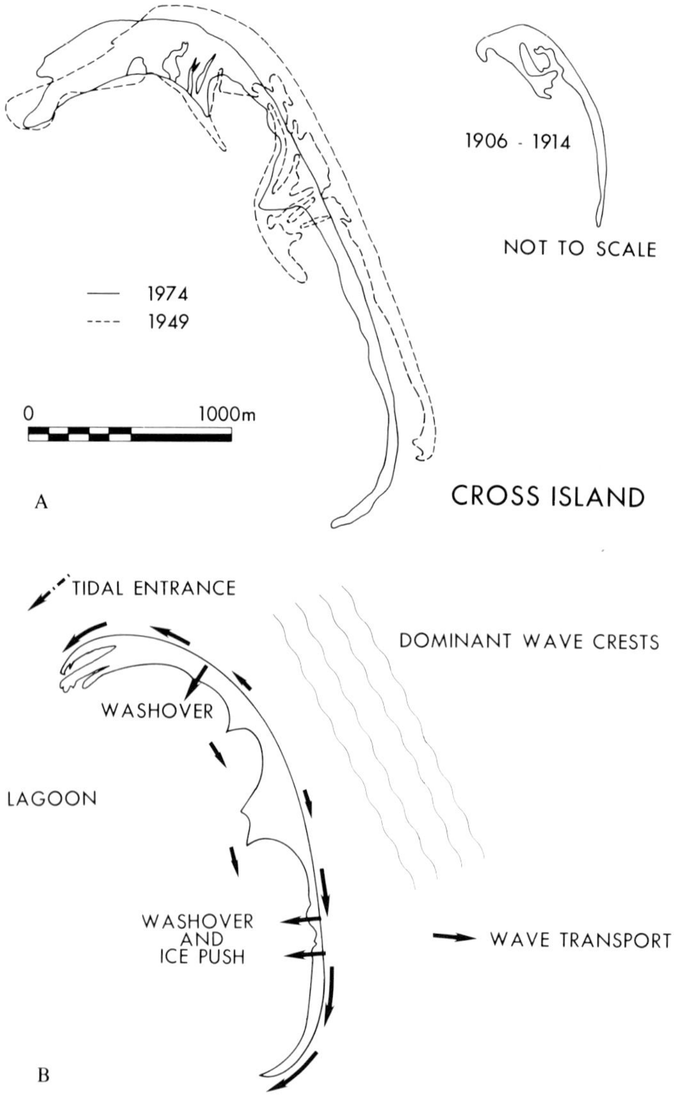

FIGURE 25. (A) Historical migration of Cross Island off Prudhoe Bay, Beaufort coast of Alaska.[86] (B) Sketch summarizing the processes of coarse-grained sediment dispersal at a wave-dominated Beaufort barrier island. (From Nummedal, D., *Proc. Port and Ocean Eng. Arctic Conditions, 5th Conf.*, Norwegian Institute of Technology, Trondheim, 1979, 845. With permission.)

With an increase in tidal prism the inlet-associated shore-normal tidal currents increase in strength. Consequently, they can support larger ebb-tidal deltas against the destructive influence of the waves. These ebb-tidal deltas, in turn, supply sediment to the downdrift beaches.[27,28] As the sediment moves from the updrift tidal delta "source" towards the ultimate "sink" of the next downdrift tidal inlet, the intervening beach will attempt to attain a profile of dynamic equilibrium. This profile is one which must be able to transmit the same amount of sediment across any shore-normal cross section. In light of the refraction of waves around the margin of the tidal delta, this transport requirement is fulfilled by the longshore profile of a concave seaward logarithmic spiral, hence the overall "drumstick" shape[38] of a tide-dominated barrier island (Figures 26A and B) (see also Walton,[102] Silvester,[92] Komar et al.,[50] and Chapter 10 of this volume for analytic treatment of related shoreline shapes).

The capacity for onshore-offshore sediment transport increases with the tide range. Furthermore, in areas of high tide range the wave energy is distributed across a wide shorezone during the tidal cycle. The result is that situations occur where tide dominance is strong enough to completely eliminate barrier islands, apparently regardless of the level of wave energy. On a worldwide basis it is found that barrier islands generally do not exist along coasts where the tide range is more than 4 m.[40,76] This corresponds to the macrotidal category of Davies[15] and is generally found only in embayments where resonance, shoaling, and funneling effects amplify the tide range to many times its open oceanic value.[18] The central part of the German Bight,[76] the Bay of Fundy, Canada,[48] and Nushagak Bay, Alaska, are examples of such macrotidal embayments. Instead of forming barrier islands, the sand in such macrotidal embayments is deposited as linear, lunate, or sigmoidal shoals with their long axes parallel to the predominant tidal currents.

Tidal Inlet Sediment Bypassing

The stability of a barrier island beach, its propensity to erode or accrete, is probably the issue of greatest concern to coastal developers, planners, and politicians. Unfortunately, this is an issue of extreme complexity that lends itself to few generalizations. One statement of general validity is that a beach will erode when faced with a rising sea level and reduced sediment supply. It also appears true that low-profile shores erode faster than high-profile ones.

Barrier island beaches are typically least stable close to the inlets. On many of the tide-dominated South Carolina barrier islands it has been found that a sudden influx of sediment from an ebb-tidal delta may cause dramatic accretion of the adjacent island shore, to be followed by equally rapid and often greater net erosion over the ensuing decades.[39,96] The Friesian Islands in West Germany have a similar pattern of behavior, and Luck[56] refers to that part of the barrier island where sand from the ebb-tidal delta makes landfall as one of a "changing" sediment budget. This zone is located between the updrift erosional beach (negative budget) and the downdrift accretionary beach (positive budget) (Figure 27). Homeier and Kramer[44] attribute this to the episodic nature of the welding of bars from the adjacent tidal delta. These bars originally derived their sand from the island updrift of this inlet through a complex pathway of sediment movement.[78] This process is generally referred to as inlet sediment bypassing (Figure 28).

Although studies of inlet sediment bypassing do not lead directly to generalized models permitting predictions of patterns of change on the adjacent shorelines it can certainly be argued that the temporal behavior of a barrier island beach can never be understood without proper knowledge of the mechanisms of inlet sediment transfer.

At the downdrift end of a barrier island the sediment in motion enters the next tidal inlet. Even though the deep channel of a tidal inlet may seem to be a formidable barrier to continued longshore transport, the fact of the matter is that most tidal inlets are very effective sediment bypassers. Inlet transfer of sediment from updrift to downdrift occurs in at least four distinctly different modes:

1. The common bypassing of large inlets in the southeastern U.S. involves wave-induced delivery of sand into the main inlet currents. The ebb-phase of the tide carries the sand seaward to the terminal lobe of the ebb-tidal delta where wave action again will remobilize it and carry it along the downdrift tidal delta margin to the next beach (Figure 26B).
2. Many inlets are subject to periodic channel switching. Over a period of some years the channel may gradually become deflected downdrift. As this happens a spill-over lobe commonly develops on the updrift side of the channel bend. Because of the inertia

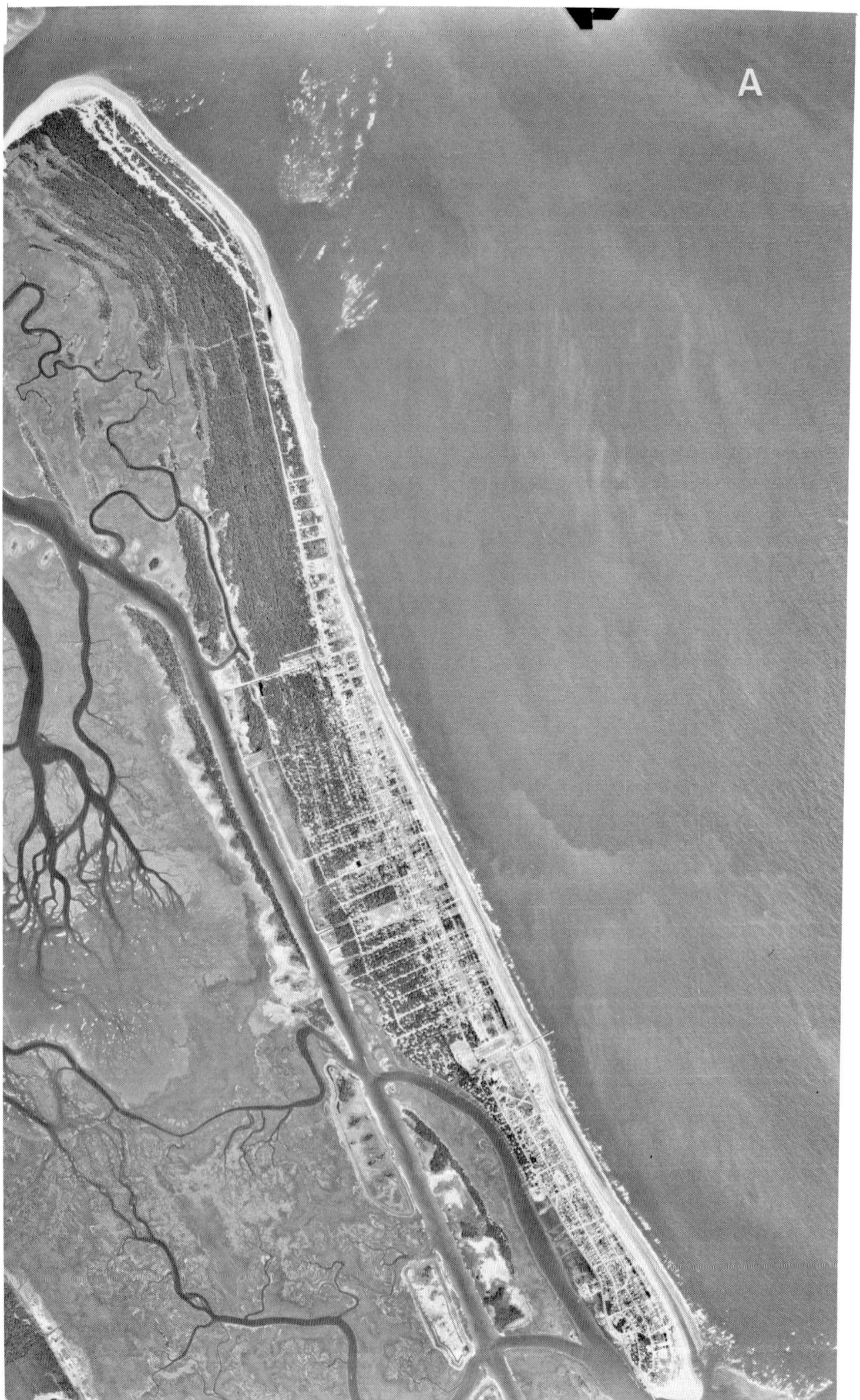
A

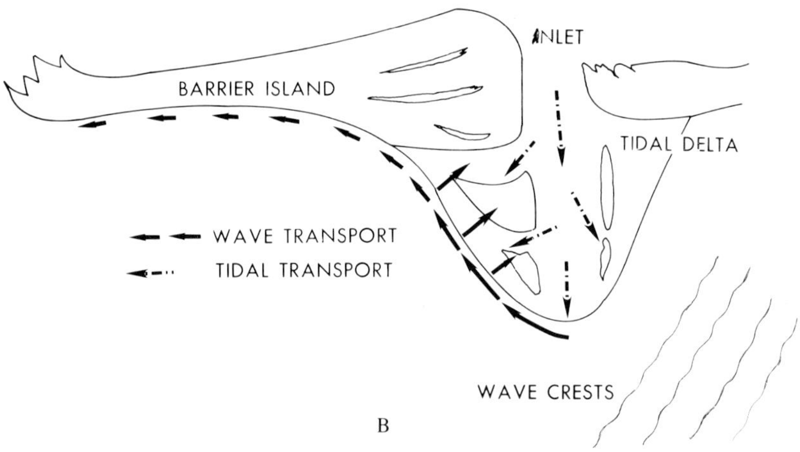

B

FIGURE 26. (A) Vertical airphoto of Isle of Palms, South Carolina. The morphology of this 10-km-long "drumstick" barrier is typical of the tide-dominated Holocene islands of the central Georgia Bight. Note the breaker zones which outline the ebb-tidal deltas at both ends of the island. (Date: March 1, 1973). (B) Generalized sediment dispersal pattern along tide-dominated barrier island shores.

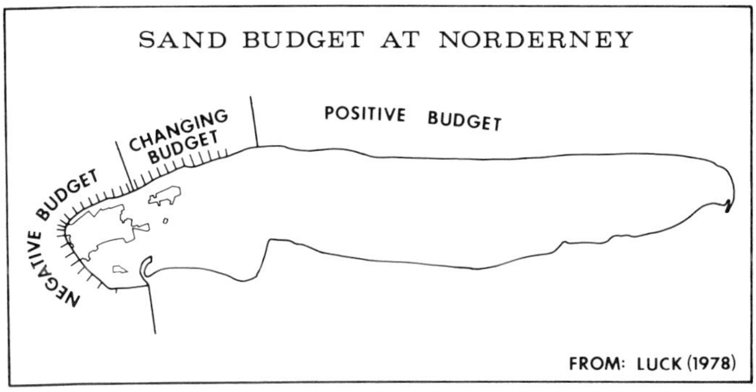

FIGURE 27. Sediment budget for Norderney, one of the Friesian Islands on the North Sea coast of West Germany. The "changing" sediment budget corresponds to the region of episodic attachment of sandbars from the ebb-tidal delta. (From Luck, G., *Die Küste,* 32, 85, 1978. With permission.)

of the ebb-tidal current, this spill-over lobe gradually increases its share of the discharge, ultimately to become the new main channel. The sand body that originally was situated between the two channels now finds itself free to migrate onto the downdrift beach under the influence of continued wave swash. This mechanism of bypassing leads to sudden, episodic delivery of large quantities of sediment onto the downdrift island beach, causing temporary rapid accretion.[27]

3. The mechanism of channel switching just described may occur on a larger scale, involving a significant portion of the entire barrier island. The downdrift end of a barrier island is often a recurved spit connected to the remainder at a narrow neck. During storms this neck might be breached, displacing the inlet to a new position updrift[27,85] (Figure 29).
4. A final mechanism of bypassing involves wave-induced bar migration in an arc around the periphery of the tidal-delta platform. This mode of bypassing is beautifully illustrated on the stormy North Sea coast of Germany (reviewed above) and is a charac-

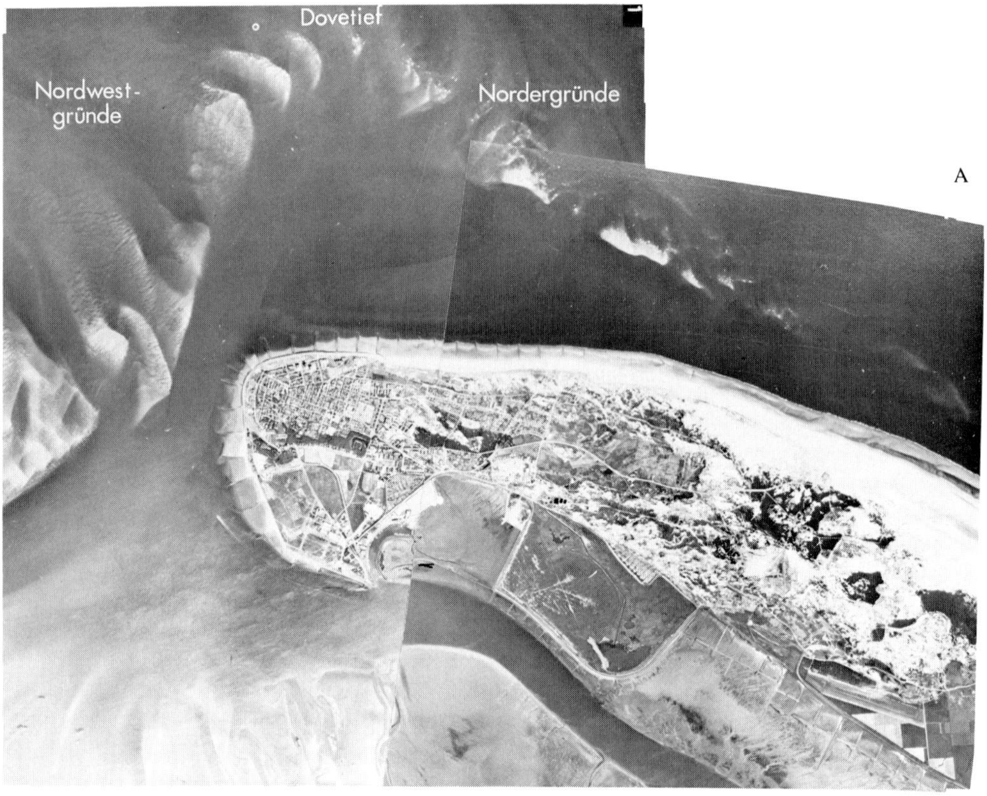

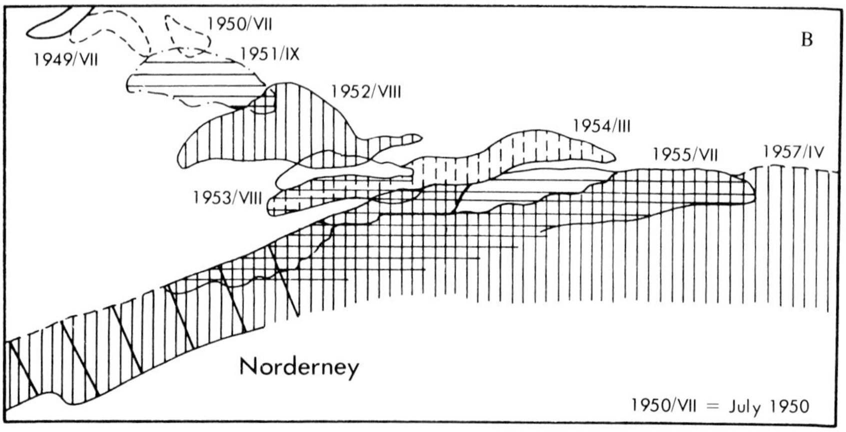

FIGURE 28. (A) Vertical airphoto of the west end of Norderney and its updrift ebb-tidal delta taken 23 June 1973. A series of intertidal bars outline the periphery of the ebb-tidal delta. Sand is supplied to this arcuate bar system from the island of Juist (outside photo to the left) and transferred to Norderney through the actual migration of individual bars through the arc. See Reference 78 for details. Airphotos obtained through the courtesy of Forschungsstelle Norderney. (B) Map of the beach attachment process of a single large swash bar at Norderney between 1949 and 1957. Note how the bar becomes gradually more shore-parallel during the attachment process. Data from Reference 44. (From Nummedal, D. and Penland, S., International Association of Sedimentologists, Spec. Publ. No. 5, Blackwell Scientific, Oxford, 1981, 187. With permission.)

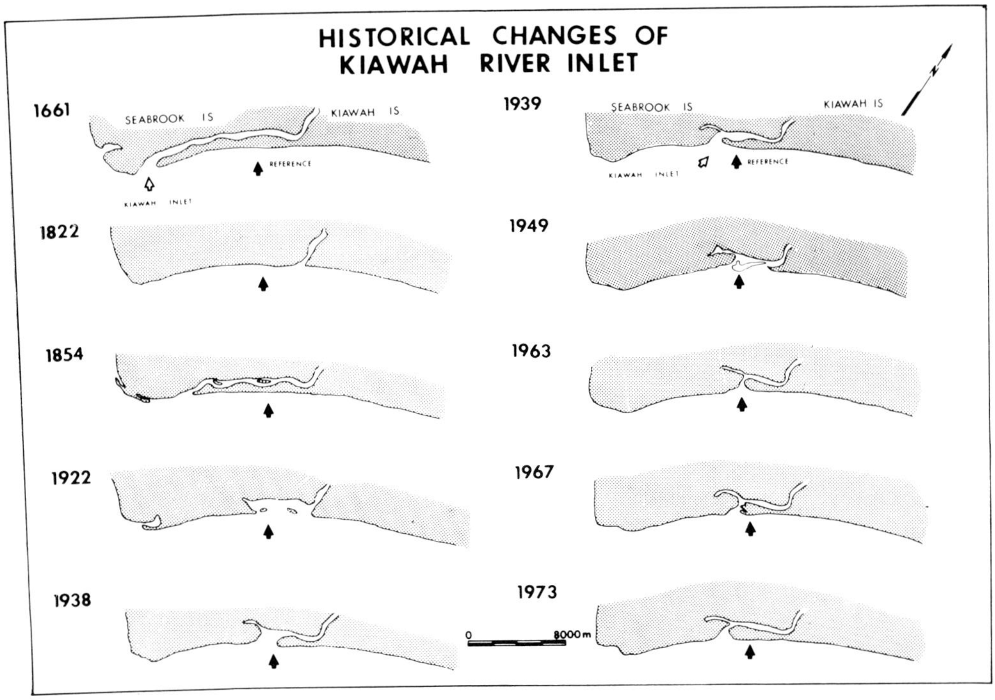

FIGURE 29. Changes in the Kiawah Island (S.C.) recurved spit in historical times. Note repeated episodes of breaching in the same area. (From FitzGerald, D. M., Hubbard, D. K., and Nummedal, D., *Proc. Coastal Zone' 78*, American Society of Civil Engineers, New York, 1978, 1973.)

teristic of inlets with higher wave energy and tide range combinations than is commonly found along the American barrier island shores.

The factors discussed so far profoundly affect the potential for safe habitation on a barrier island; their relative significance varies with the time-scales of the associated effects. The island profile is the paramount control on storm impact. Long-term changes in shoreline position, on the other hand, are influenced to an equal degree by the island profile, planimetric shape, and the pathways of inlet sediment bypassing.

THE EFFECTS OF HURRICANES

Hurricanes are a major, in many areas a dominant, agent in the development of the barrier shoreline of the U.S. Numerous hurricanes over the last two decades have provided an excellent data set on which to base generalized models for the environmental, stratigraphic, and sedimentological effects of hurricane landfalls.

Hurricanes affecting the U.S. generally are developed over the tropical Atlantic Ocean between latitudes 5° and 15°N. They cannot form farther south because of the lack of an adequate Coriolis inertial force, nor do they generally develop farther north where the sea surface temperature falls below 27°C.[94] The initial disturbance is generally found within the Intertropical Convergence Zone, and as this zone moves north into the latitude belt conducive for hurricane formation in the late summer, the hurricane season begins.

Figure 30 presents a regional breakdown in hurricane landfalls by month along the U.S. coastline. The statistics are derived from the National Climatic Center listing of all hurricanes since 1886.[71] Along the Gulf Coast there is a distinct eastward delay in the peak hurricane season. Texas has the highest probability of hurricane landfall in August; Louisiana's peak

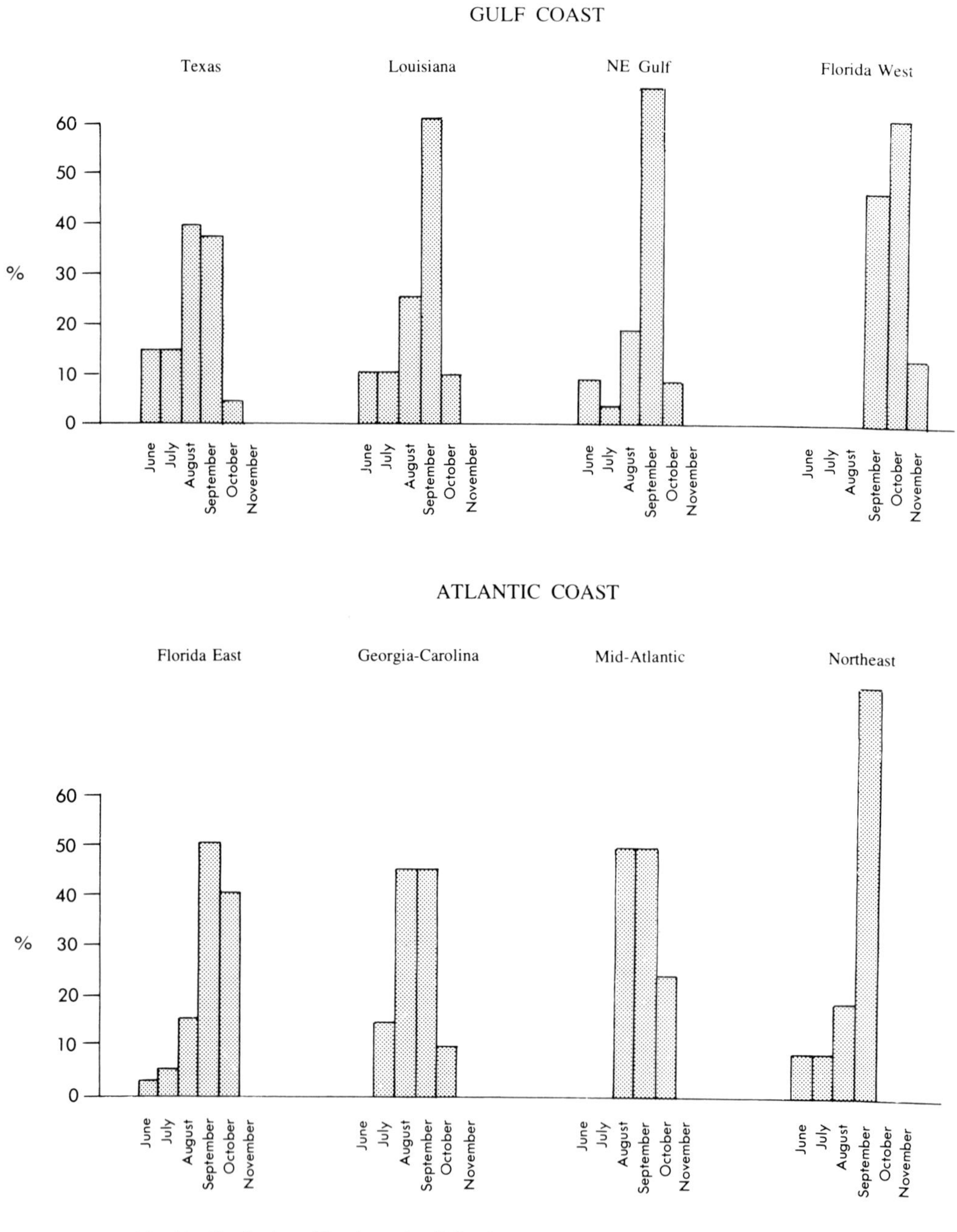

FIGURE 30. Monthly distribution of hurricane landfalls along the U.S. coast. (Compiled from data in Reference 71.)

is in September; the west coast of Florida experiences its peak season in October. August and September are generally the months of peak hurricane landfall probability along the entire east coast.

The spatial distribution of hurricane landfall probabilities is depicted in Figure 31. Based on nearly a century of hurricane records, Simpson and Lawrence[93] identified the upper Texas coast and the Mississippi delta as being the most hurricane-prone areas along the Gulf coast. Southeast Florida and North Carolina's Outer Banks have the maximum hurricane landfall probabilities along the east coast. Considering the fact that hurricane classification is con-

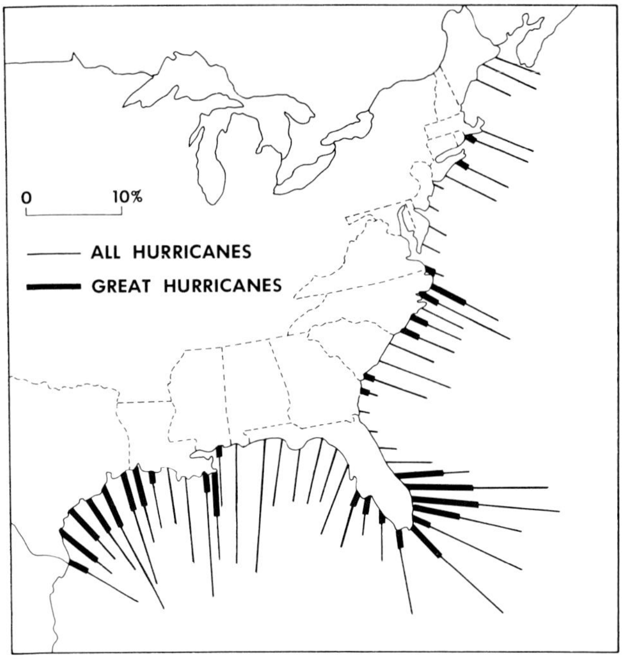

FIGURE 31. Annual hurricane landfall probabilities in the U.S. Compiled from data in Simpson and Lawrence.[93] Hurricanes refer to storms with maximum winds greater than 33 m/sec (74 mi/hr), great hurricanes have sustained winds in excess of 56 m/sec (125 mi/hr).

troversial even today, and highly suspect in pre-aircraft time, these statistics should only be considered tentative. Other estimates of hurricane landfall probabilities exist, some of which differ significantly from the distribution shown here (e.g., Ho et al.[43]).

From data in Neumann et al.[71] one can extract statistics on hurricane landfalls within any specified region in North America. For the entire Texas-Louisiana coast the average landfall frequency since the year 1900 has been about 0.64 hurricanes per year. To put this in perspective, the average number of Atlantic hurricanes per year is about 5, out of which 1.8 make landfall in the U.S.

The Neumann et al.[71] data show no strong indication of a temporal clustering in hurricane landfalls, although there is an indication of a slight peak in the 1930s and 1940s. These two decades also correspond to the warmest decades in North America during the last century.[34] Because the hurricane derives its energy from the ocean surface one would expect warmer temperatures to increase the likelihood of hurricane generation. The question is: have the temperature fluctuations over the last century been of great enough magnitude to have had a noticeable effect on tropical storm production? A preliminary analysis has yielded an affirmative answer to this question.[74] In view of the strong indication of a global warming over the next century,[34] this correlation suggests that one should be prepared for a significant increase in hurricane landfalls and storm-tide inundation levels.

As the hurricane moves across the sea it is accompanied by a superelevated water surface both because of the reduced local atmospheric pressure and the wind stress imparted to the water surface. As the hurricane approaches land the water level along the shore will rise in response to this "surge". Prior to the actual arrival of the hurricane surge water levels may already have risen a couple of feet from the effects of the long-period forerunners of the hurricane storm waves.

The wind distribution in a hurricane is generally asymmetrical with the most intense winds

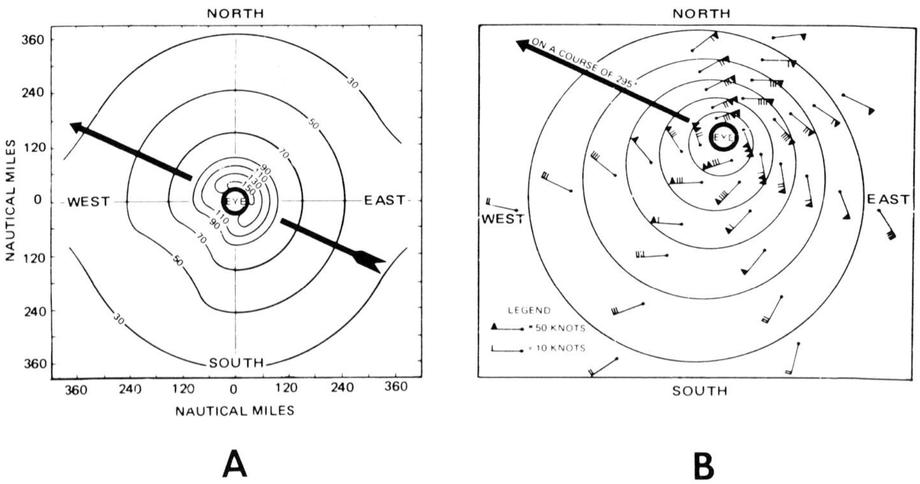

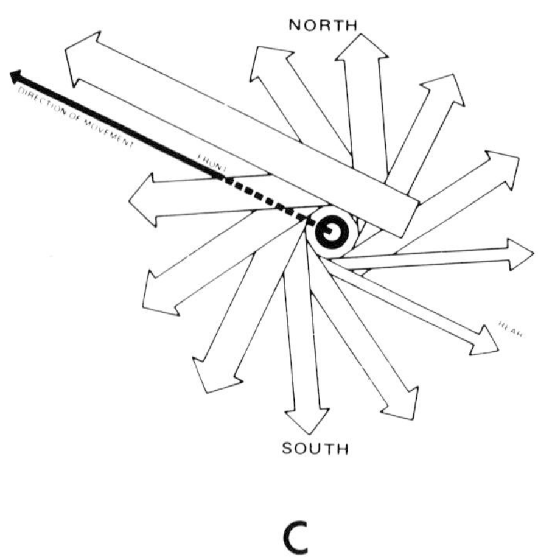

FIGURE 32. Characteristics of a tropical cyclone in the northern hemisphere. (A) Distribution of surface wind speeds (in knots) around a typical 150-kn cyclone; (B) wind circulation pattern around a typical 150-kn cyclone; (C) pattern of swell generation within a cyclone. Arrows indicate direction of travel of the swells, their widths indicate the relative heights of the generated swells. (From Crutcher, H. L., and Quayle, R. G., Mariners Worldwide Climatic Guide to Tropical Storms at Sea, Naval Weather Service Command, U.S. Government Printing Office, Washington, D.C., 1974.)

occurring to the right of the storm track because of the vectorial addition of the circulatory storm winds and the forward movement of the storm system (Figure 32). The resultant storm surge is also asymmetric with peak surge always recorded some tens of kilometers to the right of the point of landfall. The maximum wave energy is transmitted directly forward in the direction of movement of the storm (Figure 32).

The magnitude of the surge depends on the central pressure index of the hurricane (difference between pressure within the eye and that at the periphery of the storm system), the speed of travel across the water surface, the angle of the storm track relative to the local coastline, the extent of the storm, and the radius of the central vortex of maximum wind. The surge is also critically dependent on the bathymetry of the continental shelf. A wide, shallow shelf will produce a much higher surge than one which is narrow and steep.[47] The actual superelevation measured at the coastline during hurricane landfall is the resultant of the surge, the set-up due to breaking waves, and the stage of the astronomical tide. This measured water elevation is referred to as the storm tide.[10]

A series of continuously recording tide gages and surge gages (measuring only peak storm-tide elevation) have been in operation through recent hurricanes along the Gulf coast. These recorders have demonstrated the nature of the coast-wise distribution in storm-tide elevation.[35,99,100] Some of these data have been summarized for *Hurricanes Carla* (1961) and *Allen* (1980) in Figure 33. These two hurricanes were chosen for illustration because they represent the largest and most recent ones, respectively, along the northwest Gulf coast.

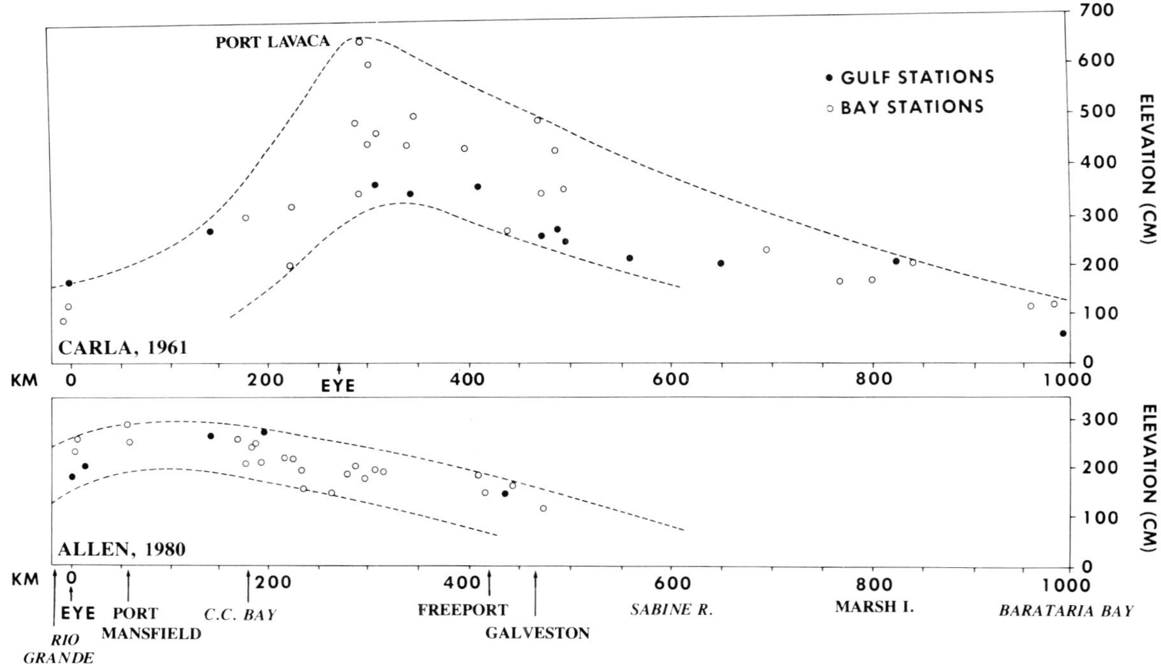

FIGURE 33. Storm tide elevations along the northwest coast of the Gulf of Mexico for hurricanes Carla and Allen. (From Nummedal, D., *Sedimentary Processes and Environments Along the Louisiana-Texas Coast,* Geological Society of America, Boulder, Colo., 1982. With permission.)

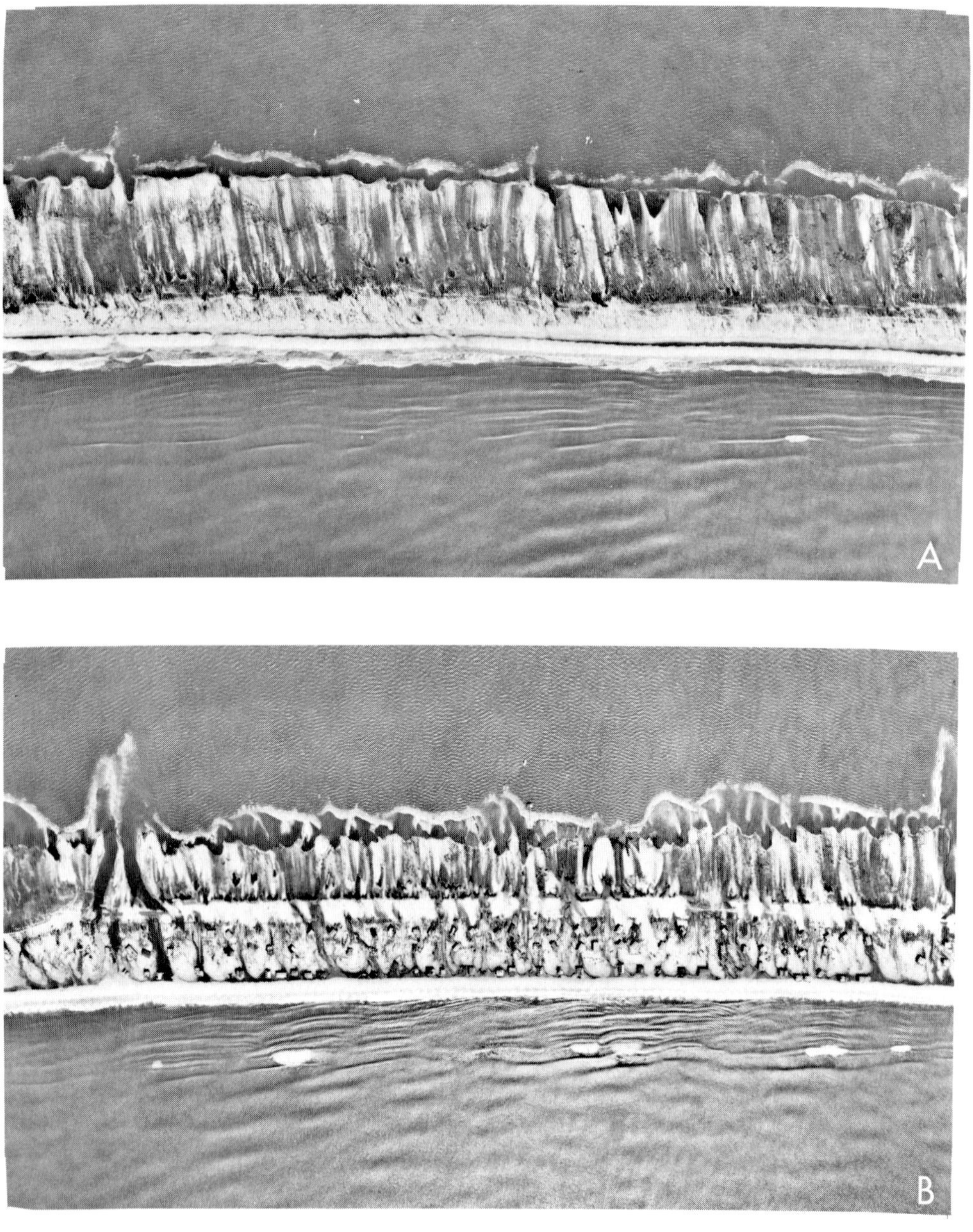

FIGURE 34. Vertical airphotos of Dauphin Island, Alabama, taken on September 22, 1979, only 9 days after the landfall of *Hurricane Frederic*. (A) Undeveloped western section of the island. (B) Developed part of the island. Note the channels formed near many of the beach-front homes, and the scour pools and sand deposits along the Mississippi Sound (top) side of the island. Airphotos obtained through the courtesy of the U.S. Army Corps of Engineers, Mobile District.

Both distributions show an asymmetry in storm-tide elevations along the coast with respect to the point of landfall of the eye. The peak tide is invariably recorded some tens of kilometers to the right of (to the north and east of) the point of landfall of the eye. Furthermore, the tide elevation as a function of distance declines much more rapidly south of (to the left of) the point of landfall than to the north (or east). Actual recorded tide elevations fall within a broad range because of local effects at the gage location.

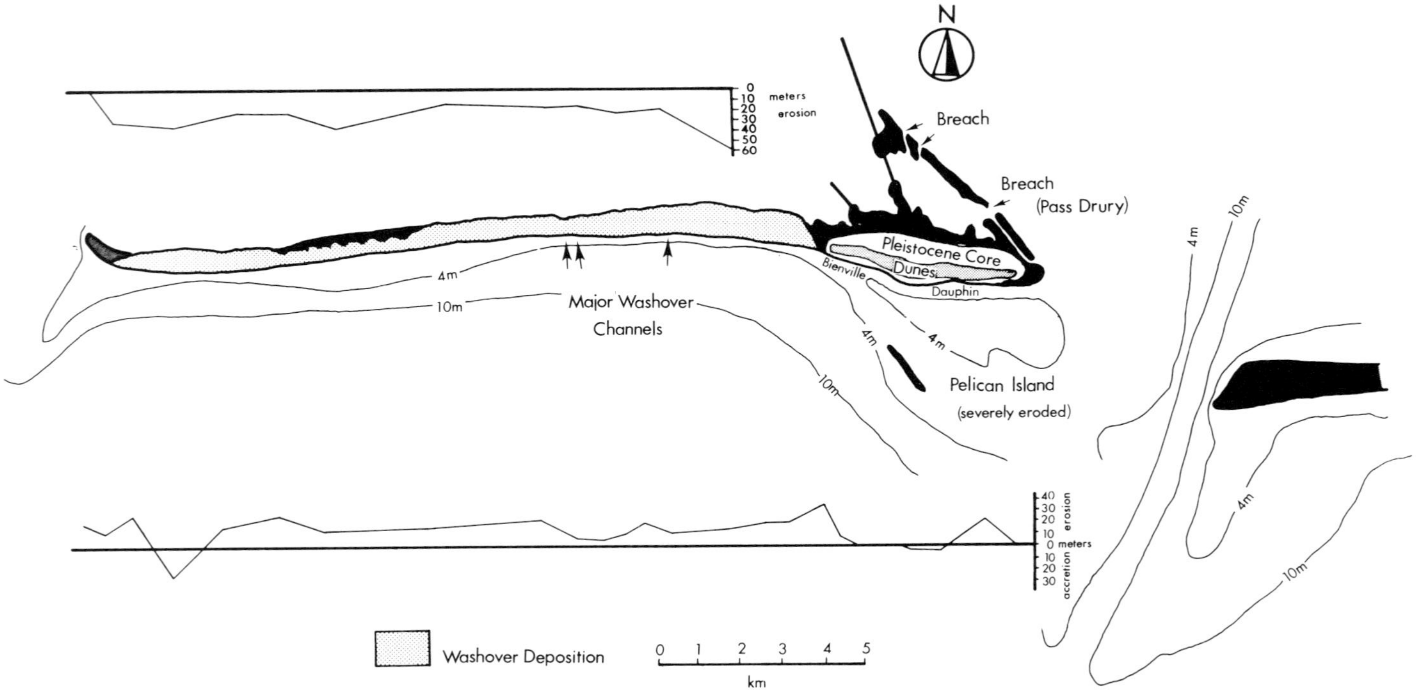

FIGURE 35. The amount of shoreline erosion on Dauphin Island by *Hurricane Frederic* was determined by comparison of pre- and poststorm aerial photographs. Note that the average amount of erosion along the Mississippi Sound side of the island (top) was about the same as that along the Gulf of Mexico beach (bottom). Minimum shoreline retreat occurred along the protected Pleistocene core at the east end of the island, maximum retreat was found immediately to the west. (From Nummedal, D., Penland, S., Gerdes, R., Schramm, W., Kahn, J., and Roberts, H., *Gulf Coast Assoc. Geol. Soc. Trans.*, 30, 183, 1980. With permission.)

The *Hurricane Carla* storm tide attained an elevation of nearly 7 m at Port Lavaca in the upper part of Matagorda Bay. This is the highest storm tide recorded along the northwest Gulf coast in recent decades. The enormous lateral extent of *Carla* is evident from Figure 33, which shows that storm tides in excess of 1 m were observed along the entire coast from the Mississippi delta to the Rio Grande. *Hurricane Allen,* which began as a major hurricane in the Caribbean, had decayed at the time of landfall and consequently produced a much smaller storm tide both in terms of maximum height (3 m at South Padre Island) and lateral extent. *Hurricane Camille* (1969) produced a 7.3-m storm tide on the coast at Bay St. Louis, Miss.[10]

Environmental effects of hurricanes, such as property damage, extensive flooding and short-term shoreline erosion are functions of wind intensity, rainfall, wave heights, and the storm tide. Effects directly attributable to the storm tide are generally the most severe. Long-term geological effects, with a distinct imprint in the stratigraphic record, appear to be directly controlled by the storm tide. Storm-tide elevations, their recurrence intervals, and the stratigraphy and morphology of the local coastline are, therefore, factors of great interest in studies of the geological effects of major storms.

The concepts and data on barrier island morphology and sediment transport together with the oceanographic data on hurricanes permit us to better understand patterns of storm-related destruction, the causes of failure of many schemes for long-term beach protection and hurricane mitigation, and to aid in the formulation of barrier land-use practices compatible with the dynamic nature of barrier islands. We will briefly review two specific examples.

Two recent Gulf coast hurricanes, *Frederic* in 1979 and *Allen* in 1980, provided ample evidence of the dramatic role of barrier island profile on the patterns and degrees of island modification. *Frederic* made landfall at Dauphin Island, Ala. at midnight September 13th, 1979, after a 2.5 day crossing of the eastern Gulf of Mexico from Cuba. At the time of landfall *Frederic* had a maximum sustained wind of about 220 km/hr and an associated storm tide of 365 cm at Gulf Shores, Ala. about 30 km east of the point of landfall of the eye. Storm tides at Dauphin Island ranged from about 300 cm at the east end to about 220 cm at the west end.[79]

Immediately following the storm an exhaustive study of island impact was conducted. The western low-profile, Holocene transgressive part of the island had been completely submerged. Strong currents had carried sediment across the island from the Gulf beach to the Mississippi Sound margin and formed large flame-shaped sand deposits in the shallow waters of the Sound (Figure 34A and B). The shoreline retreated an average of about 25 m both on the Gulf and the Sound sides (Figure 35). In fact, the already narrow Dauphin Island lost about 10% of its width in this single hurricane. Beach accretion over the months following hurricane landfall, however, restored about half of the lost land. Man-made features including driveways, paths, marina entrances, and drainage canals which were oriented perpendicular to the shore acted to locally intensify the flow and consequent scour during the washover events. All major channels cross cutting the island after the hurricane coincide with the former location of a man-made topographic low (Figure 34B).

The low-profile parts of South Padre Island, Tex. suffered in a very similar manner during *Hurricane Allen* which made landfall on August 10th, 1980. *Allen's* storm tide level was 200 cm along the Port Isabel beaches (Figure 33). Padre Island's response to *Hurricane Allen* illustrates one additional point: many preexisting hurricane-generated barrier breaches were reopened. *Hurricane Beulah* (1967) was the most recent major pre-*Allen* storm in that area. When post-*Allen* and post-*Beulah* aerial photography are compared a clear correlation is found to exist between breach locations.

That such a correlation should exist is perhaps obvious, yet most often ignored in building plans on many barrier islands. Most of the American barrier island shoreline has been hit by a hurricane or major northeast storm within the last century. Maps and aerial photography

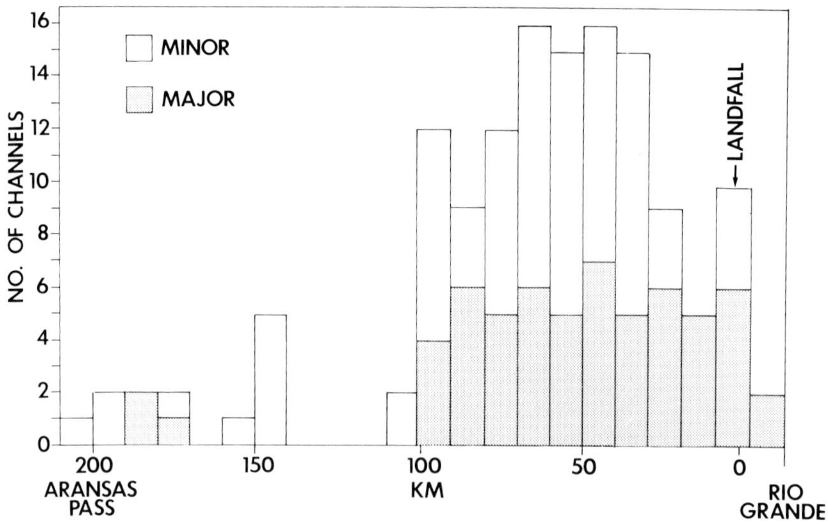

FIGURE 36. Distribution of channels across Padre and Mustang Islands, Tex. after the landfall of *Hurricane Allen* on August 10th, 1980. Major hurricane channels are cut below mean sea level and remain active until sealed by the littoral drift. Minor channels, by contrast, are not cut below sea level and are active only as long as the storm surge is present.

accurately depict the damage that was caused. This information is of great value because, as *Allen* proved, once a pattern of storm response is established future storms most probably will enhance that same pattern. This point is further demonstrated in a systematic analysis presented in Chapter 6.

The history of Dauphin Island proves this point rather dramatically. In 1916 the eastern 8 km of the Holocene low-profile part of the island was completely removed by a major hurricane. It took 20 years for the breach to heal. In 1947 a portion of this old scar was reopened by a new hurricane. This time the breach healed quickly. When the causeway linking Dauphin Island to Cedar Point on the Alabama mainland was completed in 1955 housing development began in earnest on the island. Most of this development took place within the region of that 1916 breach, largely, one would surmise, because of the unobstructed view of the ocean.

Hurricane Frederic ravaged this development. The hurricane damaged other parts of the island too, yet most of the large washover channels, the largest amount of shoreline retreat, and what by indirect evidence appears to have been the highest waves occurred within this development. There is a physical reason why the damage repeatedly has been most severe on this segment of the island. The cause is found in refraction of the incoming hurricane waves around the western margin of the ebb-tidal delta platform which is built at the entrance to Mobile Bay. This refraction focuses maximum wave energy onto the site of the housing development.[79]

In many other instances we may not yet have enough geologic and oceanographic information to identify precisely the physical mechanisms behind an established pattern of coastal storm damage. In such cases let us, at the very least, use history as a guide. Both Dauphin Island and Padre Island are part low-profile and part high-profile. On Dauphin Island the high-profile segment consists of a Pleistocene core at the east end, about 8 km long. This core is sheltered behind a 10-m-high, nearly continuous ocean-front dune ridge (Figure 35). Moreover, the core of the island, in its entirety, is located behind the large ebb-tidal delta platform of the Mobile Bay entrance. Most of the pre-causeway homes and some of the recent ones, are built behind this dune. They suffered minor damage during *Frederic*. Even

FIGURE 37. Undermining by *Hurricane Allen* of one of the units at the "Lost Colony" condominiums at Mustang Island, Tex. illustrates that dune scarp retreat is the typical storm response on a high-profile regressive barrier island. Contrast this response to that illustrated by the low-profile Dauphin Island (Figure 34). Picture taken on August 15th, 1980, 5 days after landfall of *Allen*. Mounds of sand on the right have been dumped there by truck after the storm.

the beach front of the island core suffered only slight erosion, this because the shallow tidal-delta platform caused dissipation of hurricane wave energy far offshore (Figure 35).

The large-scale effect of *Allen's* landfall was to inundate almost entirely that part of Padre Island which lies south of the Mansfield Ship Channel. That segment of the barrier was covered with an almost continuous washover terrace. The terrace was segmented by numerous channels. Figure 36 plots the coast-wise distribution of these channels based on mapping from NASA high-altitude photography obtained one week after landfall. The sudden decrease in channel density about 100 km north of the point of landfall of the hurricane eye corresponds to the equally abrupt transition from low-profile transgressive South Padre Island to high-profile stable or regressive central and north Padre.

The southern part of Mustang Island also is low-profile because of the recent migrations of Corpus Christi Pass and related tidal inlets.[67] That part of the coast responded to the landfall of *Allen* in a fashion very similar to South Padre Island. There were a lot fewer channels, however.

The high-profile, regressive barrier island segments responded to the storm surge through dune-scarp retreat (Figure 37) and the deposition of small interdune fans. Along Mustang Island, which is all high-profile to the north of the passes, the dune scarp was eroded back about 30 m by *Hurricane Allen*. A number of small interdune fans occupy former dune ridge blowouts and other small depressions. The beach in front of this dune scarp is slowly recovering.

Housing on high-profile barriers is relatively safe compared to that on low-profile islands

provided the developments are set far enough back from the beach to provide space for storm-related displacements of the dune scarp. A general reconnaissance of America's barrier islands unfortunately reveals that most developments occur on low-profile islands. It also reveals that the lessons of history are generally ignored.

The geological effects of hurricane landfall go beyond the island surface proper which has been the focus of this discussion. Hayes[37] and Brown et al.[8] present detailed models for the overall coastal geologic impact of hurricanes. Morton[65] has analyzed the sedimentary effects of hurricane wind-driven circulation on the back-barrier tidal flats, and also discusses[66] possible mechanisms for the deposition of coarse-grained sedimentary units on the Texas shelf during hurricanes.

CONCLUSIONS

This discussion of barrier island stratigraphy, morphology, and sedimentation has revealed that the knowledge now exists to provide scientific input into plans for the optimum utilization of this fragile coastal ribbon of sand.

Any barrier island is but one link in a continuously changing system which spatially involves adjacent tidal inlets, old river deltas, other barriers and the inner continental shelf. Temporally, the characteristics of a barrier island depend on its stage of evolution. The long-range dynamic equilibrium can only be maintained if the sand is left to freely migrate between all related components of the system.

The barrier islands have evolved according to physical laws. Their morphology is a predictable response to external environmental parameters (sea level history, sediment supply, tide range, wave energy). Human modifications of the barrier islands have consequences which also are predictable according to these same laws of nature. Storms will impact our barrier coastlines according to the laws of atmospheric science and oceanography. The barrier response can be predicted. Coastal storm disasters are not "acts of God", but the results of society's willful ignorance of existing knowledge.

REFERENCES

1. **Beard, J. H., Sangree, J. B., and Smith, L. A.,** Quaternary chronology, paleoclimate, depositional sequences and eustatic cycles, *Am. Assoc. Petrol. Geol. Bull.*, 66, 158, 1982.
2. **Belknap, D. F. and Kraft, J. C.,** Holocene relative sea-level changes and coastal stratigraphic units on the northwest flank of the Baltimore Canyon Trough geosyncline, *J. Sediment. Petrol.*, 47, 610, 1977.
3. **Belknap, D. F. and Kraft, J. C.,** Preservation potential of transgressive coastal lithosomes on the U.S. Atlantic shelf, *Mar. Geol.*, 42, 429, 1981.
4. **Bernard, H. A., LeBlanc, R. J., and Major, C. F.,** Recent and Pleistocene geology of southeast Texas: *Geology of the Gulf Coast and Guidebook of Excursions,* Excursion #3, Geological Society of America, Boulder, Colo., 1962, 174.
5. **Bernard, H. A. and LeBlanc, R. J.,** Resume of the Quaternary geology of the northwestern Gulf of Mexico province, in *The Quaternary of the United States,* Wright, H. E., Jr., and Frey, D. G., Eds., Princeton University Press, Princeton, N.J., 1965, 137.
6. **Blackwelder, B. W.,** Late Wisconsin and Holocene tectonic stability of the United States mid-Atlantic coastal region, *Geology,* 8, 534, 1980.
7. **Blatt, H., Middleton, G., and Murray, R.,** *Origin of Sedimentary Rocks,* Prentice-Hall, Englewood Cliffs, N.J., 1972.
8. **Brown, L. F., Jr., Morton, R. A., McGowen, J. H., Kreitler, C. W., and Fisher, W. L.,** *Natural Hazards of the Texas Coastal Zone,* Spec. Rep. University of Texas, Austin, 1974.
9. **Brooks, M. N., Colquhoun, D. J., Pardi, R. R., Newman, W., and Abbott, W. J.,** Preliminary archaeological and geological evidence for Holocene sea level fluctuations in the lower Cooper River Valley, S.C., *Fla. Anthropologist,* 32, 85, 1979.

10. **CERC,** *The Shore Protection Manual,* Coastal Engineering Research Center, Ft. Belvoir, Va., 1975.
11. **Coates, D. R., Ed.** *Coastal Geomorphology,* State University of New York, Binghamton, 1973.
12. **Crutcher, H. L. and Quayle, R. G.,** Mariners Worldwide Climatic Guide to Tropical Storms at Sea, Naval Weather Service Command, U.S. Government Printing Office, Washington, D.C., 1974.
13. **Curray, J. R.,** Transgressions and regressions, in *Papers in Marine Geology — Shepard Commemorative Volume,* Miller, R. L., Ed., Macmillan, New York, 1964, 175.
14. **Curray, J. R.,** Late Quaternary history, continental shelves of the United States, in *The Quaternary of the United States,* Wright, H. E., Jr. and Frey, D. G., Eds., Princeton University Press, Princeton, N.J., 1965, 723.
15. **Davies, J. L.,** A morphogenetic approach to world shorelines, *Z. Geomorphol.,* 8, 27, 1964.
16. **Davies, J. L.,** *Geographical Variation in Coastal Development,* Hafner Publ., New York, 1973.
17. **De Beaumont, E.,** *Lecons de Geologie Practique,* Paris, 1845, 223.
18. **Defant, A.,** *Ebb and Flow,* University of Michigan Press, Ann Arbor, 1958.
19. **Dietz, R. S. and Holden, J. C.,** The breakup of Pangea, *Sci. Am.,* 223, 30, 1970.
20. **Dillon, W. D. and Oldale, R. N.,** Late Quaternary sea level curve: reinterpretation based on glacio-eustatic influence, *Geology,* 6, 56, 1978.
21. **Fairbridge, R. W.,** Shellfish-eating preceramic indians in coastal Brazil, *Science,* 191, 353, 1976.
22. **Fairbridge, R. W. and Krebs, O. A., Jr.,** Sea-level and the southern oscillation, *Geophys. J.,* 6, 532, 1962.
23. **Field, M. E., and Duane, D. B.,** Post-Pleistocene history of the United States inner continental shelf: Significance to the origin of barrier islands, *Geol. Soc. Am. Bull.,* 87, 691, 1976.
24. **Field, M. E., Meisburger, E. P., Stanley, E. A., and Williams, S. J.,** Upper Quaternary peat deposits on the Atlantic inner shelf of the United States, *Geol. Soc. Am. Bull.,* 90, 618, 1979.
25. **Fischer, A. G.,** Stratigraphic record of transgressing seas in light of sedimentation of the Atlantic coast of New Jersey, *Am. Assoc. Petrol. Geol. Bull.,* 45, 1656, 1961.
26. **Fisk, H. N.,** Bar-finger sands of the Mississippi delta, in *Geometry of Sandstone Bodies,* Paterson, J. A. and Osmond, J. C., Eds., Spec. Publ., American Association of Petroleum Geologists, Tulsa, Okla., 1961.
27. **FitzGerald, D. M., Hubbard, D. K., and Nummedal, D.,** Shoreline changes associated with tidal inlets along the South Carolina coast, *Proc. Coastal Zone '78,* American Society of Civil Engineers, New York, 1978, 1973.
28. **FitzGerald, D. M., Nummedal, D., and Kana, T. W.,** Sand circulation pattern at Price Inlet, South Carolina, *Proc. 15th Coastal Eng. Conf.,* American Society of Civil Engineers, New York, 1976, 1868.
29. **Frazier, D. E.,** Recent deltaic deposits of the Mississippi River, their development and chronology, *Gulf Coast Assoc. Geol. Soc. Trans.,* 17, 287, 1967.
30. **Galloway, W. E., Hobday, D. K., and Magara, K.,** Frio Formation of Texas Gulf coastal plain: depositional systems, structural framework, and hydrocarbon distribution, *Am. Assoc. Petrol. Geol.,* 66, 649, 1982.
31. **Gerdes, R. G.,** Stratigraphy and History of Development of the Caminada-Moreau Beach Ridge Plain, Southeast Louisiana, M.Sc. thesis, Department of Geology, Louisiana State University, Baton Rouge, 1982.
32. **Gilbert, G. K.,** The topographic features of lake shores. U.S. Geol. Survey, 5th Ann. Rep., U.S. Department of Commerce, Washington, D.C., 1885.
33. **Gornitz, V., Lebedeff, S., and Hansen, J.,** Global sea level trend in the past century, *Science,* 215, 1611, 1982.
34. **Hansen, J., Johnson, D., Lacias, A., Lebedeff, S., Lee, P., Rind, D., and Russell, G.,** Climate impact of increasing atmospheric carbon dioxide, *Science,* 213, 957, 1981.
35. **Harris, D. L.,** Characteristics of the hurricane storm surge, Tech. Pap. no. 48, U.S. Weather Bureau, U.S. Department of Commerce, Washington, D.C., 1963.
36. **Hayes, M. O.,** Sedimentation on a Semiarid, Wave-Dominated Coast (South Texas), with emphasis on Hurricane Effects, Ph.D. dissertation, University of Texas, Austin, 1965.
37. **Hayes, M. O.,** *Hurricanes as Geological Agents: Case Studies of Hurricane Carla, 1961, and Cindy, 1963,* Bur. Econ. Geol. Rep. Inv. 61, University of Texas, Austin, 1967.
38. **Hayes, M. O.,** Barrier island morphology as a function of tidal and wave regime, in *Barrier Islands from the Gulf of St. Lawrence to the Gulf of Mexico,* Leatherman, S. P., Ed., Academic Press, New York, 1979, 1.
39. **Hayes, M. O., FitzGerald, D. M., Hulmes, L. J., and Wilson, S. J.,** Geomorphology of Kiawah Island, South Carolina, in *Terrigenous Clastic Depositional Environments,* Tech Rep. CRD-11, Hayes, M. O. and Kana, T. W., Eds., University of South Carolina, Columbia, 1976, II-80.
40. **Hayes, M. O. and Kana, T. W., Eds.,** *Terrigenous Clastic Depositional Environments,* Tech. Rep. CRD-11, University of South Carolina, Columbia, 1976.
41. **Hays, J. D. and Pitman, W. C.,** Lithospheric plate motions, sea-level changes and climatic and ecological consequences, *Nature (London),* 246, 18, 1973.

42. **Hicks, S. D.**, Long-period variations in secular sea level trends, *Shore Beach,* 36, 32, 1968.
43. **Ho, E. P., Schwerdt, R. W., and Goodyear, H. V.**, Some climatological characteristics of hurricanes and tropical storms, Gulf and East coasts of the United States, *NOAA Tech. Rep.*, NWS-15, 1975.
44. **Homeier, H. and Kramer, J.**, Verlagerung der Platen im Riffbogen von Norderney und ihre Anlandung an den Strand, *Jahr. Forsch. Norderney,* 8, 37, 1957.
45. **Hoyt, J. H.**, Barrier island formation, *Geol. Soc. Am. Bull.*, 78, 1125, 1967.
46. **Inman, D. L. and Nordstrom, C. E.**, On the tectonic and morphologic classification of coasts, *J. Geol.*, 79, 1, 1971.
47. **Jelesnianski, C. P.**, SPLASH — Special program to list amplitudes of surges from hurricanes, *NOAA Tech. Memo,* NWS TDL-46, 1972.
48. **Klein, G. de V.**, Depositional and dispersal dynamics of intertidal sand bars, *J. Sediment. Petrol.*, 40, 1095, 1970.
49. **Kolb, C. R. and Van Lopik, J. R.**, Depositional environments of the Mississippi River deltaic plain, southeastern Louisiana, in *Deltas in Their Geologic Framework,* Shirley, M. L., Ed., Houston Geological Society, Tex., 1966, 17.
50. **Komar, P. D., Lizarraya-Areiniega, J. R., and Terich, T. A.**, Oregon coast shoreline changes due to jetties, *J. Waterways, Harbors Coastal Eng.*, 102, 13, 1976.
51. **Kumar, N.**, Barrier island facies, in *The Encyclopedia of Sedimentology,* Fairbridge, R. W. and Bourgeois, J., Eds., Dowden, Hutchinson & Ross, Stroudsburg, Pa., 1978, 33.
52. **Kraft, J. C.**, Sedimentary facies patterns and geological history of a Holocene marine transgression, *Geol. Soc. Am. Bull.*, 82, 2131, 1971.
53. **Kraft, J. C., Allen, E. A., Belknap, D. F., John, C. J., and Maurmeyer, E. M.**, Delaware's Changing Shoreline, Tech. Rep. No. 1, Delaware Coastal Zone Management Program, Lewes, Del., 1976.
54. **Kraft, J. C. and John, C. J.**, Lateral and vertical facies relations of transgressive barriers, *Am. Assoc. Petrol. Geol.*, 63, 2145, 1979.
55. **Lohse, E. A.**, Dynamic geology of the modern coastal region, northwest Gulf of Mexico, in *Finding Ancient Shorelines,* Spec. Publ. 3, Society of Economic Paleontologists and Mineralogists, Tulsa, Okla., 1955, 99.
56. **Luck, G.**, Islands in front of the southern North Sea Coast, *Die Küste,* 32, 85, 1978.
57. **Manabe, S. and Stouffer, R. J.**, Sensitivity of a global climate model to an increase of CO_2 concentration in the atmosphere, *J. Geophys. Res.*, 85, 5519, 1980.
58. **Matthews, R. K. and Pore, R. Z.**, Tertiary $\delta^{18}O$ record and glacio-eustatic sea-level fluctuations, *Geology,* 8, 501, 1980.
59. **McGowen, J. H., Garner, L. E., and Wilkinson, B. H.**, *The Gulf Shoreline of Texas: Processes, Characteristics, and Factors in Use,* Geol. Circ. 77-3, University of Texas, Austin, 1977, 27.
60. **McGowen, J. H. and Scott, A. J.**, Hurricanes as geologic agents on the Texas coast, in *Estuarine Research,* Vol. 2, Cronin, L. E., Ed., Academic Press, New York, 1975, 23.
61. **Milliman, J. D. and Emery, K. O.**, Sea-levels during the past 35,000 years, *Science,* 162, 1121, 1968.
62. **Moerner, N. A.**, Eustatic changes during the last 8,000 years in view of radiocarbon calibration and new information from the Kattegat region and other northwestern European coastal areas, *Paleogeogr. Paleoclimatol. Paleoecol.*, 19, 63, 1976.
63. **Morton, R. A.**, Nearshore changes at jettied inlets, Texas Coast, *Proc. Coastal Sediments '77,* American Society of Civil Engineers, New York, 1977, 267.
64. **Morton, R. A.**, Temporal and spatial variations in shoreline changes and their implications, examples from the Texas Gulf Coast, *J. Sediment. Petrol.*, 49, 1101, 1979.
65. **Morton, R. A.**, Subaerial storm deposits formed on barrier flats by wind-driven currents, *Sediment. Geol.*, 24, 105, 1979.
66. **Morton, R. A.**, *Formation of Storm Deposits by Wind-Forced Currents in the Gulf of Mexico and North Sea,* Spec. Publ. 5, International Association of Sedimentologists, Blackwell Scientific, Oxford, 1981, 385.
67. **Morton, R. A. and McGowen, J. H.**, *Modern Depositional Environments of the Texas Coast,* Guidebook 20, University of Texas, Austin, 1980, 167.
68. **Morton, R. A. and Nummedal, D.**, Regional geology of the N.W. Gulf coastal plain, in *Sedimentary Processes and Environments along the Louisiana-Texas Coast,* Nummedal, D., Ed., Geological Society of America, Boulder, Colo., 1982.
69. **Moslow, T. E.**, Stratigraphy of Mesotidal Barrier Islands, Ph.D. dissertation, University of South Carolina, Columbia, 1980.
70. **Moslow, T. E. and Heron, S. D., Jr.**, Holocene depositional history of a microtidal cuspate foreland cape: Cape Lookout, North Carolina, *Mar. Geol.*, 41, 251, 1981.
71. **Neumann, C. J., Cry, G. W., Caso, E. L., and Jarvinen, B. R.**, Tropical cyclones of the North Atlantic Ocean, 1971-1977, National Climatic Center, Asheville, N. C., 1978.
72. **Nummedal, D.**, Coarse-grained sediment dynamics — Beaufort Sea, Alaska, *Proc. Port and Ocean Eng. Arctic Conditions, 5th Conf.*, Norwegian Institute of Technology, Trondheim, 1979, 845.

73. **Nummedal, D.,** Future sea level changes along the Louisiana coast, *Proc. Conf. Coastal Erosion Wetlands Modification in Louisiana,* Louisiana Universities Marine Consortium, Cocodrie, 1982, 164; U.S. Fish and Wildlife Service, FWS/OBS-82-59, Washington, D.C.
74. **Nummedal, D.,** Hurricane landfalls along the N.W. Gulf coast, in *Sedimentary Processes and Environments Along the Louisiana-Texas Coast,* Geological Society of America, Boulder, Colo., 1982.
75. **Nummedal, D., Ed.,** *Deltaic Sedimentation on the Louisiana Coast,* Society of Economic Paleontologists and Mineralogists, Tulsa, Okla., 1982.
76. **Nummedal, D. and Fischer, I. A.,** Process-response models for depositional shorelines: the German and the Georgia Bights, *Proc. 16th Coastal Eng. Conf.,* American Society of Civil Engineers, New York, 1978, 1215.
77. **Nummedal, D., Oertel, G. F., Hubbard, D. K., and Hine, A. C.,** Tidal inlet variability — Cape Hatteras to Cape Canaveral, *Proc. Coastal Sediments '77,* American Society of Civil Engineers, New York, 1977, 543.
78. **Nummedal, D. and Penland, S.,** *Sediment Dispersal in Norderneyer Seegat, West Germany,* Spec. Publ. No. 5, International Association of Sedimentologists, Liege, Belgium, 1981, 187.
79. **Nummedal, D., Penland, S., Gerdes, R., Schramm, W., Kahn, J., and Roberts, H.,** Geologic response to hurricane impact on low-profile Gulf Coast Barriers, *Gulf Coast Assoc. Geol. Soc. Trans.,* 30, 183, 1980.
80. **Oldale, R. N.,** Glacial deposits and features of outer Cape Cod, in *Environmental Geologic Guide to Cape Cod National Seashore,* Leatherman, S. P., Ed., Society of Economic Paleontologists and Mineralogists, Tulsa, Okla., 1979.
81. **Penland, S. and Boyd, R.,** Mississippi delta barrier systems: an overview, in *Deltaic Sedimentation of the Louisiana Coast,* Nummedal, D., Ed., Society of Economic Paleontologists and Mineralogists, Tulsa, Okla., 1982, 71.
82. **Penland, P. S., Boyd, R., Nummedal, D., and Roberts, H.,** Deltaic barrier development on the Louisiana Coast, *Gulf Coast Assoc. Geol. Soc. Trans.,* 31, 471, 1981.
83. **Pilkey, O. H., Blackwelder, B. W., Knebel, H. J., and Ayers, M. W.,** The Georgia Embayment continental shelf: stratigraphy of a submergence, *Geol. Soc. Am. Bull.,* 92, 52, 1981.
84. **Pitman, W. C.,** The relationship between eustacy and stratigraphic sequences of passive margins, *Geol. Soc. Am. Bull.,* 89, 1389, 1978.
85. **Price, W. A. and Parker, R. H.,** Origins of permanent inlets separating barrier islands and influence of drowned valleys on tidal records along the Gulf coast of Texas, *Gulf Coast Assoc. Geol. Soc. Trans.,* 29, 371, 1979.
86. **Reimnitz, E., Barnes, P. W., and Melchior, J.,** Changes in barrier island morphology — 1949 to 1975, Cross Island, Beaufort Sea, Alaska, Open-file rep. 77-477, U.S. Geological Survey, part F, U.S. Department of Commerce, Washington, D.C., 1977.
87. **Rhode, H.,** The history of the German coastal area, *Die Küste,* 32, 7, 1978.
88. **Saxena, R. S.,** *Modern Mississippi Delta — Depositional Environments and Processes,* American Association of Petroleum Geologists, Tulsa, Okla., 1976.
89. **Schwartz, M. L.,** The multiple causality of barrier islands, *J. Geol.,* 79, 91, 1971.
90. **Scruton, P. C.,** Delta building and the deltaic sequence, in *Recent Sediments, Northwest Gulf of Mexico,* Shepard, F. P., Phleger, F. B., and van Andel, T. H., Eds., American Association of Petroleum Geologists, Tulsa, Okla., 1960, 82.
91. **Shepard, F. P.,** *Submarine Geology,* 3rd ed., Harper & Row, New York, 1973.
92. **Silvester, R.,** *Coastal Engineering,* Vol. 2, Elsevier, Amsterdam, 1974.
93. **Simpson, R. H. and Lawrence, M. B.,** Atlantic hurricane frequencies, *NOAA Tech. Mem.,* NWS SR-58, 1971.
94. **Simpson, R. H. and Riehl, H.,** *The Hurricane and Its Impact,* Louisiana State University Press, Baton Rouge, 1981.
95. **Stephen, M. F.,** Effects of Seawall Construction on Beach and Inlet Morphology and Dynamics at Caxambas Pass, Florida, Ph.D. dissertation, University of South Carolina, Columbia, 1981.
96. **Stephen, M. F., Brown, P. J., FitzGerald, D. M., and Hubbard, D. K.,** Beach erosion inventory of Charleston County, South Carolina: a preliminary report, Sea Grant Tech. Rep. no 4, Columbia, S. C., 1975.
97. **Swift, D. J. P.,** Barrier island genesis: Evidence from the central Atlantic shelf, eastern U.S.A., *Sediment. Geol.,* 14, 1, 1975.
98. **Swift, D. J. P.,** Continental shelf sedimentation, in *Marine Sediment Transport and Environmental Management,* Stanley, D. J. and Swift, D. J. P., Eds., John Wiley & Sons, New York, 1976, 311.
99. **USACE,** Galveston District, Report on Hurricane Allen, 3-10 August, 1980, U.S. Army Corps of Engineers, Washington, D.C., 1981.
100. **USACE,** Mobile District, Hurricane Frederic, Post Disaster Report, 30 August-14 September, 1979, U.S. Army Corps of Engineers, Washington, D.C., 1981.

101. **van Heerden, I. Ll.,** Deltaic sedimentation in Atchafalaya Bay (south-central Louisiana), in *Deltaic Sedimentation on the Louisiana Coast,* Nummedal, D., Ed., Society of Economic Paleontologists and Mineralogists, Tulsa, Okla., 1982, 40.
102. **Walton, T. L., Jr.,** Equilibrium shores and coastal design, *Proc. Coastal Sediment '77,* American Society of Civil Engineers, New York, 1977, 1.
103. **White, W. A., Morton, R. A., Kerr, R. S., Kuenzi, W. D., and Brogden, W. B.,** *Land and Water Resources, Historical Changes, and Dune Criticality, Mustang and North Padre Islands, Texas,* Rep. Invest., 92, University of Texas, Austin, 1978.
104. **Wilkinson, B. H.,** Matagorda Island, Texas: the evolution of a Gulf coast barrier complex, *Geol. Soc. Am. Bull.,* 86, 959, 1975.
105. **Winker, C. D.,** Late Pleistocene Fluvial-Deltaic Deposítion, Texas Coastal Plain and Shelf, M.Sc. thesis, University of Texas, Austin, 1979.
106. **Winker, C. D. and Howard, J. D.,** Correlation of tectonically deformed shorelines on the southern Atlantic coastal plain, *Geology,* 5, 123, 1977.

Chapter 6

PATTERNS AND PREDICTION OF SHORELINE CHANGE

Robert Dolan and Bruce Hayden

INTRODUCTION

Inspection of satellite photographs of the Atlantic coast reveals many large-scale, crescent-shaped embayments. Some of these patterns are the result of variations in the rates of shoreline processes. For example, the rate of erosion on the 11 Virginia barrier islands varies with the orientation of the shoreline, with erosion greatest where the islands are oriented near 28° east of north, and less at smaller and larger angles. At some places the shorelines are concave and at other locations the shorelines are convex, forming what we call "false capes".[3]

The largest of the crescentic shoreline landforms are the broad arcs of the North Carolina Outer Banks, which span distances of approximately 100 km. Within these arcs are lesser arcs ranging in size from a few kilometers to several tens of kilometers.[2] Smaller crescentic forms occur within these large arcs, including beach cusps (10 to 30 m) and giant beach cusps (100 to 200 m).[3] Inshore bars and troughs may also assume crescentic and rhythmic configurations in response to sea states, tides, and sea level. Smaller forms appear, disappear, and may migrate along the shoreline, while larger ones are fixed at least for the time scales (four decades) of our observations.[9] The larger forms establish the spatial context for shoreline distribution of erosion and storm damage (Figure 1).

Crescentic coastal landforms must result from differential and periodic shorezone processes. Several investigators have described the processes responsible for smaller forms, such as beach cusps,[1,11,15-17] and we have showed that along-the coast periodicities exist in shoreline rates of change at scales ranging from a few kilometers to tens of kilometers, based on analyses of rates of shoreline change (recession or progradation) and the landward limit of overwash penetration.[4,5,13] We have suggested that these patterns may be associated with standing edge waves or some related process, although we cannot make a definite statement on the process-response relationship at this time. In this chapter we will summarize our research on the patterns in long-term rates of shoreline change and overwash penetration for a 1300-km section of the Atlantic coast.

THE DATA BASE

At the University of Virginia we have developed a common-scale aerial photographic mapping system (called the orthogonal grid mapping system or OGMS) to produce data on changes in the shoreline and overwash penetration position at 100-m intervals along the coast.[4] To date, we have completed measurements at this resolution for 1300 km of the Atlantic coast between New Jersey and South Carolina. The method provides for the rapid and systematic acquisition of shoreline and storm penetration data from sequences of historical aerial photographs. A 100-m sampling interval is used because it is smaller than the crescentic shoreline forms found along the Atlantic coast other than beach-face cusps.

The simple definition of a shoreline is the physical interface of land and water (Figure 2). The position of the shoreline on the beach face is, however, variable because water levels change with tides and waves. The slopes or gradients of the beaches along the Atlantic and Gulf coasts vary from 1:10 to 1:50. With a tidal range of approximately 1.0 to 1.25 m,

FIGURE 1. Storm damage along the mid-Atlantic coast due to the Ash Wednesday Storm of March 1962.

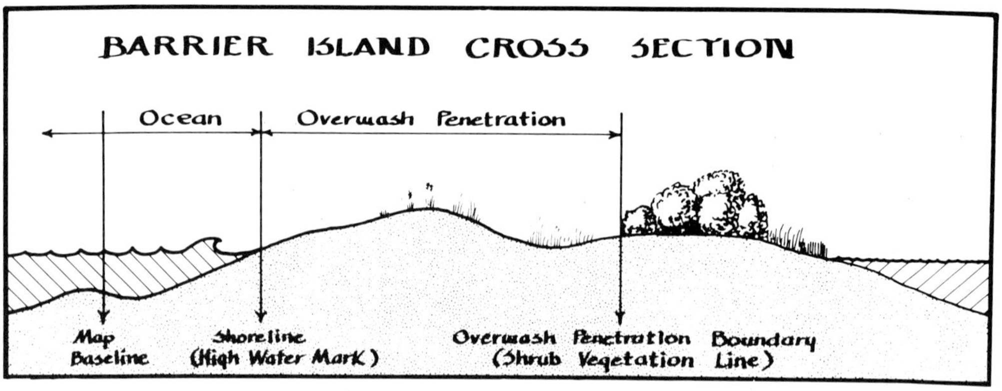

FIGURE 2. Definition of the shoreline (SL) and the landward limit of overwashed lands.

intersection of the beach and ocean has a horizontal range during a tidal cycle of about 10 to 60 m for 1:10 and 1:50 slopes, respectively. This range in shoreline positions is unacceptable in our mapping program. Two alternatives are available: (1) correct all the photos for the tidal stage and the beach slopes present at the time the photos were taken and map the mean water line, or (2) define some other more stable marker of the shoreline which is less dependent on tidal stage.

It was determined early in our research that although it would be possible to reposition the shoreline as shown in the historical aerial photography to account for tidal stage, secondary factors such as beach slopes, sediment size, and wind and wave conditions at the time of the photography would be almost impossible to determine. These secondary factors are necessary for computation of the mean high water line. In addition, scheduling future flights to coincide with ideal tide, wave, and wind conditions could also be difficult or impossible. We decided, therefore, to explore the use of the high water line (HWL) as a substitute for the mean high water line.

High Water Line (HWL)

The HWL is reestablished with each high tide as the upper beach is wetted. The resulting boundary between moist and dry sand is evident on both color and black and white aerial photography as a distinct tonal change (Figure 3). Variations in position of the HWL during the lunar tidal cycle, as recognized on aerial photographs, are of two types: (1) up-beach (landward) movement associated with the rising tide and (2) down-beach (seaward) movement forced by drying of the sand surface. At high tide, the swash terminus and HWL are the same. As the high tide falls both the swash terminus and HWL move seaward. The HWL, however, moves much more slowly than the swash terminus. The extent of the horizontal displacement of the swash terminus is a function of the slope of the beach, the tidal range, roughness of the beach face, and the wave height and period at the time of wave runup. In addition, occasional variations in water level due to longer-period tidal components, storm surges, and wind set-up and set-down may further increase the horizontal displacement. Horizontal displacement of the HWL seaward during the falling tide has the additional factor of drying of the beach sand. The drying process is slow except when winds are fast and temperatures are high. Thus, the HWL has a smaller horizontal displacement than the swash terminus and is, therefore, a more suitable choice for measuring shoreline changes.

We conducted a series of field tests along the Outer Banks of N. C. and at Virginia Beach, Va. to determine the variability of the HWL over the course of several tidal cycles.[8] The results indicate that on medium-size sand beaches with slopes in the 1:10 to 1:50 range, the average movement of the HWL over a tidal cycle is 1 to 2 m. These tests were carried out

FIGURE 3. Tonal changes at the point of the high water line (HWL) are evident in aerial photographs.

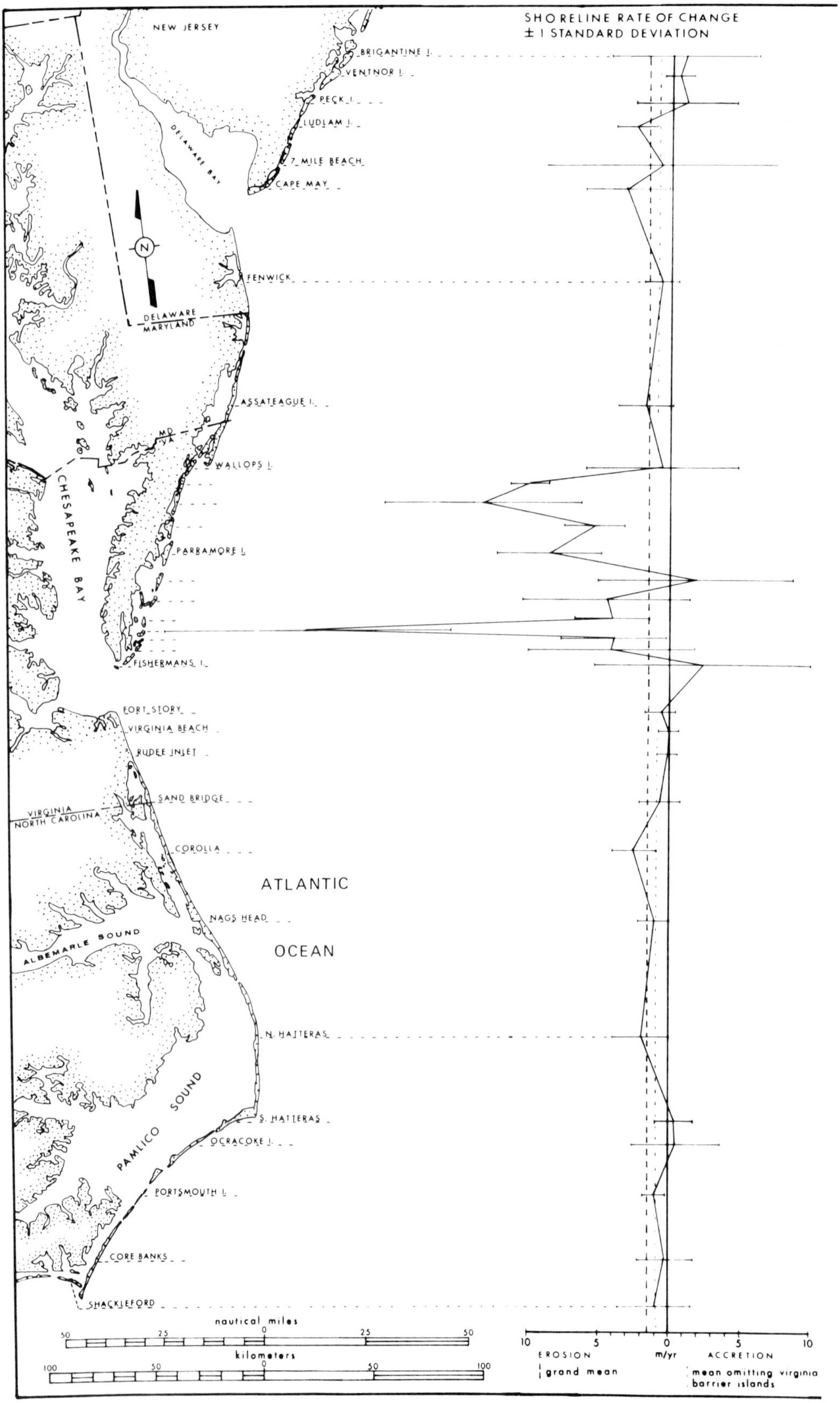
SHORELINE RATE OF CHANGE
± 1 STANDARD DEVIATION
NEW JERSEY
BRIGANTINE I.
VENTNOR I.
PECK I.
LUDLAM I.
7 MILE BEACH
CAPE MAY
DELAWARE BAY
N
FENWICK
DELAWARE
MARYLAND
ASSATEAGUE I.
MD
VA
WALLOPS I.
CHESAPEAKE BAY
PARRAMORE I.
FISHERMANS I.
FORT STORY
VIRGINIA BEACH
RUDEE INLET
SAND BRIDGE
VIRGINIA
NORTH CAROLINA
COROLLA
ATLANTIC
OCEAN
NAGS HEAD
ALBEMARLE SOUND
N. HATTERAS
PAMLICO SOUND
S. HATTERAS
OCRACOKE I.
PORTSMOUTH I.
CORE BANKS
SHACKLEFORD
nautical miles
50 25 0 25 50
kilometers
100 50 0 50 100
10 5 0 5 10
EROSION m/yr ACCRETION
grand mean
mean omitting virginia barrier islands

in sunny, breezy conditions which would maximize any seaward movement of the HWL associated with the drying process. Little such movement was observed. Additionally, it was apparent in the field that the movement of the HWL position was concentrated in the period of the rising tide. Little or no movement was noticed during the period from high tide to the next low tide. Therefore, measurements made from photographs taken on the falling tide will contain negligible error associated with the tidal stage, while the measure of error introduced by fluctuations in the HWL over the entire tidal cycle is less than the resolution of the photography at 1:5000 scale.

Storm Surge Penetration

During periods of extreme storm activity (high waves and/or wind surge) low-lying coastal areas may be flooded and overwashed by sea waters. The overwash is a bore of highly turbulent, sediment-laden water which moves across the beach and onto subaerial portions of the coast. As this flow moves inland, its velocity is reduced so that at some point the flow of water can no longer transport sediments. Thus a zone of sediment transport is produced between the beach and the line of inland penetration of the bore: the zone of overwash deposit (Hayden et al.[14]).

In the months immediately following an overwash event the newly deposited sand is clearly evident against the contrast of vegetated surfaces (Figure 3). As time passes, vegetation encroaches on the sand deposit and the overwash zone begins to narrow. Thus there are two separate processes that give definition to the width of the overwash penetration zone: (1) overwash events that widen the zone, and (2) vegetation regrowth that narrows the zone.

The width of the overwash penetration zone is defined as the distance between the shoreline and the line of encroaching vegetation on the overwash deposit. The width of the overwash zone varies over time and changes along the coast. Along the Atlantic coast, areas that experienced deep overwash penetration in the 1940s usually experienced deep penetration in the 1950s, 1960s, and 1970s.

Orthogonal Grid Mapping System (OGMS)

Using the OGMS, shoreline erosion and overwash information can be acquired rapidly and systematically from aerial photographs. Prior to interpretation of the aerial photographs, standard 1:5000 scale base maps of the study region are prepared. These maps are produced by photo enlargement of 7–1/2 minute series USGS maps (1:24,000). Each base map represents an area 3500 m by 2100 m. The frame of each base map is oriented with the long side parallel to the coastline and positioned over the active portion of the coast. The long axis, lying entirely over the ocean, is the baseline from which all measurements are made. The aerial photographs are then enlarged to the exact scale of the base map by projection onto the base map. On each projected image, a stable-base, transparent rectilinear grid with 100-m spacings is laid. Any coastal location is thus specified by base map number and coordinates. The shoreline (HWL) and the overwash penetration line are then digitized from the projected image with a Numonics® 1224 Electronic Graphics Calculator. These traces constitute the raw cartographic data base from which subsequent measurements are extracted.

Errors in the OGMS

The OGMS is subject to limitations of accuracy and precision inherent in the maps and photographs available for use in studies of shoreline change. The smallest unit that can be measured easily by this method, using an electronic digitizer is 0.25 mm. This produces a potential ground measurement error of 1.25 m at a scale of 1:5000. To determine rates of

FIGURE 4. Average rates of shoreline changes for the mid-Atlantic coast. The error bars indicate the spatial standard deviations for the within-island erosion rates determined at 100-m intervals.

FIGURE 5. Patterns of overwash immediately to the north of the village of Buxton on the Hatteras coast (A), and Indian Inlet on the Fenwick coast (B). Arrows indicate sites of major overwash fans. White bar indicates a distance of 1 km.

FIGURE 5B.

change, two position measurements are required, resulting, therefore, in a combined potential ground measurement error of as much as 2.5 m.[8]

The process of matching the projected photographs to the base maps is another source of error which is difficult to assess quantitatively. Much depends upon the skill of the photointerpreter, but this error is no greater than that involved in determining the scale of a photograph by comparing measurements between points on the photo and the same points on a map. In either case the error can be substantial if there are few stable man-made or natural features available for matching or for making scale measurements. Under good conditions, it is difficult to match ground features on photos and maps to less than a 10 m ground distance. The combined maximum potential error in measurements from two photographs enlarged to a 1:5000 scale is 12.5 m. This error is substantially reduced when the area in the vicinity of the principal point (intersection of the optic axis with the plane of the image) is utilized. To restrict measurement to the area near the principal point, 60% overlap between adjacent photos is used whenever possible. Further improvement of measurement is achieved with a zoom transfer scope which permits adjustments for optic amorphic stretch. The 12.5-m potential error is the sum of three separate errors (each with 0.5 probability), so the joint probability of such an error is 0.125. If the error is normally distributed about zero error, than the standard deviation should be about 6.3. With quality imagery and use of an electronic digitizer, the error of measurement is within 6.3 m. In certain cases, such as the Cape Hatteras area, high-quality imagery, orthophotos, and the use of the electronic digitizer have permitted us to make measurements of known ground distances with an error of less than 1 m. However, this mapping was carried out under exceptional circumstances and a 5-m error is a more realistic overall estimate.

In summary, the total measurement error is potentially as much as 12.5 m but usually within 6.3 m. When annual rates of change are calculated the potential error must be normalized for time, i.e., a 5-m error with 10 years between photos converts to an error of 0.5 m/year in terms of rates. Measurements of long-term erosion rates to a resolution as high as ±0.05 m/year are readily feasible if the time interval between photos is large enough and the shoreline position changes between photos are as large as they are in most open-coast exposures.

Measurements in our OGMS are commonly made over a 40-year period to determine shoreline rates of change. A 12.85-m ground measurement error over 40 years results in a potential error of ±0.32 m/year for shoreline rate-of-change information. However, the standard deviation of the shoreline rates of change along the Atlantic coast varies from 1.18 to 16.4 m/year; therefore, the precision of the OGMS method is well within the year-to-year variance of the natural beach system.

One can improve precision through development of better photogrammetric techniques and reduction of subjectivity (people) in each step of the process. Improving precision results in a more efficient data collection process and, therefore, lower cost per unit of information. In contrast, improving accuracy requires more accurate measurement of the natural variation of the beach system — i.e., corrections for tide, waves, beach slopes, etc. Such improvements must be traded off against significant and continuing increases in the cost per unit of information. Our work indicates that the aerial photographic method provides statistically reliable measures of erosion along the Atlantic and Gulf coasts which appear to be adequate for long-term planning. Currently, our data are used in determining setback lines in North Carolina and in general, regional-scale planning in Louisiana. Therefore, increased costs to improve the accuracy seem unwarranted at this time. We believe that greater gains can be achieved through development of systems designed to measure shoreline change at close intervals along the coast with frequent updating of the data base.

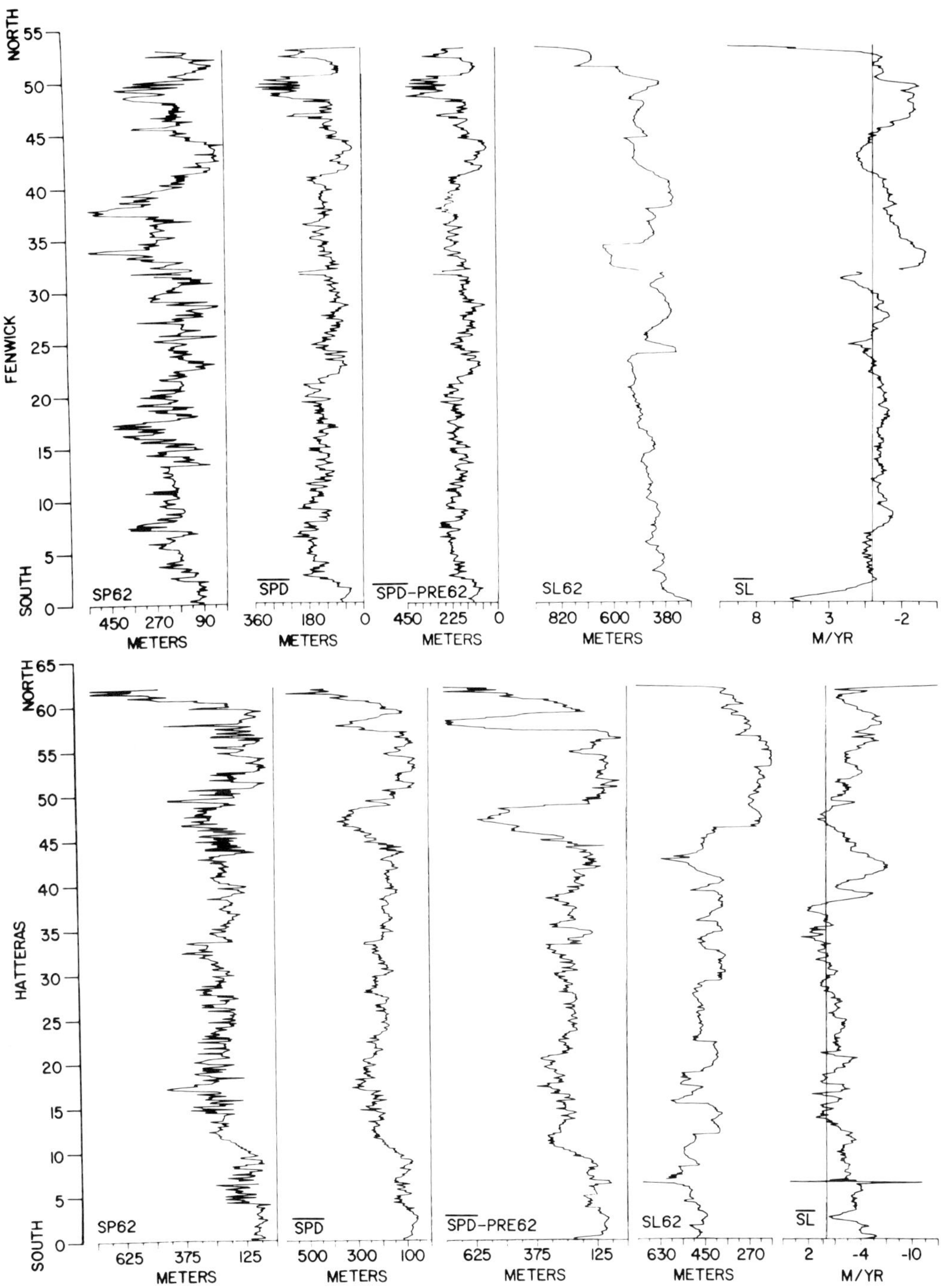

FIGURE 6. Along-the-coast variations in shoreline and storm-surge penetration limits and their rates of change. SP62 = the inland limit of storm-surge penetration due to the March 1962 storm. $\overline{SPD}$ = the long-term average distance between the shoreline and the storm surge penetration limit. SL62 = the distance between the shoreline and offshore parallel bar lines following the 1962 storm. $\overline{SL}$ = the long-term rates of change of the shoreline.

Table 1

FENWICK

	SL62	$\overline{SL}$	SP62	$\overline{SPD}$	$\overline{SPD\text{-}PRE62}$
SL62	—	−0.0660	−0.0843	0.0036	0.0104
$\overline{SL}$		—	−0.3943[a]	−0.2751	−0.2299[a]
SP62			—	0.6974[a]	0.4175[a]
$\overline{SPD}$				—	0.9241[a]
$\overline{SPD\text{-}PRE62}$					—

HATTERAS

	SL62	$\overline{SL}$	SP62	$\overline{SPD}$	$\overline{SPD\text{-}PRE62}$
SL62	—	−0.0184	0.0134	−0.0333	−0.1211
$\overline{SL}$		—	0.1628[a]	0.2068[a]	−0.0020
SP62			—	0.8313[a]	0.5381[a]
$\overline{SPD}$				—	0.8599[a]
$\overline{SPD\text{-}PRE62}$					—

[a] Statistically significant at the 0.01 level.

RESULTS

From OGMS, three measurements are available for analysis: (1) mean rate of shoreline change in m/year, (2) mean overwash penetration distance in meters, and (3) the mean rate of change of the landward limit of overwash penetration. Figure 4 shows the rates of shoreline change for the mid-Atlantic coast based on OGMS-measured rates of change.[6] The calculation of mean rates of change for entire islands does not give an accurate picture of the dynamics, as most of the islands have higher rates of change at one end than at the other. If the mean value is used, the overall shoreline erosion rate for this reach of the Atlantic coast is 1.5 m/year. Islands with more southerly exposures have lower rates of erosion and islands with northern exposures have higher rates. These rates are a function of the direction of storm tracks and wave approach, and of the orientation of the shoreline.

Aerial photographs taken soon after the Ash Wednesday storm in 1962 show the spatial pattern of overwash deposits along the North Carolina coast (Figure 5). All sections of the mid-Atlantic barrier islands showed some degree of overwash resulting from the 1962 storm; however, the distance that sand was transported inland varied markedly (Figure 6). Spectral analysis of these storm patterns and of the 40-year averages of erosion rates along the same reach suggest that along-the-coast periodicities (that is, regular, recurring patterns) exist for both the long-term average shoreline erosion rates and the inland penetration of overwash during a single storm.[9] Along the North Carolina coast from Cape Henry to Cape Hatteras, these rhythmic patterns in erosion have wavelengths of 31, 15.5, 5.0, and 2.5 km. In the vicinity of the U.S. Army Corps of Engineers Coastal Engineering Research Center Pier at Duck, N.C., erosion rates are periodic at 3.5- and 2.3-km wavelengths.

Fiqure 6 shows raw data plots of the positions of shoreline and overwash penetration lines for Fenwick Island and Hatteras Island at the time of the 1962 Ash Wednesday storm. The historic mean rates of change of the shoreline prior to the storm are also included. The position of the shoreline following the March 7, 1962 storm was further inland at places with the highest long-term mean erosion rates (SL) for the period *before the storm*. The shoreline was driven inland shorter distances at places where the erosion rates were low before the storm. The March 7, 1962 storm penetrated (overwashed) the coast further in areas that had historically been areas of maximum overwash penetration.[9]

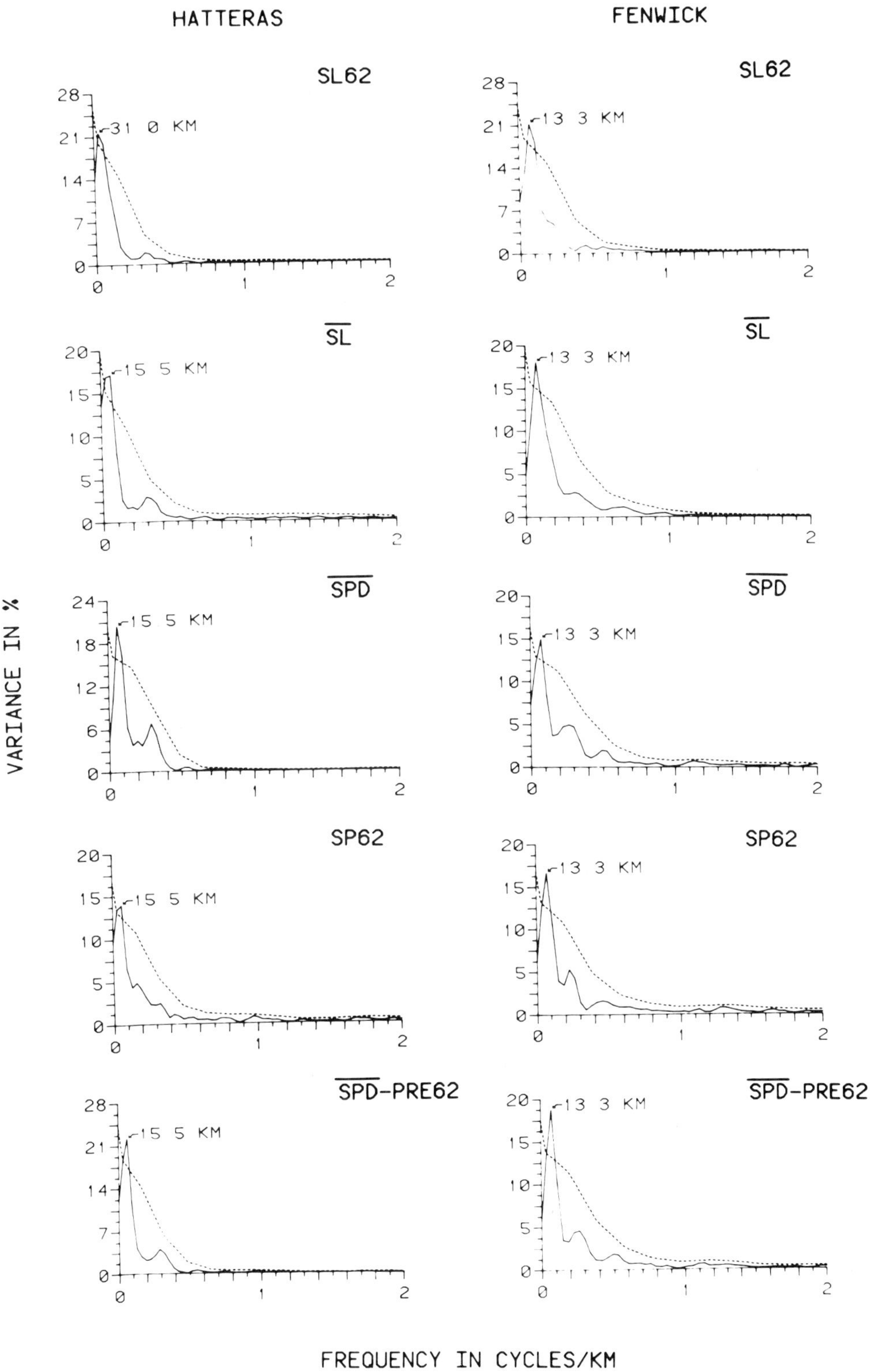

FIGURE 7. Power spectral estimates for each of the data series shown in Figure 6.

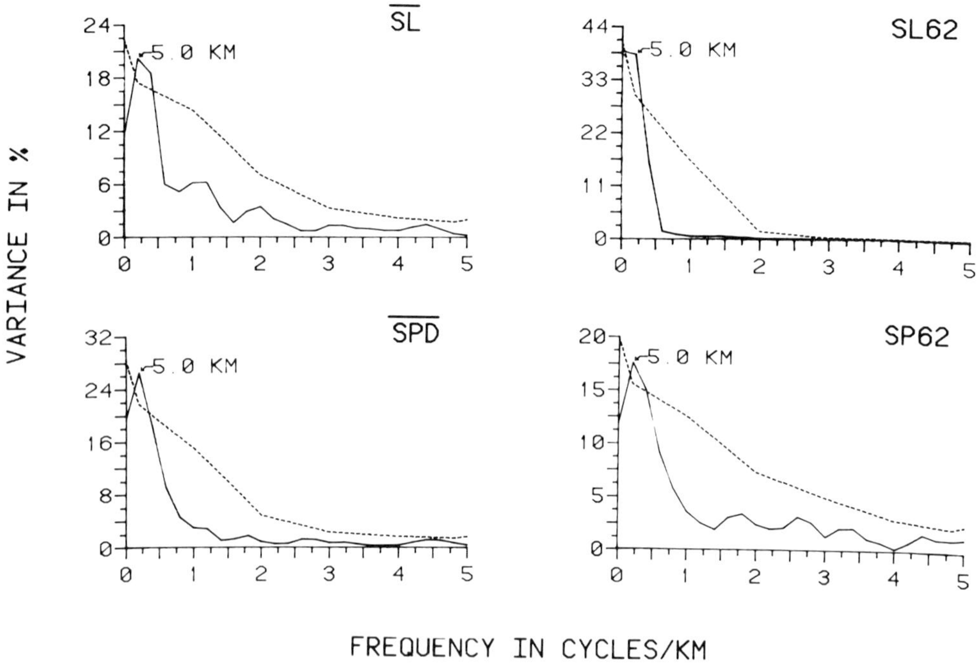

FIGURE 8. Typical power spectral estimates selected from 10-km reaches along the Hatteras coastline.

Table 1 gives the correlation coefficients between (1) the distance from a fixed base line to the shoreline following the 1962 storm, (2) the average distance between the fixed base line and the shoreline for the years 1934 to 1962, (3) the distance from a fixed base line to the overwash penetration line following the 1962 storm, and (4) the average distance between the fixed base line and the overwash penetration line for the 1934 to 1962 and the 1934 to 1979 time periods. The correlation coefficients in Table 1 were calculated for both Fenwick Island and Hatteras Island. High-positive correlations are found between the overwash penetration distance before and after the 1962 storm. These correlations indicate that regions with high historical rates of change of overwash penetration experienced great change at the time of the March 1962 storm. Regions that experienced large shoreline changes during the 1962 storm were not correlated with either the earlier history of shoreline change, the overwash penetration distance following the 1962 storm, or the mean width storm penetration for either the 1934 to 1979 period or the pre-storm period (1934 to 1962). The mean rates of change of the shoreline and overwash penetration line are synchronous in their along-the-coast variability as indicated by the modest but significant correlation (Table 1). This along-the-coast variability is further shown to be spatially periodic.

Figure 7 shows the results of spectral analyses of shoreline and overwash penetration after the 1962 storm and the long-term averages of both the shoreline and overwash penetration positions. The dominant wavelength for the 62.3-km reach along the Outer Banks is 15.5 km, and for the 53.7-km reach of Fenwick Island is 13.3 km. Although not statistically significant at the 0.05 level, there is spectral power along the Cape Hatteras and Fenwick reaches near 4.5 km. We therefore examined individual 10 km-reaches at both sites in order to filter out the longer (15 km) wavelengths that dominate the spectral patterns in the entire data set.[9] Figure 8 gives an example of the 21 separate 10-km analyses which are further summarized in Table 2. Smaller-scale regularities apparently exist within the large wavelengths.

Based on these analyses, we conclude that the natural configuration of sedimentary coastlines, as determined by shorezone processes, is sinuous and periodic rather than straight. The larger-wavelength crescentic forms are less apparent because their curvature is smaller

Table 2
STATISTICALLY SIGNIFICANT (0.01 LEVEL) WAVELENGTHS AS DETERMINED FROM SPECTRAL ANALYSES FOR 10-km REACHES TO NORTH OF CAPE HATTERAS AND OCEAN CITY INLET

	$\overline{\text{SL}}$	SL62	$\overline{\text{SPD}}$	SP62
Cape Hatteras				
0—10 km	5.0	5.0	5.0	5.0; .4
5—15	5.0	5.0; 2.5	5.0	5.0
10—20	5.0	5.0	5.0; 2.5	2.5
15—25	2.5	5.0	5.0	2.5
20—30	5.0; 2.5	5.0	5.0	—
25—35	5.0; 2.5	5.0	5.0	5.0; 2.5
30—40	5.0	5.0; 2.5	5.0	5.0
35—45	5.0	5.0	2.5	2.5
40—50	5.0	5.0	5.0	0.26; 0.25
45—55	5.0	5.0	5.0	2.3
50—60	5.0; 2.5	5.0	5.0; 2.5	2.5
55—	3.6	3.4	3.4	3.6
Oregon Inlet				
		Fenwick		
Ocean City inlet				
0—10 km	5.0	5.0	5.0	2.5; 1.3
5—15	5.0	5.0	2.5; 1.7	2.5; 1.7
10—20	2.5	5.0	0.22	5.0
15—25	5.0	5.0	5.0	5.0; 2.5
20—30	5.0	5.0; 2.5	2.5	2.5
25—35	5.0	5.0	2.5	5.0; 2.5
30—40	5.0	5.0	4.6	5.0; 2.5
35—45	5.0	5.0	5.0	5.0
40—50	5.0	5.0	5.0	5.0
45—	4.2	5.0	4.2	5.0
Cape Henlopen				

and thus their relative amplitude is lower. However, in analyzing long reaches, one sees that the larger features dominate the variance because their absolute amplitudes are greater than those of smaller crescentic features. Nevertheless, it is difficult to recognize the larger crescentic forms while carrying out site level field work.

PREDICTION

If indeed there exists, over the time scales of the last four decades, an along-the-coast stationarity (as in the sense of a stationary time series) in the positions and the rates of change of the shoreline and the line indicating the inland penetration of overwash, then detailed statistical properties of each line and its temporal changes permit a stochastic prediction of future conditions. That the shorezone is periodic in the along-the-coast direction provides clues as to the cause of the attribute of stationarity. In this section we detail the stochastic approach to prediction of shorezone changes.[5]

Among the several information sets generated by using OGMS is a graph of the mean rate of change of shoreline calculated over all time periods and plotted with an envelope of ± 1 standard deviation (Figure 9). This is a useful graph for determining areas along the coast which have been subjected to the greatest change. It visually presents a measure of

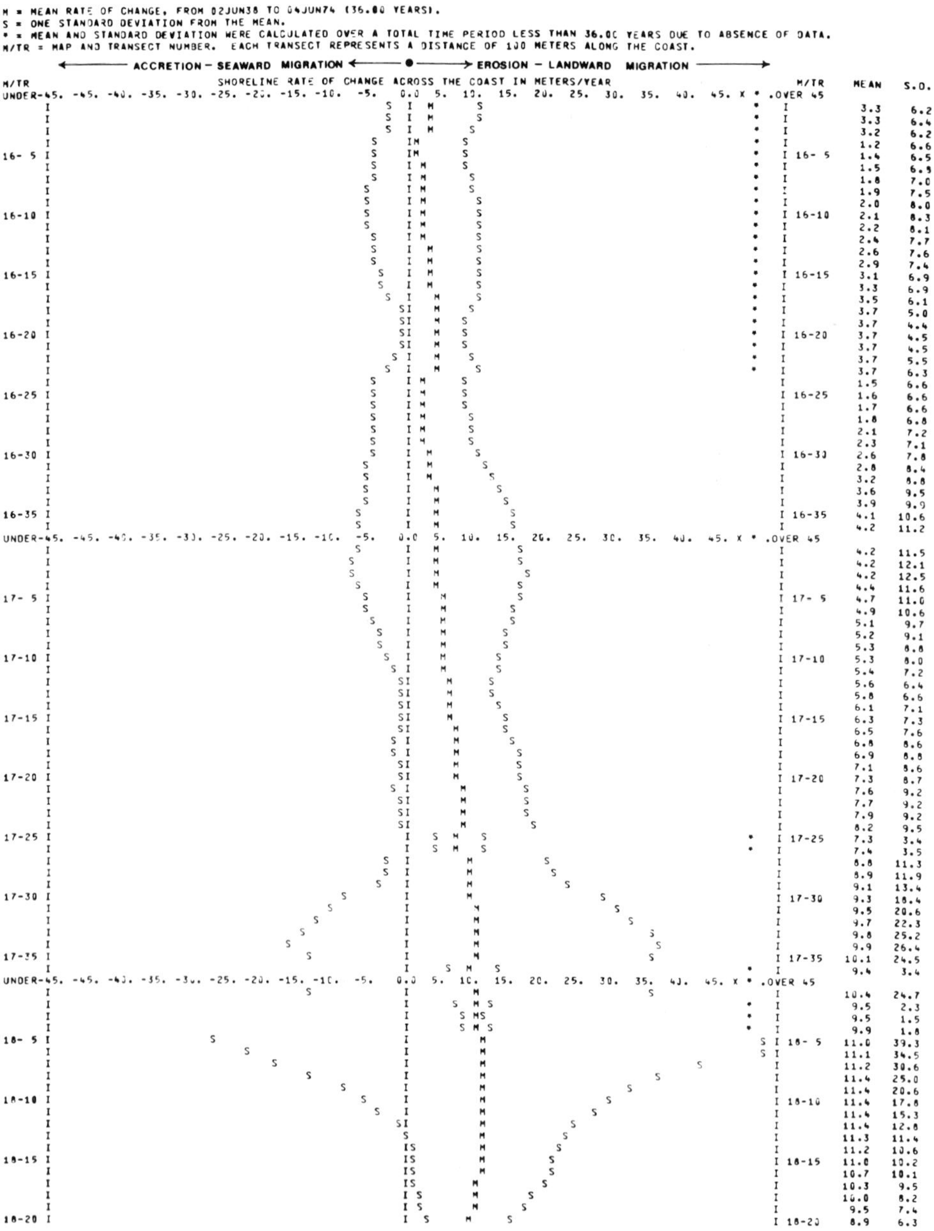

FIGURE 9. Sample printout of shoreline erosion rates and standard deviations (100-m intervals along the coast).

the erosion rates, stable areas, and areas along the coast that are more vulnerable to erosion. Numerical values for the mean and standard deviation at each of the 100-m transects are listed. The standard deviation lines are of additional value because they couple the historical trend with episodic changes associated with extreme storm events.

These statistics, like any statistics of natural systems, must be applied with caution. Although there is geological information to confirm that shoreline change has been underway for many coasts for centuries, aerial photography has been available for only three or four decades. The resulting statistics are based upon samples of short periods within a much longer trend. One must assume that the past 30 to 40 years of shoreline change is representative of the longer trend, and thus indicative of the future. Complementary information from other disciplines, including oceanography, climatology, and geomorphology can be important in establishing confidence in the statistics.

In addition, the selection of individual photographic flights can bias the statistics. In our analysis, for example we included among our five sets of photography a flight taken soon after the Ash Wednesday storm of 1962 (125-year return interval). This storm alone is responsible for a considerable amount of statistical variance in our data set.

Three separate lines are related to the question of risks and hazards in the shorezone (Figure 10): (1) the shoreline, (2) the landward limit of the major destruction zone, and (3) the landward limit of the surge damage zone. Since each of these lines varies in position over time, the risk or probability that a place or point will fall within one of the zones defined by these lines likewise varies over time. These probabilities can be estimated from the data base provided by the OGMS program.

Future positions of the shoreline and the landward limits of the major destruction and surge damage zones may be calculated for specified time intervals and desired probability levels.[5] To simplify the equations that follow, "s" is used to designate the shoreline, and "v" the vegetation line that marks the inland penetration of overwash. Data required for these calculations are (1) the rate of shoreline change (ds/dt), (2) the standard deviation of shoreline change rates (σ_s), (3) the rate of change of the overwash penetration (vegetation) line (dv/dt), and (4) the standard deviation of the rate of change of overwash penetration (σ_v).

The landward limit of the shoreline (Δs) for a design time interval (Δt) specified for a given probability level (p) is given by:

$$\Delta S \Big]_{\Delta t, p} = \Delta t \left(\frac{ds}{dt} + k\sigma_s \right) \tag{1}$$

where k is the number of standard deviations appropriate to the specified probability level.

The landward limit of the major damage zone (ΔMDZ) for a design time interval (Δt) specified for a given probability level (p) is given by:

$$\Delta \mathrm{MDZ} \Big]_{\Delta t, p} = \mathrm{ABZ} + \Delta t \left(\frac{ds}{dt} + k\sigma_s \right) \tag{2}$$

where ABZ is the width of the existing active beach zone. In other applications, and if data are available, temporal variation of ABZ may be incorporated into Equation 2.

The landward limit of the surge damage zone (ΔSDZ) for a design time interval (Δt) specified for a given probability level (p) is given by:

$$\Delta \mathrm{SDZ} \Big]_{\Delta t, p} = \Delta t \left(\frac{dv}{dt} + k\sigma_v \right) \tag{3}$$

These relationships may also be solved for k and thus p, the probability that a given geographic location will fall within a given zone. For example, the equation for the major damage zone (ΔMDZ) may be solved for k:

$$k = \frac{\Delta \mathrm{MDZ} - \mathrm{ABZ} - \Delta t \dfrac{ds}{dt}}{\Delta t(\sigma_s)} \tag{4}$$

NAGS HEAD, NORTH CAROLINA
SCALE 1:5,000
N
DEC. 18, 1962
1974 SHORELINE
1962 OVERWASH PENETRATION
PHOTO SOURCE: U.S. AIR FORCE
ROBERT DOLAN, COASTAL RESEARCH ASSOCIATES

FIGURE 10A

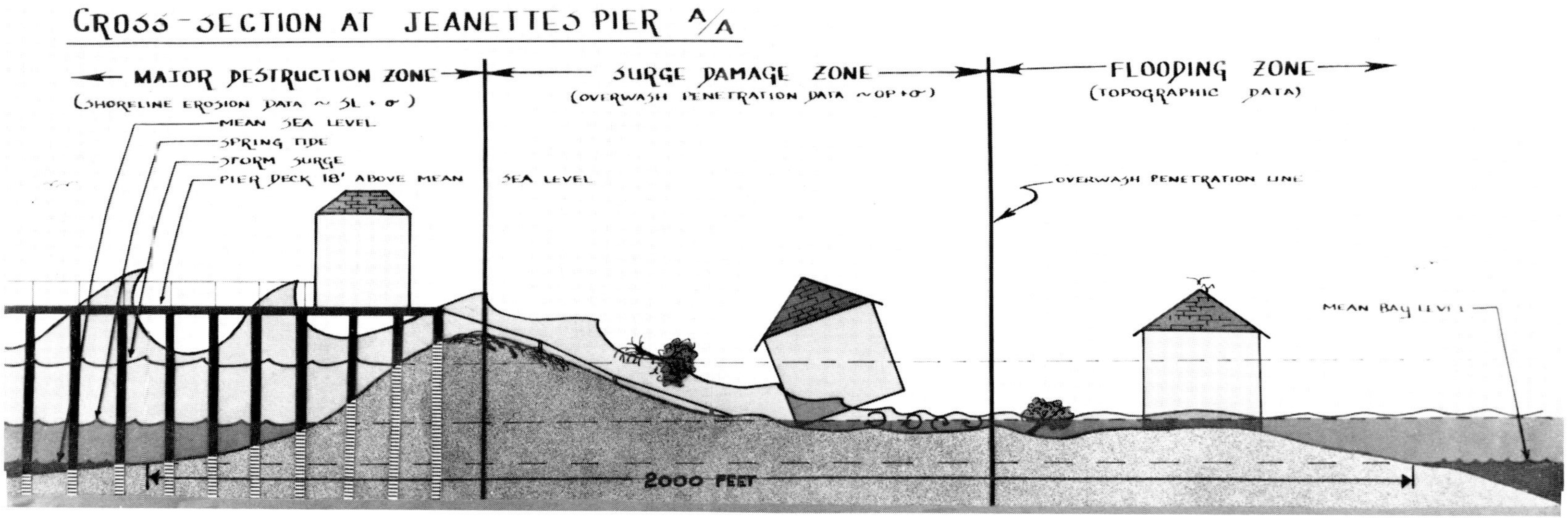

B

FIGURE 10. Definition of risk zones in the direction normal to the coast.

Table 3

Temporal:	$\overline{SL}$		σSL		$\overline{SP}$		σSP	
Spatial:	μ	σ	μ	σ	μ	σ	μ	σ
Shackleford N_i = 123	−0.97	2.74	2.87	1.21	+5.36	4.08	19.79	12.07
Core Banks N_i = 392	−0.22	2.02	4.85	3.04	−0.26	2.06	7.64	5.31
Portsmouth N_i = 220	−0.96	0.80	16.40	8.14	+3.04	4.82	18.34	10.27
Ocracoke N_i = 239	+0.59	3.11	8.36	5.65	+24.16	23.63	32.46	27.44
S. Hatteras N_i = 175	+0.37	1.33	4.85	2.93	+4.71	2.15	6.36	3.20
N. Hatteras N_i = 600	−1.94	1.96	7.22	4.27	+4.96	7.04	16.08	10.58
Assateague N_i = 498	−1.63	1.96	5.67	3.55	+2.47	6.72	15.34	14.43
Fenwick N_i = 524	−0.58	1.20	3.32	0.95	+2.55	2.43	12.08	6.92
Cape May N_i = 90	−2.97	2.88	2.74	1.64	−2.72	2.91	10.06	7.09
Seven Mile B. N_i = 230	−0.60	7.92	11.23	15.52	+0.47	7.56	14.04	12.90
Ludlam N_i = 102	−2.28	1.40	2.85	1.45	−1.98	1.22	20.82	10.06
Peck Beach N_i = 123	+0.79	3.57	4.35	5.13	+0.60	4.15	13.11	5.41
Ventnor N_i = 135	+0.65	1.03	3.58	2.21	+0.01	0.81	8.32	12.11
Brigantine N_i = 101	+1.05	5.10	7.91	5.98	+1.70	6.90	9.50	0.75
Grand means	−0.62	3.01	6.16	4.83	+3.22	5.95	14.57	9.95

Note: Spatial means (μ) and standard deviations (σ) of the long-term mean rates of change of the shoreline ($\overline{SL}$), the standard deviations of SL, the long-term mean rates of change of the storm-surge penetration line ($\overline{SP}$), and the standard deviations of SP (m/year). Negative sign indicates recession and a positive sign accretion. N_i indicates the number of transects for each island.

where ΔMDZ is the difference between the current position of landward limit of the major damage zone and the coastal location of concern; ABZ is the active beach zone, Δt is the design time interval, ds/dt the rate of change of the shoreline and σ_s the standard deviation of the shoreline rate of change. Once k is calculated, the probability may be found. Using solutions of this type the probability for any location may be calculated.

The relationship between probability levels (p) and equivalent k values is

p:	0.99	0.95	0.90	0.85	0.80	0.75	0.70	0.65	0.60	0.55	0.50
k:	2.33	1.65	1.28	1.04	0.84	0.67	0.52	0.39	0.25	0.13	0.00

Since k is merely the number of σ_s desired, p is accordingly specified.

The problems of risks and hazards in coastal environments differ from the equivalent flood plain problems because of secular variation due to shoreline change. This precludes estimates of storm-surge return intervals for assigning hazard probabilities. Accordingly, given a location on the coast, the hazard due to a "100-year storm" increases systematically

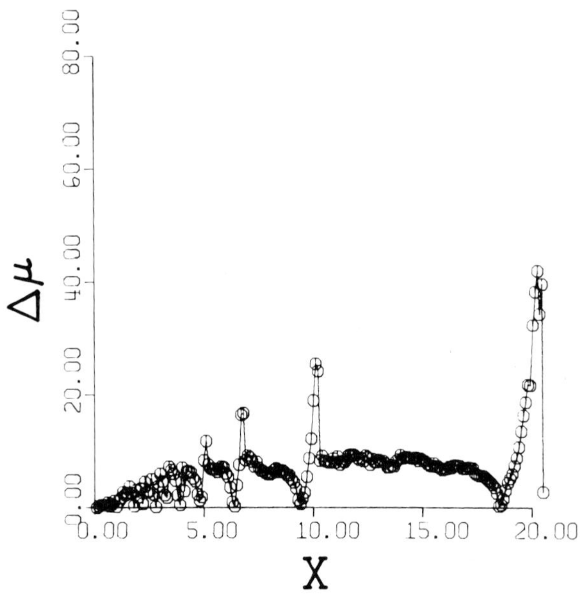

FIGURE 11. A persistence diagram for the stretch of coast from Cape Hatteras to Ocracoke Inlet. The input data is the standard deviation of shoreline erosion. The mean (m/year) for all 204 survey sites at 100-m intervals is normalized to zero. Sampling intervals greater than 100 m result in a departure (m/year) from the high resolution mean. X indicates sampling interval.

with time on an eroding coast. Since there may be several causes for shoreline change, the problem cannot be resolved by further investigation of storm frequencies and magnitudes. The only alternative is empirical evaluations of historical data. To insure that an adequate information base is available, a systematic photo reconnaissance of the coast should be initiated. As the sample size in erosion studies increases, confidence levels in the derived statistics will improve.

Our investigation also resulted in this final observation. In those areas along the New Jersey coast that have been engineered to stabilize the shoreline (groins) and to prevent overwash penetration (seawalls), the mean rates of change have been greatly reduced — 1m or less per year in many areas; however, in these same areas, the standard deviations of the rates of change are high.[13] This suggests that although the engineering works have succeeded in stabilizing the shoreline system, when extreme storms do occur damage is often greater within the stabilized areas. Thus, the hazard of systematic erosion damage is decreased, but the risk of episodic storm damage is increased.

The OGMS data base provides an opportunity to answer two long-standing sampling questions concerning shoreline variation.

1. What are the relative magnitudes of the spatial and temporal variations of shoreline changes?
2. What is the effect of different sampling intervals on the statistical stability of measurements of shoreline change?

To answer these questions, we used the along-the-coast means and standard deviations of the temporal mean rate of shoreline change (SL), the standard deviation of the rate of shoreline change, (σSL), the mean rate of change of the inland limit of overwash penetration (SP), and the standard deviation of this rate (σSP) for 14 barrier islands along the Atlantic

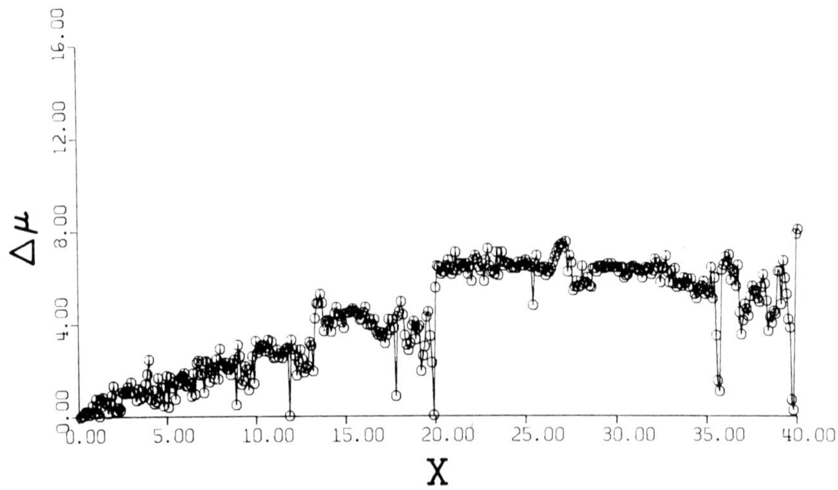

FIGURE 12. Persistence diagram as in Figure 11 for the mean rate of change for the limit of storm-surge penetration on the Core Banks, N.C.

coast.[13] In addition, we determined the departures from the island-wide means and standard deviations which resulted with sampling intervals greater than 100 m.

The means and standard deviations of the four variables (SL, σSL, SP, and σSP) for the 14 barrier islands comprising the 428-km study are shown in Table 3. For the average island within this reach, the long-term mean rate of shoreline change is −0.62 m/year. The highest erosion rate was found at Cape May, N.J. (2.97 m/year) and the highest accretion rate was on Brigantine Island, N.J. (1.05 m/year). Variations in shoreline change rates within islands were in general as large, and sometimes an order of magnitude larger than the island-wide mean rates, with an average value of ±3.01 m/year. Temporal variations of shoreline change rates average ±6.16 m/year, or an order of magnitude greater than the average rate of change.

Overwash penetrations over the last five decades have, in general, trended toward less penetration as indicated by the average "accretion" rate of this line (3.22 m/year Table 3). This accretion results not from an overall reduction in extreme storms,[12] but rather from the construction of barrier dunes and dikes and thus the retardation of overwash penetration.

SAMPLING INTERVALS

Statistics presented in Table 3 are based on data with a sampling interval of 100 m. Larger sampling intervals should result in means which differ from the high-resolution means. This difference, (Δμ), has been calculated for each of the islands studied for all possible sampling intervals which are integer multiples of the original 100-m interval.[13] Figure 11 shows the magnitudes of Δμ with various sampling intervals, for the standard deviations of the long-term means of shoreline change (σSL) on the South Hatteras reach of the coast. In Figure 12 a similar plot is given for the mean rate of change of the overwash penetration line (SP) on Core Banks. With the exception of four spikes in the curve for South Hatteras, which arise due to along-the-coast periodicities in σSL, the upper limit of Δμ defines a curve of the form

$$\Delta\mu = b\tanh kx$$

where b is the upper asymptotic bound of Δμ, k determines the slope of the curve in the

Table 4
PERSISTENCE PARAMETERS *b* AND *k* FOR BARRIER ISLANDS OF THE MID-ATLANTIC COAST

	$\overline{SL}$		σSL		$\overline{SP}$		σSP	
Island	***b***	***(k)***	***b***	***(k)***	***b***	***(k)***	***b***	***(k)***
Shackleford	8.0	(0.09)	8.0	(0.12)	6.0	(0.20)		
Core Banks	1.0	(0.20)	4.0	(0.15)	6.0	(0.10)	5.5	(0.15)
Portsmouth	1.0	(0.70)	8.0	(0.22)	11.0	(0.20)	8.0	(0.20)
Ocracoke	3.0	(0.25)	6.0	(0.15)	20.0	(0.10)	25.0	(0.05)
South Hatteras	2.0	(0.40)	8.0	(0.40)	4.0	(0.20)	6.0	(0.20)
North Hatteras	0.5	(0.90)	3.0	(0.08)	5.0	(0.10)	9.0	(0.12)
Assateague	1.0	(0.10)	2.0	(0.40)	4.0	(0.30)	20.5	(0.40)
South Fenwick	2.5	(0.08)	0.4	(0.50)	0.6	(0.50)	6.0	(0.10)
North Fenwick	1.5	(0.20)	0.5	(0.50)	2.0	(0.40)	7.0	(0.20)
Cape May	[a]	[a]	[a]	[a]	[a]	[a]	5.5	(0.40)
Seven Mile Bch.	2.0	(0.90)	6.0	(0.35)	2.0	(0.50)	10.0	(0.35)
Ludlam	5.0	(0.20)	0.7	(0.90)	1.8	(0.20)	5.0	(0.25)
Peck Beach	9.0	(0.20)	4.0	(0.20)	1.5	(0.60)	3.0	(0.25)
Ventnor	1.4	(0.25)	2.5	(0.20)	1.0	(0.50)	4.5	(0.30)
Brigantine	[a]	[a]	[a]	[a]	[a]	[a]	5.0	(0.20)

[a] A $\Delta\mu$ value $\geq 0.25\sigma$ does not occur for any sampling interval; thus *b* and *k* could not be estimated.

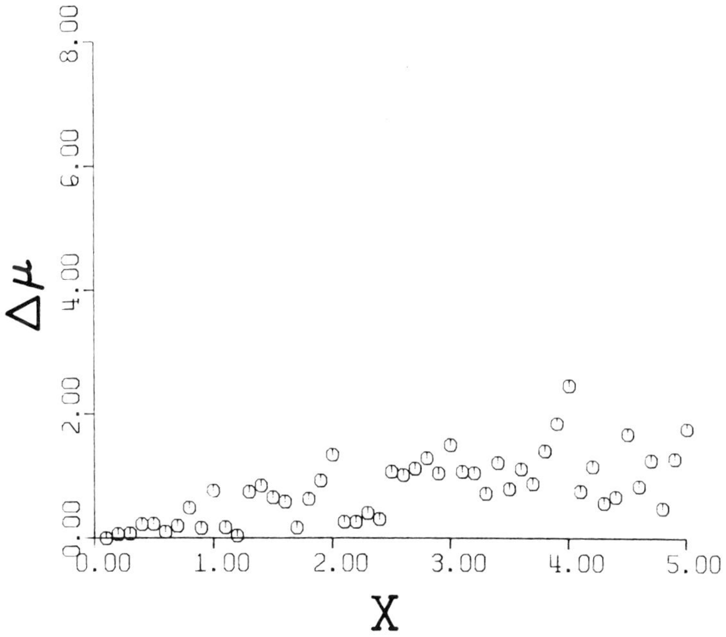

FIGURE 13. An enlargement of the left portion of Figure 12.

region near x = 0, and x is the sampling interval. In the case of the South Hatteras plot, b = 8.0 m/year and k = 0.40. The Core Banks SP data indicate a value for b of 6.0 m/year and a value for k of 0.10. Plots of $\Delta\mu$ prepared for each variable and each island are presented in the form of a table of b and k values for which the curves can be reconstructed (Table 4).

Table 5
SAMPLING INTERVAL x' FOR WHICH $\Delta\mu = 0.25\sigma$

Island	$\overline{SL}$	σSL	$\overline{SP}$	σSP
Shackleford	0.95	0.32	0.85	1.50
Core Banks	2.74	1.27	0.86	1.64
Portsmouth	0.29	1.18	0.55	1.66
Ocracoke	1.06	1.60	3.04	2.06
S. Hatteras	0.42	0.23	0.67	0.67
N. Hatteras	2.16	4.63	3.66	1.51
Assateague	5.29	1.18	1.49	0.44
S. Fenwick	1.50	1.32	∞	2.96
N. Fenwick	1.00	1.02	3.29	1.26
Cape May	2.50	1.00	1.50	0.83
Seven Mile Beach	2.72	2.20	3.48	0.95
Ludlam	0.35	0.63	0.85	2.21
Peck Beach	0.50	1.66	1.41	1.94
Ventnor	0.74	1.11	0.41	2.71
Brigantine				0.19

Since most coastal investigations call for sampling intervals smaller than 5 km, the portions of Figures 11 and 12 for sampling intervals x <5 km are enlarged and reproduced in Figure 13. In the case of the temporal standard deviation of shoreline change rates along South Hatteras (Figure 5), which averages ± 2.93 m/year, a 1-km sampling interval would provide an estimate of this variable within 1σ of the high-resolution mean. Along most reaches of the coast studied, variances are large and may be an order of magnitude greater than the mean. Therefore, sampling interval choices may be more appropriately made on the basis of a small fraction of a standard deviation, for example, 0.25σ. The sampling interval for which $\Delta\mu = 0.25\sigma$ is here referred to as x'. The parameters b, k, and x' thus define the along-the-coast statistical persistence of the data. Accordingly, b, k, and x' values associated with the four variables are summarized in Tables 4 and 5 for each barrier island studied. Given the functional (hyperbolic tangent) relationship between b and k, the standard deviation data given in Table 1, and the value of x', adequate information for efficient sampling design is available in Table 5 for the mid-Atlantic coast, and should provide guidelines for other areas.

The representations given in Figures 11, 12, and 13 and the summarizations in Tables 3, 4, and 5 also provide the information needed to address the question of how representative an estimate of erosion is. Is the estimate representative of adjacent reaches? In general we find the erosion measurements to be representative for reaches less than 1 km distance along the coast.

Variations in the position of the shoreline and in the position of the limit of overwash penetration are largely the result of extreme meteorological events. Accordingly, the statistics associated with change data are characterized by large variances. The major component of these variances is customarily thought of as fixed at the time-scale of synoptic weather conditions and resides not so much in the frequency of these events as in their magnitude. While estimates of temporal variance are usually available, they appear to be three to four times larger than the spatial variance for a given reach of coast. Departures from this rule notwithstanding, it seems to have general applicability. In view of the fact that temporal variances are found to exceed those along the coast, decadal averages of shoreline rates of change must be used with caution and with a wide margin for error. This result is illustrated in Table 6 where the ratios of temporal to spatial variance for each of the islands studied are presented. In most cases this ratio exceeds unity by a substantial margin.[13]

Table 6
RATIOS OF THE TEMPORAL TO SPATIAL STANDARD DEVIATIONS OF RATES OF CHANGE OF THE SHORELINE AND THE STORM-SURGE PENETRATION LINE

Island	$\mu\ (\sigma_{SL})/\sigma\ (\overline{SL})$	$\mu\ (\sigma_{SP})/\sigma\ (\overline{SP})$
Shackleford	1.05	4.86
Core Banks	2.40	3.71
Portsmouth	20.63	3.80
Ocracoke	2.69	1.34
S. Hatteras	3.66	2.96
N. Hatteras	3.68	2.28
Assateague	2.85	2.28
Fenwick	2.75	4.96
Cape May	0.95	3.46
Seven Mile Beach	1.42	1.86
Ludlam	2.04	17.17
Peck Beach	1.22	3.16
Ventnor	3.47	10.28
Brigantine	1.55	1.38
Grand means	3.59	4.52

PERIODICITIES AND SAMPLING DESIGN

The $\Delta\mu$ vs. sampling interval plot for Seven Mile Beach, N.J. (Figure 14), clearly illustrates the problems encountered with sampling design for coastal reaches which have spikes in the $\Delta\mu$ curve. The mean rate of shoreline change (spatial) along this reach of coast is only 0.60 m/year; however, the standard deviation of this mean is large (7.92 m/year). If a sampling interval of 7.8 km was used along this reach, the mean shoreline rate of change measured would be 21.95 m/year, or nearly 40 times greater than the 100-m interval mean. A slightly larger sampling interval of 8.0 km would result in a calculated mean of 0.70 m/year, a good estimate.

In an earlier report,[5] we suggested that periodicities in shoreline rates of change, overwash and overwash penetration data for the coastal reach between Cape Hatteras and Cape Lookout are the result of standing edge waves trapped between the two capes. When the coastal dynamics data are viewed as a data series of climatological form, the attribute of statistical stationarity applies. In the absence of knowledge of the underlying periodicities, a random sampling scheme is recommended.

Based on our analyses of rate-of-change data for both the shoreline and the limit of overwash penetration for more than 1300 km of coast at 100-m intervals, we conclude the following:

1. Shoreline change rates using an island-wide average vary from 3.0 m/year erosion to 1.0 m/year accretion with a mean for the study area of 0.6 m/year.
2. Along-the-coast variations in shoreline and overwash penetration line change are as much as an order of magnitude larger than the mean rates.
3. Temporal variance of shoreline erosion is three to four times larger than the along-the-coast spatial variance, while temporal variance of overwash penetration change rates is four to five times larger than spatial variance.
4. For the islands studied, sampling at sites farther apart than 1.0 km provides little more precision than one sampling site for an entire island.

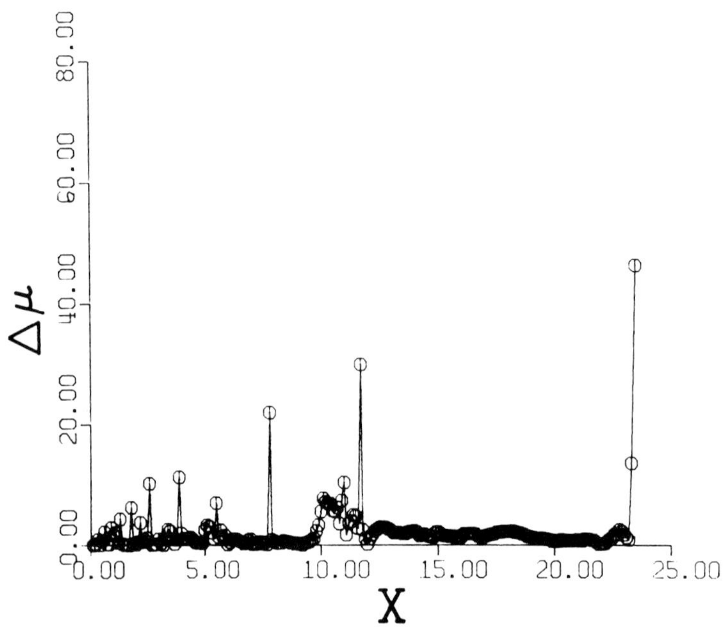

FIGURE 14. Persistence diagram for the long-term mean rate of shoreline change for Seven Mile Beach, N.J.

5. In general, sampling intervals of 2.5 km or less are required in order to specify island-wide means within ±0.25 of the high resolution mean.
6. Site-specific (transect or profile) measures of shoreline or overwash penetration rates are representative of approximately ±500 m along the coast.
7. Mid-Atlantic coast shorelines are characterized by a complex series of along-the-coast periodicities in shoreline dynamics and, in general, unless the objective is to define the periodicities, regular sampling intervals should be avoided.

CONCLUSIONS

In the 1950s, some homes on the barrier islands of North Carolina were constructed on concrete slabs. Some are still there today, having weathered hundreds of storms, including the great Ash Wednesday northeaster. Other houses nearby have long since disappeared. Is the vulnerability of some places along the coast simply a matter of chance or is there a pattern to the hazards? The research we have summarized in this chapter suggests that even at site-level scales (hundreds of meters), shorezone processes and shorezone landforms assume systematic patterns both along and across the coast. This is a departure from the more common conception that coastal change and coastal hazards are random or happenstance events.

If storm hazard zones along the coast are systematically distributed and stationary over time, then they should be predictable. The problem is that detailed historical information for establishing patterns is not always available. Our research, however, suggests that sections of sedimentary coasts which have experienced storm damage and serious erosion in the past, are likely to experience more of the same in the future. We believe there is a natural "template of change", a stationarity of periodic coastal landforms and processes that is governed by the coastal configuration, and the location of headlands, capes, and shoals.

The inhabitants of sandy coasts face a continual need for assessment of environmental hazards and their associated risks. The assessment problem is complex because there exist century-long trends in rising sea level, storm frequencies, and increasing residential densities. In addition to confronting the problems of long-term environmental trends, we must gauge the hazards of individual storms. There is no way yet to predict exactly when or where storms will occur. However, this does not mean that we are limited to general assessments of along-the-coast variations in hazard probabilities. Our research clearly indicates that the impact of large and small storms differs more in intensity than in the geography of their impact. The Ash Wednesday storm caused severe erosion and property destruction in the same locations as previous storms of lesser severity. Adjacent standing and destroyed beachfront homes reflect this variation in hazard intensity. The great storms offshore provide the energy for coastal change, but the application of this energy is determined by the regional morphology of the coast. Our "template of change" can be identified from the historical record[9] and it can be applied in evaluating the risks of living near the sea.

REFERENCES

1. **Bowen, A. J. and Inman, D. L.,** Edge waves and crescentic bars, *J. Geophys. Res.*, 8662, 1971.
2. **Dolan, R. and Ferm, J.,** Crescentic landforms along the Atlantic coast of the United States, *Science*, 159(3815), 627, 1968.
3. **Dolan, R., Vincent, L., and Hayden, B.,** Crescentic coastal landforms, *Z. Geomorphol.*, 18(1), 1, 1974.
4. **Dolan, R., Hayden, B., and Heywood, J.,** A new photogrammetric method for determining shoreline erosion, *Coastal Eng.*, 2, 31, 1978.
5. **Dolan, R., Hayden, B., and Heywood, J.,** Analysis of coastal erosion and storm surge hazards, *Coastal Eng.*, 2, 41, 1978.
6. **Dolan, R., Hayden, B., Rea, C., and Heywood, J.,** Shoreline erosion along the middle Atlantic coast, *Geology*, 7, 602, 1979.
7. **Dolan, R., Hayden, B., and Felder, W.,** Shoreline periodicities and edge waves, *J. Geol.*, 87, 175, 1979.
8. **Dolan, R., Hayden, B., May, P., and May, S.,** The reliability of shoreline change measurements from aerial photographs, *Shore Beach*, 48(4), 22, 1980.
9. **Dolan, R. and Hayden, B.,** Templates of change: storms and shoreline hazards, *Oceanus*, 23(4), 32, 1981.
10. **Dolan, R. and Hayden, B.,** Storms and shoreline configuration, *J. Sediment. Petrol.*, 51(3), 737, 1981.
11. **Guza, R. T. and Inman, D. L.,** Edge waves and beach cusps, *J. Geophys. Res.*, 80, 2997, 1975.
12. **Hayden, B.,** Storm wave climates at Cape Hatteras, North Carolina: recent secular variations, *Science*, 190 (4218), 981, 1976.
13. **Hayden, B., Dolan, R., and Felder, W.,** Spatial and temporal analysis of shoreline variations, *Coastal Eng.*, 2, 351, 1979.
14. **Hayden B., Dolan, R., and Ross, P.,** Barrier island migration, in *Thresholds in Geomorphology*, Coates, D. R. and Vitek, J. D., Eds., George Allen and Unwin, Ltd., London, 1980, 364.
15. **Komar, P. D.,** Nearshore cell circulation and the formation of giant cusps, *Geol. Soc. Am. Bull.*, 82, 2643, 1971.
16. **Komar, P. D.,** Rhythmic shoreline features and their origins, in *Large-scale Geomorphology*, Gardener, R. et al., Oxford University Press, London, 1982.
17. **Sonu, C.,** Field observations of nearshore circulation and meandering currents, *J. Geophys. Res.*, 77(18), 3232, 1972.

Chapter 7

MODELS FOR BEACH PROFILE RESPONSE

R. G. Dean and E. M. Maurmeyer

INTRODUCTION

As the forces which affect the equilibrium of the nearshore zone change, the beach profile will respond to these changes in a manner which tends to restore equilibrium. The relevant forces are operative on short- and long-term time scales and include: mean water level, wave characteristics (height and period), longshore sediment transport, and in some cases human interaction, such as sand addition to or removal from the surf zone.

In order to develop a basis for predicting future shoreline positions and to interpret shoreline history, substantial efforts have been devoted to the idealization of shoreline behavior and to the formalization of simple models. In general, these models include both kinematic and dynamic elements. The kinematic element requires conservation of sediment and the dynamic element relates the beach shape and/or position to some characteristics of the affecting forces.

This chapter reviews the characteristics of the available models and, where possible, attempts to provide an evaluation of their validity. Eustatic sea level change rates will not be reviewed in this chapter, as they have been examined in Chapter 1.

GENERAL CHARACTERISTICS OF MODELS

As noted, most models include a kinematic element (which is exact in conceptual form) and a dynamic element (which is very empirical, approximate, and based on observations or intuitive concepts).

The Kinematic Element

The kinematic element of a shoreline change model simply ensures conservation of volume of sediment. The associated equation can be expressed in a very compact mathematical form; however, the form of greatest use here will be one which relates the time rate of change in sand volume $\forall$ in a beach profile cross section to the spatial gradients in the components (Q_{sx}, Q_{sy}) of sediment transport as

$$\frac{\partial \forall}{\partial t} = -\left(\frac{\partial Q_{s_x}}{\partial x} + Q_{s_{y_o}} - Q_{s_{y_s}}\right) \qquad (1)$$

where the variables are presented in Figure 1. Equation 1 is exact and the transport components into and out of the control volume can be due to any cause, including wind-blown sediment to the dunes, removal from or addition of sediment to the system, etc. An often-used version of Equation 1 for numerical modeling purposes is obtained by regarding $Q_{sys} = Q_{syo} = 0$, and by assuming that the profile is simply translated landward or seaward without change of form out to some depth, h_*, to ensure conservation of sediment. For this case

$$\Delta \forall = \Delta y \ (B + h_*) \qquad (2)$$

in which B is the berm height, and Equation 1 becomes

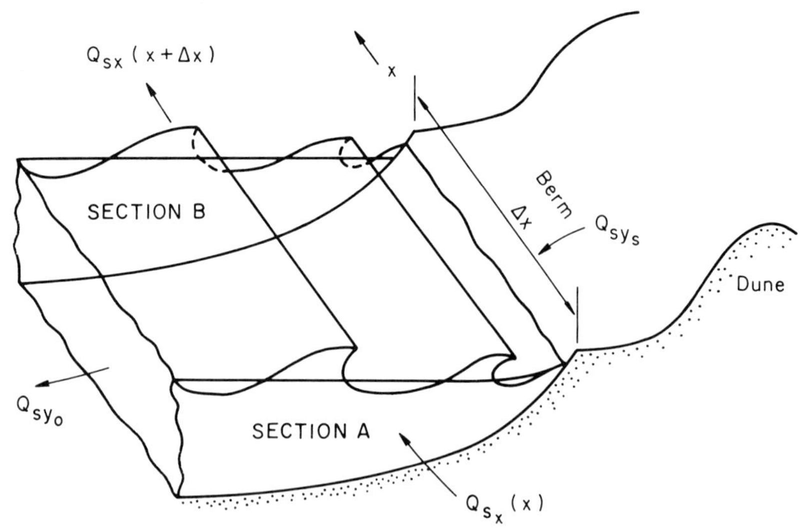

FIGURE 1. Volumetric changes through gradients in longshore transport.

$$\frac{\partial y}{\partial t} = -\frac{1}{(B + h_*)}\left(\frac{\partial Q_{sx}}{\partial x}\right) \qquad (3)$$

The Dynamic Element

Usually the dynamic element of a model is coupled in some manner to the change in force that is taking place. The most common type of model attempts to address the changes that occur due to storms, i.e., due to elevated mean water levels and storm waves or due to long-term sea level rise. The most simple model assumes an equilibrium profile which keeps pace with the sea-level rise. Clearly, all dynamic models are idealizations and simplifications of the natural process.

PARTICULAR BEACH PROFILE RESPONSE MODELS

The Bruun Model

Bruun[2] considered the shoreline retreat, R, that would occur due to an increase in local sea level, S, assuming that the beach profile remained the same relative to the actual sea level, that the active profile depth was h_*, and that the gradient in longshore sediment transport, $\frac{\partial Q_{sx}}{\partial x}$ was not important. The volume of sand yielded, ΔV_+, by a shoreline retreat, is given by Equation 2

$$\Delta \text{V}_+ = (B + h_*)R \qquad (4)$$

and the volume required, V_-, to elevate the active width of the profile, L, a vertical distance, S (see Figure 2), is

$$\Delta \text{V} = L\,S \qquad (5)$$

Equating Equations 4 and 5

$$R = \frac{L}{(B + h_*)} S \qquad (6)$$

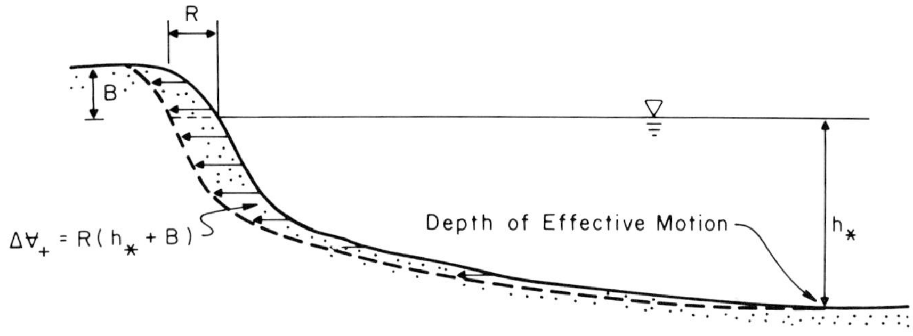

(a) Volume of Sand "Generated" by Horizontal Retreat, R, of Equilibrium Profile Over Vertical Distance (h_* + B)

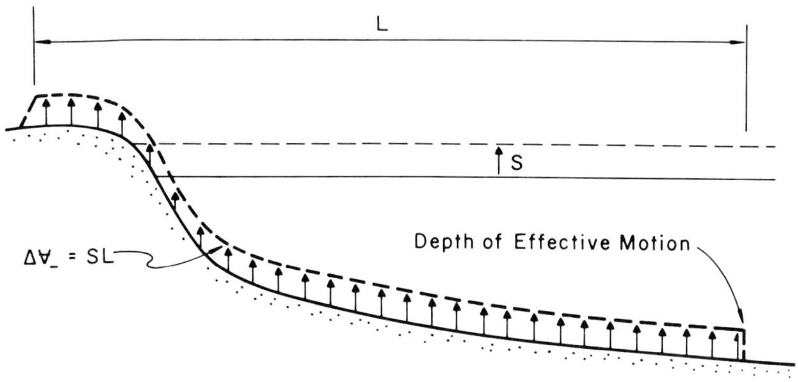

(b) Volume of Sand Required to Maintain An Equilibrium Profile of Active Width, L, Due to a Rise, S, in Mean Water Level.

FIGURE 2. Components of sand volume balance due to sea level rise and associated profile retreat.

or

$$R = \frac{1}{\tan \theta} S \tag{7}$$

where tan θ is the average beach slope over the active beach width, L. Equation 7 indicates that the horizontal recession, R, is proportional to and an amplified version of the change in sea level, S. Thus beaches with mild slopes will retreat more rapidly for a unit of sea level rise than beaches with steep slopes.

Edelman Model I

Edelman in 1968,[9] formulated an early beach response model based on numerous surveys of pre- and poststorm beach profiles along the coast of Holland. Edelman concluded that the equilibrium profile relative to the instantaneous sea level was approximately a constant and was represented by a uniform slope which depends on sediment characteristics. Edelman also considers the depth of effective motion to be given by the breaking depth. The method is then carried out by balancing the material deposited out to the breaking depth on the uniform slope (Edelman found a slope of approximately 1:40 for his measurements) with the material eroded to a landward limit of the instantaneous water level on the eroded profile. Figure 3 presents this method for an actual profile. Edelman also applied his method to an idealized profile in which the dunes were represented by a vertical face and a dune crest of

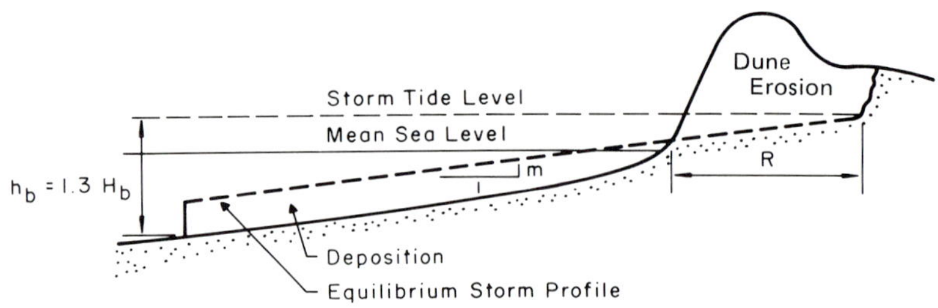

FIGURE 3. Edelman's Method I applied to actual beach profile.

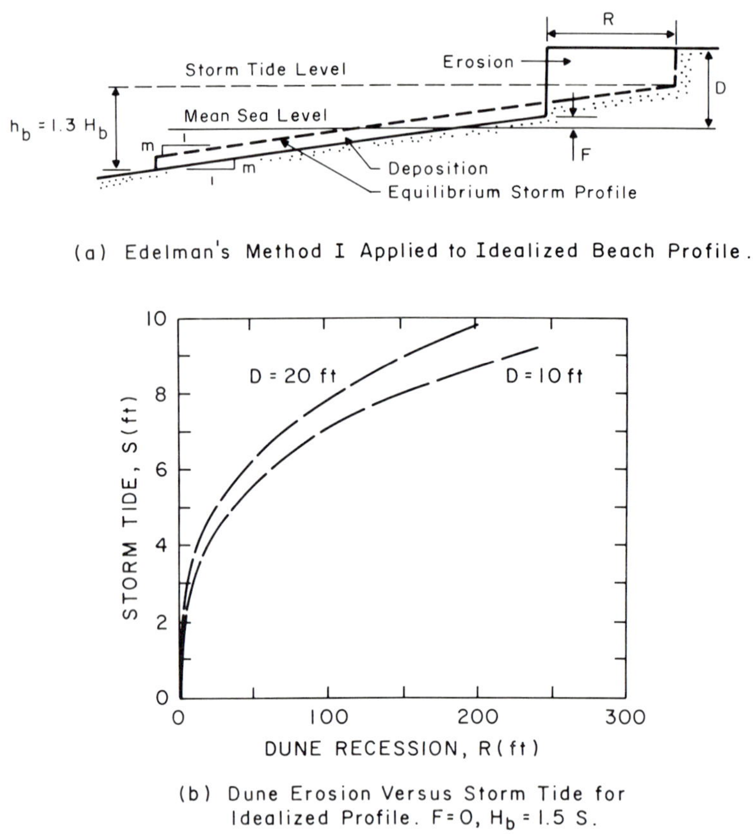

FIGURE 4. Edelman Method I applied to idealized beach profile response.

uniform elevation. In this case it is possible to develop somewhat generalized graphs which depict the effect of different profile and storm (surge and wave) variables. Figure 4 presents the dune erosion as a function of storm surge level and dune height. In this graph, the breaking wave height has been taken to be 1.5 times the storm surge.

Edelman Model II

This model (1970)[10] is much more sophisticated than that described in the preceding section and provides an analytical solution to the problem. It is assumed that the dune face is vertical with height (relative to storm level) h_D and that the profile changes keep pace with the rising storm water level, Figure 5. The original dune height and breaking depths are h_{Do} and h_{bo}, respectively. At some time after a water-level rise of magnitude, S, the

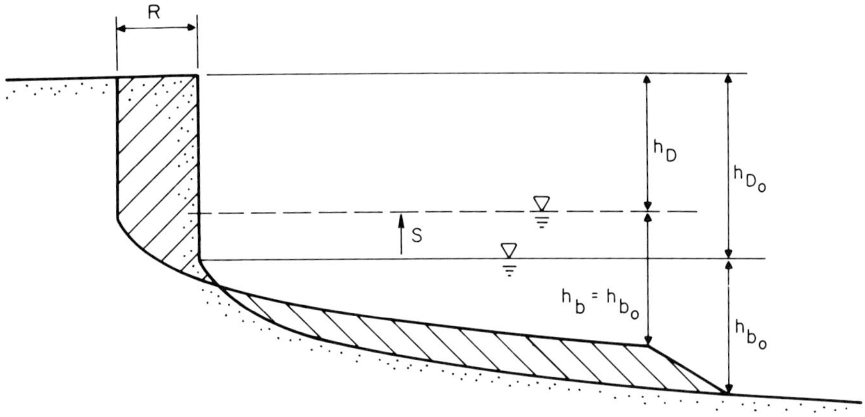

FIGURE 5. Characteristics of the Edelman II response model.

dune height and breaking depth are h_D and h_b. Considering the equilibrium profile to translate upward and landward with the rising sea level and denoting the horizontal and vertical rates of change (velocities) of this profile as U and V, respectively, the conservation of sand, counterpart to Equation 6, is

$$U(h_D + h_b) = VL \tag{8}$$

in which L again represents the active width of the profile, and it is noted that since the speed at which the profile moves upward is the same as the rate of sea level rise, then

$$V = \frac{dS}{dt} \tag{9}$$

and, by definition,

$$U = \frac{dR}{de} \tag{10}$$

Moreover, since the wave conditions are assumed to remain constant with time ($h_b = h_{bo}$), inspection of Figure 5 will show

$$h_D + h_b = h_{D_o} + h_{b_o} + (h_D - h_{D_o}) = h_{D_o} + h_{b_o} - S \tag{11}$$

Combining Equations 8, 9, 10, and 11 yields

$$dR = L\frac{dS}{(h_{D_o} + h_{b_o} - S)} \tag{12}$$

which can be integrated from S = 0 to S = S to yield

$$R = L \ln\left(\frac{h_{b_o} + h_{D_o}}{h_{b_o} + h_{D_o} - S}\right) \tag{13}$$

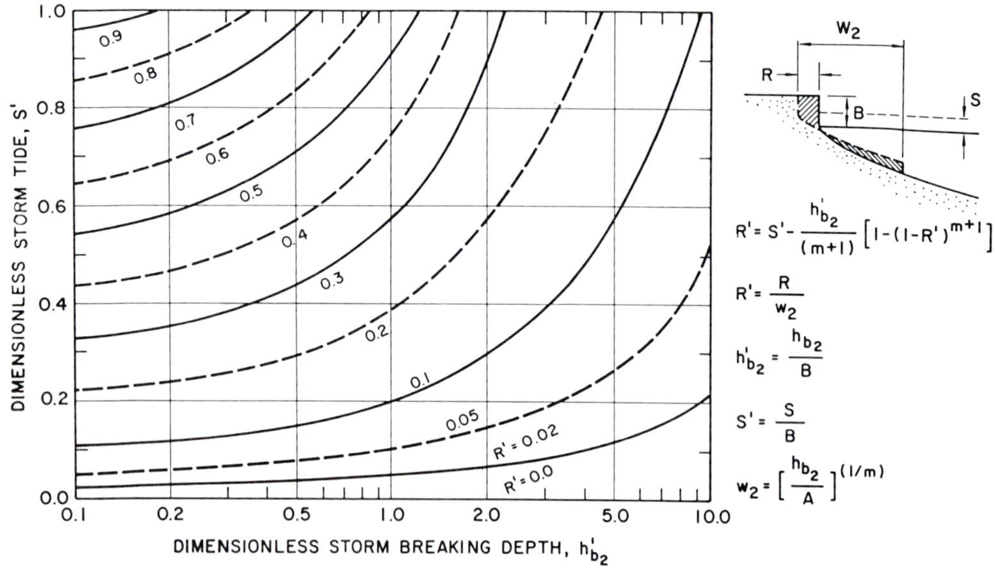

FIGURE 6. Isolines of dimensionless berm recession, R', vs. dimensionless storm breaking depth, h'_{b_2}, and dimensionless storm tide, S', $m = 2/3$.

It is noted that the only difference between Equations 7 and 13 is that the latter accounts for the diminishing dune height relative to the water level as the profile moves upward. Thus it is expected that Equation 13 would predict a somewhat greater beach recession than Equation 7. A later intercomparison of results obtained by various methods will confirm this expectation.

Equilibrium Profile of the Form: h = Ax$^{2/3}$

Bruun,[1] and later Dean[3] have investigated natural profiles and found a reasonably good fit to the form

$$h(x) = Ax^m \tag{14}$$

in which A is a scale factor primarily dependent on sediment characteristics and m is a shape factor, found to be 2/3. Equation 14 is clearly an idealization, representing a monotonic profile, so that bar effects cannot be represented. Dean[5] has shown that the dimensionless recession, R', can be represented in the following implicit form

$$R' = S' - \frac{3}{5} h'_{b_2} [1 - (1 - R')^{5/3}] \tag{15}$$

which is shown in Figure 6 along with the definition of dimensionless terms. The primary difference between Equations 13 and 15 is that the latter considers an abrupt rise in sea level with the equilibrium occurring over a long time period after the sea-level rise. Thus, the offshore terminus of the profile predicted by Equation 15 would be vertical, whereas that predicted by Equation 13 would have a slope:

$$\frac{dh}{dx} = \frac{h_{D_o} + h_b - S}{L} \tag{16}$$

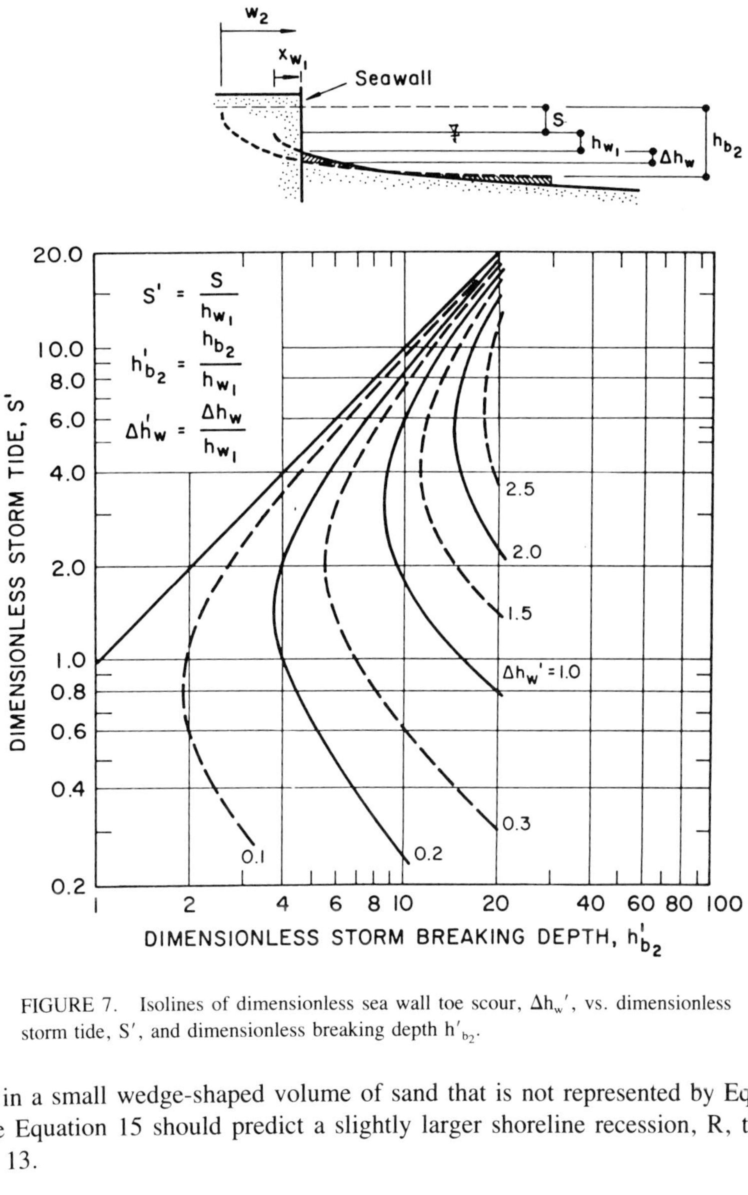

FIGURE 7. Isolines of dimensionless sea wall toe scour, $\Delta h_w'$, vs. dimensionless storm tide, S', and dimensionless breaking depth h'_{b_2}.

resulting in a small wedge-shaped volume of sand that is not represented by Equation 15. Therefore Equation 15 should predict a slightly larger shoreline recession, R, than would Equation 13.

Equilibrium Profile of the Form: $h = Ax^{2/3}$ (With Seawall Present)

Dean[5] has examined the profile changes that result due to a sea-level rise, S, for an equilibrium profile of the form of Equation 14. In this case, since the seawall prevents sand from being drawn from the subaerial portion of the profile, there must be a volumetric balance seaward of the seawall. Again the results are presented in graphical dimensionless form (Figure 7) as

$$\Delta h'_w\ (h'_{b_2}, S') \tag{17}$$

in which $\Delta h_w'$ represents the dimensionless increase in water depth at the toe of the seawall. With the depth at the wall known for the elevated water level,

$$h_w = h_{w_0} + S + \Delta h_w \tag{18}$$

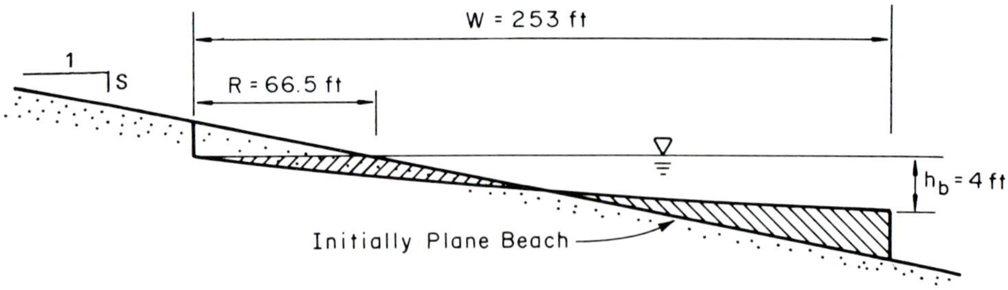

FIGURE 8. Evolution of a plane beach to a profile of the form: $h = Ax^{2/3}$. For this example, $h_b = 4$ ft, $s = 1{:}25$, $R = 66.5$ ft, and $W = 253$ ft.

the remainder of the profile can be computed from the equation

$$h(x) = Ax^{2/3} \tag{19}$$

by recognizing that there is a virtual origin, x_w, Equation 14, where

$$x_w = \left(\frac{h_w}{A}\right)^{3/2} \tag{20}$$

It can be shown that one effect of a seawall is to cause a greater lowering of the profile at the location of the seawall than would occur if no seawall were present. Thus in nature, it can be expected that during storm conditions, the presence of a seawall will cause some additional stress and associated erosion on the adjacent unarmored beaches.

Equilibrium Profile of the Form: $h = Ax^{2/3}$, Initial Uniform Slope

Consider the case in which a beach fill is placed at a slope of s and later the waves configure this material to a profile of the form of Equation 14 (see Figure 8). It can be shown that the shoreline recession, R, from the initial water level is given by

$$R = \frac{W}{2} - \frac{3}{5}\frac{h_b}{s} \tag{21}$$

Generalized Bruun Rule

Maurmeyer and Dean[13] have generalized the Bruun Rule to the case of the landward and upward migration of an entire barrier island system, Figure 9. The resulting equation is

$$R = \frac{S(L_o + W + L_L)}{(B_o + h_{bo}) - (B_L + h_{bL})} \tag{22}$$

in which L_o and L_w represent the widths of the "active" nearshore zones on the ocean and lagoon sides, respectively, and h_{bo} and h_{bL} are the associated breaking depths, and B_o and B_L denote the berm heights on the ocean and lagoon sides, respectively. W is the width of the barrier island, which is considered to remain constant in time but to accrete vertically at the same (average) rate as sea level rise. It is noted that this equation reduces to Bruun's Rule[2] for the case of no deposition on the barrier island or in the lagoon, i.e., for W, L_L, B_L, $h_{BL} = 0$, thus

$$R = \frac{SL_o}{(B_o + h_{bo})} \tag{23}$$

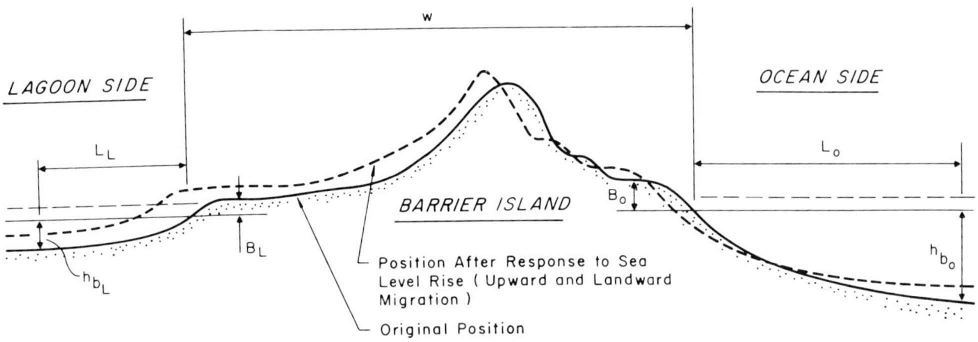

FIGURE 9. Generalized shoreline response model due to sea level rise. Applicable for a barrier island system which maintains its form relative to the adjacent ocean and lagoon.

It is also of interest to note that Equation 22 always predicts a greater retreat than does the Bruun Rule. The interpretation, of course, is (1) sand is being added to the barrier island width to maintain its vertical position relative to seal level, and (2) sand is being added to the lagoon side of the barrier island to maintain the width of the barrier island.

There is some evidence[17] of barrier islands "drowning" in place rather than surviving by landward migration during sea-level rise. It is of interest that one interpretation of Equation 22 is that if the denominator is zero, $(B_L + h_{bL}) = (B_o + h_{bo})$, then there would be no "yield" of sediment due to landward migration and any rise in sea level would cause the barrier island to become narrower and it would, in effect, "drown in place".

EVALUATION OF MODEL CONCEPTS

There are a number of conceptual elements of the various models that, although expedient, do not provide an accurate representation of natural processes or time scales.

Breaking Conditions

It is usually assumed in most models that waves break in a water depth which is proportional to the wave height, i.e.,

$$H_b = \kappa h_b \tag{24}$$

Equation 24 is often termed the "spilling breaker" assumption and κ is taken as approximately 0.8. In reality, a finite beach slope causes an increase in the breaking height, H_b, and other factors such as wave height and length can modify the character of the breaking wave. Thus although Equation 24 is a convenient form for idealized conditions, it does not effectively represent all wave conditions in nature.

Depth of Effective Motion

Inherent in all models of profile response is the requirement for defining some limit of offshore adjustment. In many models in which short-term response is of greatest interest, the assumption is made that the breaking depth defines the limit of effective motion. On a longer term basis, the "rule of thumb" on exposed shorelines is approximately 20 to 25 ft *(Shore Protection Manual, 1977)*. In reality, the depth of effective motion depends on the time scale of interest. At some locations on the east coast of the U.S. the effective depth associated with periods of interest of 1 year, 5 years, and 20 years may be 15 ft, 20 ft, and 25 ft, respectively. Hallermeier[11] has developed a procedure in which wave data are employed to determine the annual depth of effective sand motion. The detailed method was calibrated with data from a number of laboratory and field cases. Hallermeier concludes that the annual depth of effective sand motion is approximately equal to twice the significant wave height which is exceeded 12 hr/year.

Constancy of Profile

There are two characteristics of profiles in nature that differ from the constancy assumption. The first is that the equilibrium profile of a particular type sediment definitely depends on wave conditions, changing from a relatively monotonic steep beach profile for mild (i.e., low steepness) waves to a barred profile of relatively mild average slope for storm waves. Secondly, the evolution of a beach profile with time from one state of equilibrium to another state of equilibrium does not occur with a constant profile form, even if the initial and final equilibrium forms are the same relative to the associated water levels. The nature of the variation from constancy of profile is that the changes are initiated where the profile is the most out of equilibrium. With rising water level, this greatest deviation appears to occur in the shallow portions of the profile, hence, the greatest initial changes occur in this region. With falling water level, there is frequently a bar offshore which is greatly out of equilibrium relative to the reduced water level and wave conditions. This feature then responds first, resulting in the familiar ridge and runnel system associated with the recovery process of an eroded beach.

Response Time

Many of the models consider the profile response to a change in conditions to be essentially instantaneous. In many cases this can limit severely the validity of the resulting predictions. In reality, the profile response time may be much longer than the time scale of certain affecting hydrologic factors. For example, the time scale for peak storm tides due to small intense hurricanes is generally on the order of hours, whereas the response time of beach profiles is an order of magnitude longer. The results of applying any of the methods described earlier should be regarded as an upper limit or an erosion potential that would result if the storm conditions were held constant indefinitely. Kriebel[12] has shown by detailed numerical modeling of beach profile response due to hurricanes that the erosion usually reaches only a fraction (on the order of 25%) of the erosion potential.

COMPARISON OF MODELS WITH LABORATORY AND FIELD DATA

Bruun,[2] Schwartz,[18,19] Dubois,[6-8] and Rosen[15] have evaluated the Bruun Rule for appropriate conditions. Bruun has compared his results with measured shoreline retreat rates of the eastern and "panhandle" (western) areas of the Florida shoreline. For the southeast shoreline, it was found that the long-term rate of sea level rise of 1.2 mm/year yielded a shoreline recession rate of approximately 11 cm/year, considerably less than the observed rate of approximately 60 cm/year. Bruun reasoned that the shoreline retreat is due to the short-term rates of sea-level rise of 6 mm/year which results in predicted rates more in accord with observations. On the upper east coast of Florida (Cape Canaveral to Daytona Beach), the long-term sea level rise of 1.2 mm/year yielded recession rates of approximately 40 cm/year, which is reasonable; however, the results are quite sensitive to the width, L, of the active zone. In the panhandle area of Florida, Bruun found the observed shoreline retreat rates were in better accord with prediction based on the long-term sea-level rise rates (1.2 mm/year).

Schwartz[18] carried out small-scale laboratory model studies and field investigations to determine whether the increase in water level caused an offshore deposition equal to the rise in water level, as predicted by the Bruun Rule. The laboratory studies were conducted in a very small wave basin (approximately 80 × 115 cm in plan) in which waves of 0.33

sec were generated. Medium sand, with an average diameter of approximately 0.2 mm was used. The tests consisted of generating waves until an equilibrium profile was approximated, a phase which usually required approximately one half hour. Following this equilibration phase the resulting beach profiles were measured, the water level was raised 1.0 cm, and the waves were generated for an additional half hour. The beach profiles were then measured again. It was found that the sand level increased by 0.9 cm, which Schwartz regarded as substantiation of Bruun's hypothesis. Schwartz[19] also carried out a second series of experiments at somewhat larger scale with a wave basin of plan dimensions 100 × 232 cm. In this series, the wave period, wave height, and manner in which the water level was increased varied from test to test. Water level increases of 3 and 6 cm were used. The mean deviation between the increase in sand level offshore and that predicted by the Bruun hypothesis was only 0.4%. In the laboratory tests, Schwartz did not attempt to substantiate Bruun's prediction of shoreline recession. This, of course, would have required determination of the offshore distance to which the eroded sand was deposited, a value not reported by Schwartz. In addition to the laboratory studies, Schwartz conducted two field experiments on Cape Cod, Mass. The hypothesis was that during the period between spring and neap tidal occurrences and between neap and spring tidal occurrences, the beach would respond due to mean tidal range decrease and increase and that the response would be qualitatively similar to a decrease and increase, respectively, in mean sea level. For each of the two field sites a total of five profiles were reported including three at spring tide and two at neap tide. Although Schwartz concluded that "a recognizable upward and landward translation of the profile was noted in the intervals between neap and spring tides", the data are not absolutely convincing as there are some intervals between spring and neap tides during which the profile moves landward. Moreover, the argument is not clear that the effect of a transition from neap to spring tidal range should be qualitatively similar to an increase in mean sea level. The profiles did not extend seaward to the depth of closure, and it was therefore impossible to state conclusively that the beach profiles examined were in longshore equilibrium. Regardless, Schwartz concluded that both his laboratory and field measurements corroborated Bruun's predictions and proposed that the relationship be known as "Bruun's Rule", an appellation that is now in general use.

Dubois[6] reported on a field study at Terry Andrae State Park, Mich. in which characteristics of two beach profiles were measured on a weekly basis over a 35-week period during which the mean lake level increased by approximately 1 ft. It was found that the shoreline recession due to the 1-ft rise was 23 ft, of which 6 ft could be attributed to the direct effect of inundation of the sloped profile. Dubois conducted a correlation analysis between the water level increase, wave conditions, and various features of the two profiles. It was found that various elements of the beach profiles, e.g., the base and crest positions of the foreshore, maintained their position relative to the rising water level. Thus Dubois' study provided a qualitative validation of the Bruun Rule. Dubois suggested that certain refinements be made to the Bruun Rule; however, it appears to this writer that they were inherently present, but perhaps not explicitly stated, in the original version by Bruun.[2] Later, Dubois[8] used the same field data in an attempt to provide a quantitative verification to the Bruun Rule. It was concluded that the rule in the following form was substantiated by the field data

$$R = \frac{1}{\tan \theta_B} S \tag{25}$$

which is the form of Equation 7, except that Dubois defines θ_B as "the angle of the nearshore slope seaward from breaking waves." As derived in the earlier portion of this paper, the relevant slope, θ, is clearly the average slope (= Effective Vertical Dimension of Active Profile/Effective Horizontal Dimension of Active Profile). Although Figure 2 does not portray any bars, Equation 7 applies to an arbitrary form of beach profile, regardless of the number of bars. Rosen[16] has discussed the Dubois paper,[8] with particular reference to the concept of "dynamic equilibrium" of the beach profile, as assumed by Dubois.

Rosen[15] has evaluated the Bruun Rule on the Virginia shoreline of Chesapeake Bay, an area where the local sea level rise varied from 0.7 to 6.6 mm/year. There was also considerable variability in the bluff elevations. The break in offshore slope at 3.6 m was used to define the actual width of the beach profile and the depth of effective motion. A total of 14 beach profiles were included in the evaluation. Rosen found that the errors in erosion rates on the eastern and western shores of the bay were +58% and −7% where the positive percentage indicates that the predicted erosion exceeds measured. Locally, for smaller groups of profiles, the errors were found to be considerably larger.

Maurmeyer and Dean[13] have evaluated the generalized model for shoreline retreat (Equation 22) which allows for landward migration of a barrier island system which maintains equilibrium relative to sea level. This model was applied to six barrier island locations ranging from North Carolina to New Jersey. The limiting depths were taken to be 10 m and 2 m on the ocean and lagoon sides, respectively. The other variables in Equation 22 were determined directly from charts and knowledge of the area, including published retreat rates. Comparisons were made of the measured and predicted retreat rates with the latter based on both Bruun's Rule and the modifications for barrier island migration as represented by Equation 22. It was found that the Bruun Rule underpredicted the retreat rate by approximately 60% wherease Equation 22 predicted the average within 3%. The underprediction by Equation 7 is not surprising since it was not developed for barrier island systems.

INTERCOMPARISON OF BEACH RESPONSE MODELS

It is of interest to intercompare the Bruun (1962),[2] Edelman II (1970),[10] and Dean (1975) models of beach profile response. To reduce the number of variables, these models are cast into dimensionless form as follows.

Model I: Bruun (1962)

$$R' = \frac{S'}{(h'_b + 1)} \tag{26}$$

Model II: Edelman (1970)

$$R' = \ln\left[\frac{h'_b + 1}{h'_b + 1 - S}\right] \tag{27}$$

Model III: (Valid for $h = Ax^{2/3}$), Dean (1975)

$$R' = S' - \frac{3}{5} h'_{b2} \left[1 - (1 - R')^{5/3}\right] \tag{28}$$

in which

$$S' = S/B \tag{29}$$

$$h'_b = h_b/B \tag{30}$$

$$R' = R/W \tag{31}$$

Table 1 presents dimensionless shoreline recessions, R', for particular combinations of

Table 1
COMPARISON OF VARIOUS METHODS OF PREDICTING PROFILE RESPONSE

		R′		
h'_b	S′	Method I (Bruun)	Method II (Edelman)	Method III (Dean)
0.2	0.2	0.167	0.182	0.168
0.2	0.4	0.333	0.405	0.340
0.2	0.6	0.500	0.693	0.516
0.2	0.8	0.667	1.099	0.696
1.0	0.2	0.100	0.105	0.102
1.0	0.4	0.200	0.223	0.207
1.0	0.6	0.300	0.357	0.318
1.0	0.8	.0.400	0.511	0.433
5.0	0.2	0.033	0.034	0.033
5.0	0.4	0.067	0.069	0.067
5.0	0.6	0.100	0.105	0.103
5.0	0.8	0.133	0.143	0.139

dimensionless breaking depths and sea-level rise, S′. It is seen that the Bruun and Dean methods yield very similar results, with the Dean method being slightly higher for the larger values of S′. The Edelman method predicts considerably larger values of shoreline recession. As discussed earlier, the reason is that this method considers the response to be instantaneous and continuous, thus keeping pace with the rising sea level. Therefore, associated with the equilibrium profile would be a wedge volume of sand at the offshore terminus. The Bruun and Dean methods consider a rise in sea level followed by a beach profile response. It is seen that the differences are significant only for large sea-level increases relative to the berm height.

USEFUL EXTENSIONS TO SHORELINE RESPONSE MODELS

Several extensions to the existing models would serve to improve their validity.

1. In general, when a profile erodes, a proportion of the eroded material may be vegetative matter, muds, or otherwise incompatible with the sands residing in the surf zone. Denoting P as that decimal percentage of eroded material *which is compatible* with the surf-zone sands, the Bruun Rule (Equation 7) would be modified to

$$R = \frac{1}{P \tan \theta} S \tag{32}$$

and it is seen that if all the material is compatible (P = 1), Equation 7 reverts to the usual Bruun Rule.

2. The existing models could be extended to include the effects of longshore disequilibrium, i.e., $\frac{\partial Q_{sx}}{\partial x} \neq 0$ For the Bruun Rule, this modification would yield

$$R = \frac{LS + \frac{\partial Q_{sx}}{\partial x} \Delta t}{P(h_* + B)} \tag{33}$$

where Δt is the time interval over which the sea-level rise, S, occurs. Again it is seen that if $\frac{\partial Q_{sx}}{\partial x}$ and P are equal to 0 and 1, respectively, Equation 33 reverts to the Bruun Rule. One of the difficulties in applying Equation 33 is that the gradient of longshore transport is not usually adequately known.

3. The dynamics of shoreline response need to be better represented. This is particularly true for applications of short-term events such as storm times due to hurricanes, but also applies to cases of shoreline response due to long impermeable structures. To the writer's knowledge, Kriebel[12] has developed the most comprehensive approach to this problem to date.

SUMMARY

The following summarizes the material presented in this chapter.

1. Existing shore response models are useful for predicting long-term evolution due to relative sea-level rise. Better methods and field data are required to improve the capability of predicting depth of effective sand motion and the associated width of this zone.
2. The Bruun Rule has been validated qualitatively and, to the limit of our knowledge of the relevant processes, quantitatively for the case of nonbarrier island systems. For barrier island systems which migrate landward, the model by Maurmeyer and Dean[13] is more appropriate.
3. Of the existing models of the Bruun type, the Edelman II model represents profile evolution as a continuing process and is therefore probably more representative of long-term response.
4. There is a need for application of improved profile response models which incorporate the effects of noncompatible sediment eroded and gradients in longshore sediment transport.
5. There is a need for improved definition of the detailed dynamics of beach profile response. This will probably require both laboratory and field measurements under long- and short-term (storm) events.

REFERENCES

1. **Bruun, P.,** Coast erosion and the development of beach profiles, *U.S. Army Beach Erosion Board Tech. Memo.*, No. 44, 1954.
2. **Bruun, P.,** Sea level rise as a cause of shore erosion, *Proc. ASCE J. Waterways Harbors*, 88, 117, 1962.
3. **Dean, R. G.,** *Equilibrium Beach Profiles: U.S. Atlantic and Gulf Coasts*, Tech. Rep. No. 12, University of Delaware, Newark, 1977.
4. **Dean, R.G.,** Beach erosion: causes, processes and remedial measures, *CRC Crit. Rev. Environ. Control*, 6(3), 259, 1976.
5. **Dean, R.G.,** *Models For Beach Profile Response*, Tech. Rep. No. 30, University of Delaware, Newark, 1982.
6. **Dubois, R. N.,** Support and refinement of the Bruun rule on beach erosion, *J. Geol.*, 83, 651, 1975.
7. **Dubois, R. N.,** Nearshore evidence in support of the Bruun rule on shore erosion, *J. Geol.*, 84, 485, 1976.
8. **Dubois, R. N.,** Predicting beach erosion as a function of rising sea level, *J. Geol.*, 77, 470, 1977.
9. **Edelman, T.,** Dune erosion during storm conditions, *Proc. 11th Int. Conf. Coastal Eng.*, American Society of Civil Engineers, New York, 1968, 719.
10. **Edelman, T.,** Dune Erosion During Storm Conditions, *Proc. 12th Int. Conf. Coastal Eng.*, American Society of Civil Engineers, New York, 1970, 1305.

11. **Hallermeier, R. J.,** Calculating a yearly limit depth to the active beach profile, *U.S. Army Coastal Eng. Res. Cent. Tech. Pap.*, No. 77-9, 1977.
12. **Kriebel, D.,** Beach and Dune Response due to Hurricanes, M.S. thesis, Department of Civil Engineering, University of Delaware, Newark, 1982.
13. **Maurmeyer, E.M. and Dean, R. G.,** Sea-level rise and barrier island migration: an extension of the Bruun Rule to account for landward sediment transport, Abstr. Northeastern-Southeastern Meet., Geological Society of America, Washington, D.C., March, 1982.
14. **Rampino, M. R.,** Holocene submergence of southern Long Island, New York, *Nature (London)*, 280(5718), 132, 1979.
15. **Rosen, P. S.,** A regional test of the Bruun Rule of shoreline erosion, *Mar. Geol.*, 26, M7, 1978.
16. **Rosen, P.,** Predicting beach erosion as a function of rising water level: discussion, *J. Geol.*, 86, 763, 1978.
17. **Sanders, J. E. and Kumar, N.,** Evidence of shoreface retreat and in-place 'drowning' during Holocene submergence of barriers, shelf off Fire Island, New York, *Geol. Soc. Am. Bull.*, 86, 65, 1975.
18. **Schwartz, M. L.,** Laboratory study of sea-level rise as a cause of shore erosion, *J. Geol.*, 73, 528, 1965.
19. **Schwartz, M. L.,** The Bruun theory of sea-level rise as a cause of shore erosion, *J. Geol.*, 75, 76, 1967.

Chapter 8

THE GREAT LAKES AS A TEST MODEL FOR PROFILE RESPONSES TO SEA LEVEL CHANGES

Edward B. Hands

INTRODUCTION

The Laurentian Great Lakes system, composed of five large lakes and their interconnecting waterways (Figure 1), constitutes the largest single mass of fresh water on the surface of the earth. Regional variations and climatic factors cause long-term water level fluctuations on the Great Lakes that are uncharacteristic of ocean sites (Figure 2). The cumulative effect of persistent changes in lake levels frequently shifts monthly and annual mean surface elevations by as much as a meter in less than a decade (Table 1). Although water level changes of the type indicated in Table 1 may not appear large relative to tidal ranges at many ocean beaches, the long-term gradual nature of the changes increases their effect on shore erosion. The long duration of high water allows time for a relatively broad area of the beach and nearshore zone to adjust to the elevated water surface. This adjustment involves offshore transport of large volumes of beach material and, as a consequence, substantial shore retreat. When lake levels decline sufficiently, conditions reverse and waves transport much of the material from offshore back onto the beach. The purpose of this chapter is to summarize results of a recently completed study that has demonstrated the ability of a simple sediment budget model to accurately predict the ultimate shore response to a long-term change in water level.

Publications discussing wider aspects of Great Lakes sedimentation are far too numerous to review here, but a compilation of published and unpublished data on erosion of the U.S. shoreline was prepared by Armstrong et al.[1] An atlas by Haras[20] presented data on land use, historic flood and erosion damage, ownership, property values, and physical characteristics for all the erodible Canadian shoreline of the Great Lakes. Birkemeier[3] reported on the effects of structures on shore erosion in a southern county of Lake Michigan. Temporal changes in the rate of erosion on the eastern shore of Lake Michigan have been shown to correlate strongly with periods of storminess.[4] Seasonal and irregular variations in storminess tend to obscure the effect of long-term variations in lake level in data taken over a period of months or even a few years.

In his review of nearshore research, Hails[10] states "the biggest problem is simply the collection of data for a sufficiently long period to gain a representative picture of the changes taking place in the coastal zone". Certainly, this has been the case in attempts to document the effects of unpredictable water level changes that may fissile or reverse in a year, or either persistently rise or fall for a decade. The long-term commitment, funds, and continuity of personnel to monitor changes through-out a lake level cycle are not readily available. The long-term effects of lake level changes are obscured in compilations of time and site scrambled data that span periods of both rising and falling lake levels (Figure 3). This chapter is therefore, based on a single U.S. Army Corps of Engineers study that monitored beach and nearshore changes on a 50 km stretch of the eastern shore of Lake Michigan over a nine-year period of persistently rising and then stable annual mean lake levels.[18]

THE GREAT LAKES AS A MORE MANAGEABLE MODEL OF THE OCEANS

In many ways the Great Lakes offer a more manageable and hospitable environment than

FIGURE 1. Location map for the Great Lakes.

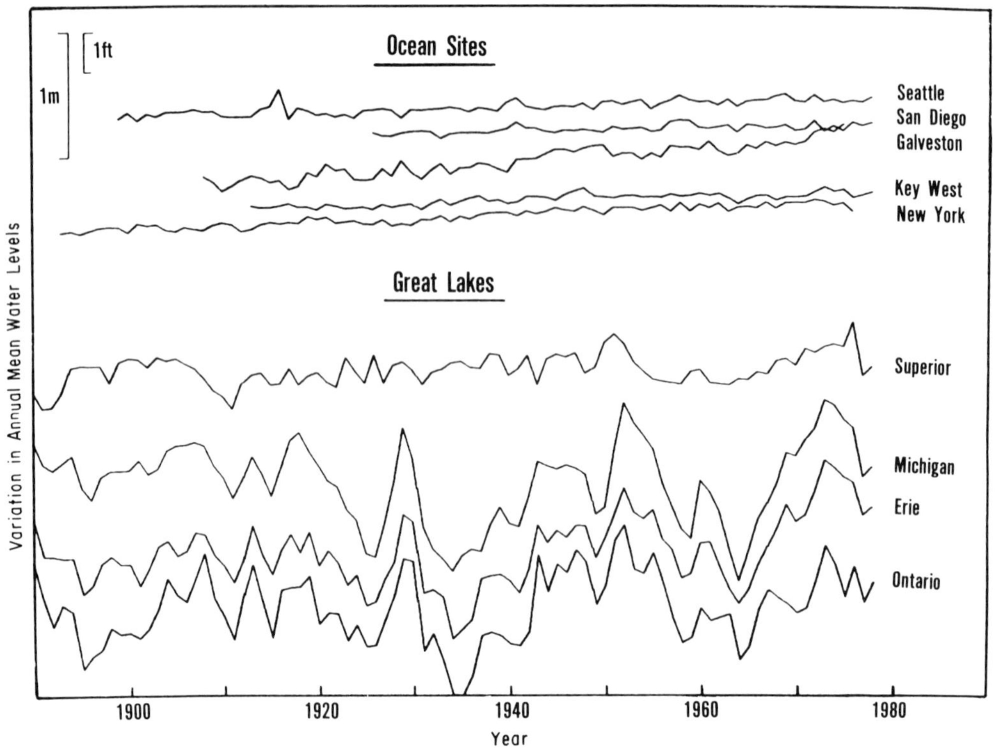

FIGURE 2. Comparison of annual mean water levels at ocean and Great Lakes sites. During rising lake levels, the shores of the Great Lakes may be submerged more in a 5 to 10 year period than most ocean sites are in a century. Reversals in trend have reduced the longer term erosive effects on the lakes, while ocean sites have been exposed to a slower, but inexorable submergence.[18]

do the oceans for the study of certain shore erosion processes. The low concentration of dissolved salt in the Lakes makes for a less corrosive environment in which to deploy instruments, while having little effect on the erosion of sandy shores. Seawater does differ from fresh water in having a lower freezing temperature and higher density. In addition, seawater increases its density as the temperature drops all the way to the freezing point,

Table 1
MAJOR INCREASES IN ANNUAL MEAN LAKE LEVEL

Lake	1925—29 (m)	1925—29 (ft)	1949—52 (m)	1949—52 (ft)	1964—73 (m)	1964—73 (ft)
Ontario	0.70	2.3	0.63	2.1	0.91	3.0
Erie	0.71	2.3	0.55	1.8	1.14	3.8
Michigan-Huron	0.99	3.3	0.88	2.9	1.45	4.8
Superior	0.47	1.6	0.19	0.6	0.26	0.9

while fresh water reaches its maximum density at about 4°C above freezing. So as fresh water cools from 4° to 0°C, it expands and floats on a column of water of nearly uniform temperature near 4°C. Seawater always sinks as it cools because the temperature of maximum density is below the freezing point. Thus, there is no winter time stratification in the oceans analogous to that in the Great Lakes. In the summer, warm surface waters are usually separated from deeper cool waters by a sharp thermal and density gradient (thermocline) in both the Lakes and the ocean. The addition of a winter stratification and spring turnover on the Lakes has no known effect on shore erosion processes.

Periods of ice cover on the Lakes are primarily times of inactivity so far as shore erosion is concerned. The ice does encourage settling of fine particles nearshore, but along sand shores these deposits are usually dispersed during spring storms.

The Great Lakes system embraces a broad range of shore types: from high rocky cliffs (e.g., much of Lake Superior, the Door Peninsula of Lake Michigan, Georgian Peninsula of Lake Huron), to low gradient marshy shores (Green Bay on Lake Michigan and sections of Lake Erie), to some of the largest coastal sand dunes in the United States (eastern shore of Lake Michigan). This chapter is limited to the response of sandy shores (loose substrates) to water level changes.

The sandy shores of the Great Lakes differ principally from those on ocean coasts in a greater development of longshore bars. Sequences of longshore bars commonly extend for tens of kilometers along sandy sections of Great Lake shores (Figure 4). In this respect, the Great Lakes are more similar in nearshore bathymetry to other enclosed bodies of water, such as the Mediterranean and the Baltic Seas, than they are to most ocean beaches. Longshore bars are persistent features throughout the year in the Great Lakes while they commonly appear and disappear seasonally off ocean beaches. Because multiple bars are so well developed on the Lakes, Hands[14] was able to demonstrate a consistent increase in all bar dimensions in an offshore sequence (Figure 5). The length of bar base, bar spacing, and crest depth all increased at a fixed rate from one Lake Michigan bar to its deeper neighbor. The rate of increase in size may vary regionally depending on the overall nearshore slope, and is less where the slope is flatter. In the Lake Michigan study area where there was a 60% rate of increase in size, the overall slope through the barred field was about 0.01 (1:100).

If the better developed bars on the Great Lakes were smoothed out, the range of slopes between sandy beaches and the 10 m contour would be comparable with the range of slopes encountered off ocean beaches. In general, nearshore slopes on Pacific beaches tend to be flatter off the stormier northwest coast and steepen towards southern California. In Figure 6, a typical profile from the eastern shore of Lake Michigan can be compared with profiles from Washington, Oregon, and California. The Lake Michigan profiles have an intermediate slope similar to that found off sandy beaches in northern California. Note that longshore bars in the same depths do have similar dimensions off both lake and ocean beaches, the ratio of bar relief to depth tending to be constant for the different environments.[14,26]

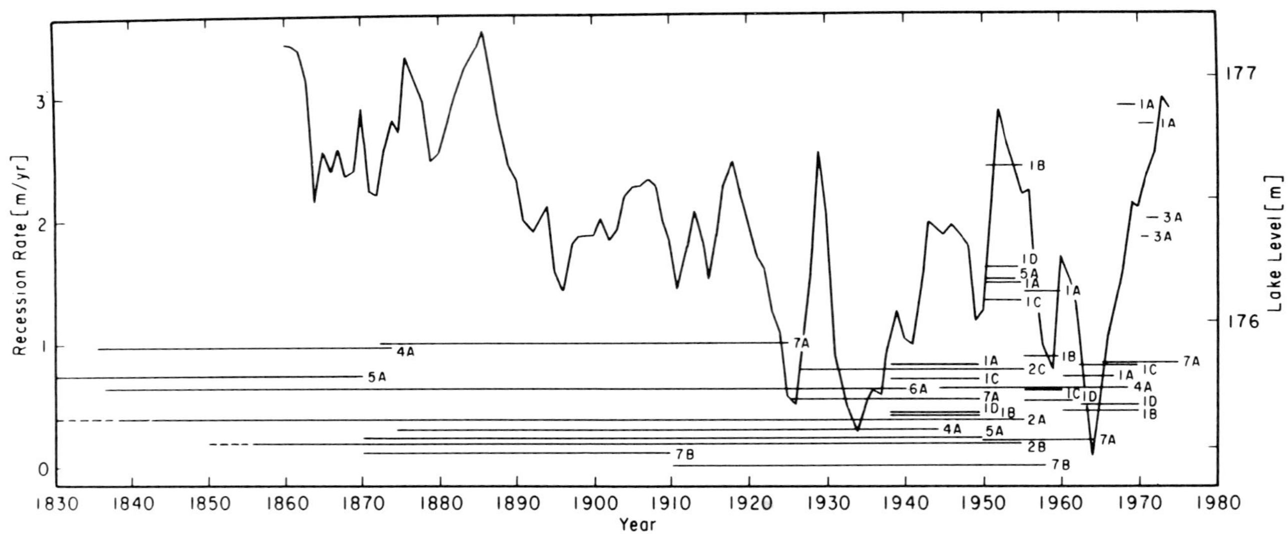

FIGURE 3. Annual mean lake level fluctuations (continuous jagged line) compared with mean bluff recession rates (straight lines). Recession rates are averages of a variable number of measurements in different sites on Lake Michigan spanning different time periods indicated by the length of each line. To examine the change in recession with time while holding the exposure and shore resistance constant, follow the rates by number/letter code. Numbers indicate sources: (1) Seibel,[34] (2) Powers,[30] (3) Davis et al.,[7] (4) Wisconsin Department of Natural Resources,[41] (6) U.S. Congress,[39] and (7) Berg and Collinson.[2] Letters indicate locations. From Hands.[14]

FIGURE 4. Longshore continuity of bars on the eastern shore of Lake Michigan. The inner two to three bars are usually visible when viewed from high bluffs and dunes along the shore.[18]

Storms erode the lake beaches and create another type of submerged sand ridge between the beach and the first persistent longshore bar. This storm ridge and its associated landward trough are sometimes referred to as the "ridge and runnel". In contrast to persistent longshore bars, the ridge is a smaller, temporary feature that may be linear, or doubly shoretied. After storms, these ridges characteristically migrate landward and weld to the beach. Davis et al.[6] compared ocean and Great Lake ridge and runnels, concluding that the overall morphology and internal structure are quite similar. Great Lakes ridges are, however, smaller in scale and migrate much more rapidly. Davis et al.[6] attribute these differences to the lack of tides on the Lakes.

The presence of a sequence of longshore bars also effects the sorting of Great Lakes nearshore sediments (Figure 7). Well-sorted sands become progressively finer as one moves from one bar crest to the next offshore. The intervening troughs, however, are lined with much coarser material. The coarsest and most poorly sorted sands are usually in the outer or next to outer trough.

Both lake and ocean waters respond to an onshore wind stress by piling up water on the beach and creating a temporarily inclined water surface. However, when the wind stress dies down on the Lakes, the water surface begins to oscillate as a function of the length and depth of the basin. These oscillations or seiches may affect onshore-offshore movement of sediment, but measurements of this phenomenon have not been obtained. The Lake Michigan study sites, upon which analogies in this chapter will be drawn, are not subject to extreme set-up or seiches typical of other sections of the Great Lakes. Therefore, these phenomenon introduce no difference between ocean and lake erosion processes that need to be considered here.

The major difference in erosion on the lakes and oceans results from the contrasting wave climates caused by restrictions in fetch on the Lakes (Chapter 1). A restriction in fetch results in a ceiling on both the wave height and wave period. The restricted size of the Great Lakes also limits the transfer of energy from short period seas to longer period swells. This shift of energy to longer period swell is characteristic of oceans where the waves travel

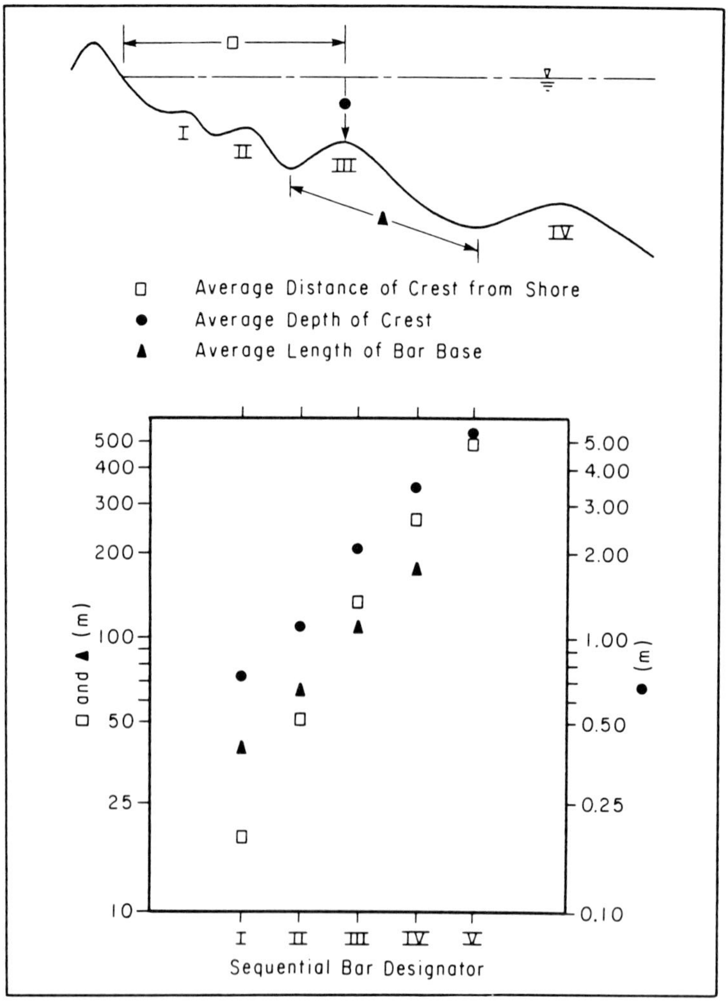

FIGURE 5. Average bar dimensions vs. position in an offshore sequence. Bar dimensions increase offshore across multiple bar fields, forming regular geometric progressions. Both the depth of the bar crest and the length of bar base appear to increase at the same rate. The rate may vary from site to site, and will be less where nearshore slopes are flatter. The above data are from a site on Lake Michigan where the slope is 1 in 100. Successive bars are about 60% deeper and 60% longer in cross-section than their landward neighbor.[14]

perhaps thousands of miles between the storm generation area and the beach. The distinction between sea and swell may be ignored on the Great Lakes.[31] In fact, wave periods rarely exceed 10 sec whereas periods in excess of 20 sec can occur on the oceans.[35] Even though restricted fetch does limit maximum wave development, the sinking of seasoned ocean vessels during their first winter voyages on the Lakes gives testimony to the frequent underestimation of Great Lakes storms by those not familiar with this region.

Insofar as shore erosion and the development of beach profiles are concerned, the surface waves interact with sediment in three ways: (1) waves impinging obliquely on the shore create longshore currents and a longshore sand transport, (2) breaking waves throw bottom material into suspension, and (3) the orbital velocities of water wave motion sufficient to lift and move sediment to-and-fro in a wide zone offshore of the breaker zone. The further offshore, the lower the velocity of these bottom orbital motions. By linear theory, at a depth

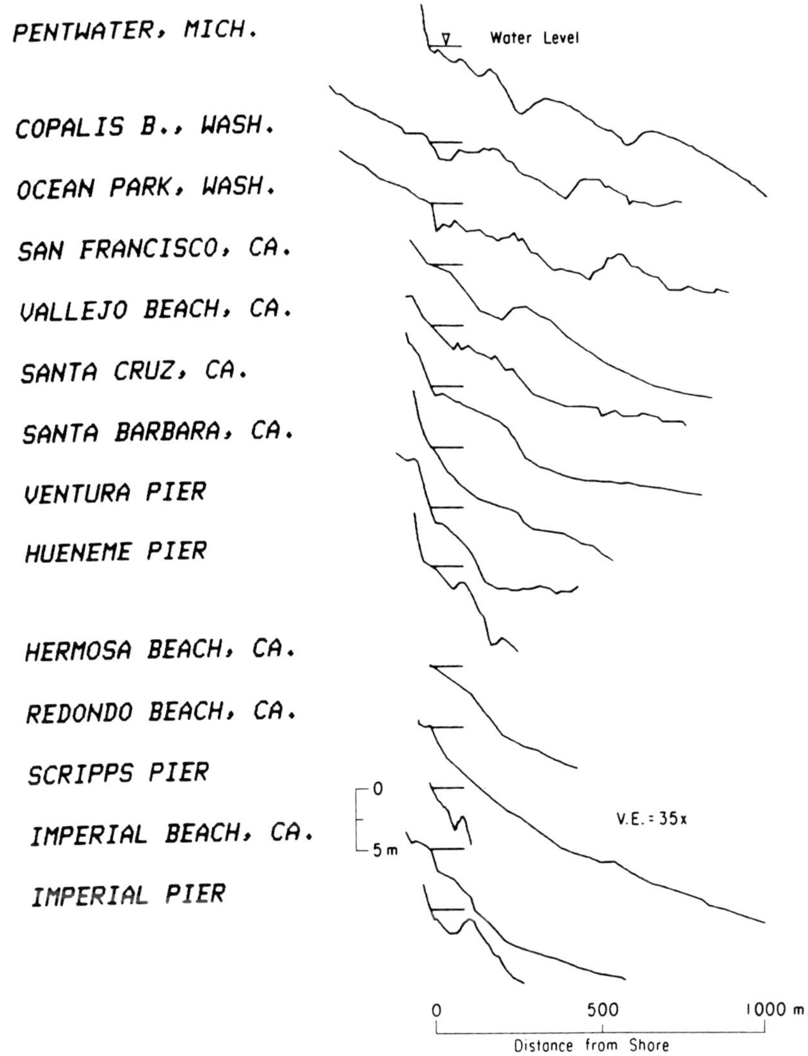

FIGURE 6. The smooth, well-developed bar pattern of the top profile is typical of the Lake Michigan study area. Pacific coast profiles (arranged from north at the top to south at the bottom) sometimes have one or two bars of a size similar to those at the same depths on the Lakes, but ocean coast profiles generally lack the orderly progression of a multiple-bar field which is typical of enclosed seas. The overall slope of the Pacific nearshore profiles increases from north to south. Smoothing out the bars, the Lake Michigan profiles have an intermediate slope similar to that of sandy sections of northern California.

equal to one half the wave length the bottom velocity should be about 4% of that at the surface. For most purposes it is assumed that the wave is not affected by the bottom before reaching this depth. Detailed analyses of wave conditions leading to motion of bottom sediments have been given by Komar and Miller[24] and Hallermeier.[12] For simplicity here, the depth at which the waves affect the bottom is taken to be the depth at which orbital velocities reach 15 cm/sec. This depth of initial bed motion will depend on both the height and period of the waves and assumes a rough cohesionless sand bottom. The depth of initial bed motion presented in Figure 8 was obtained by predicting fully-developed and fetch-limited deep-water wave characteristics using an empirical forecasting procedure based on the JONSWAP spectrum. The other formulas that were used in the construction of Figure 8 are also well established.

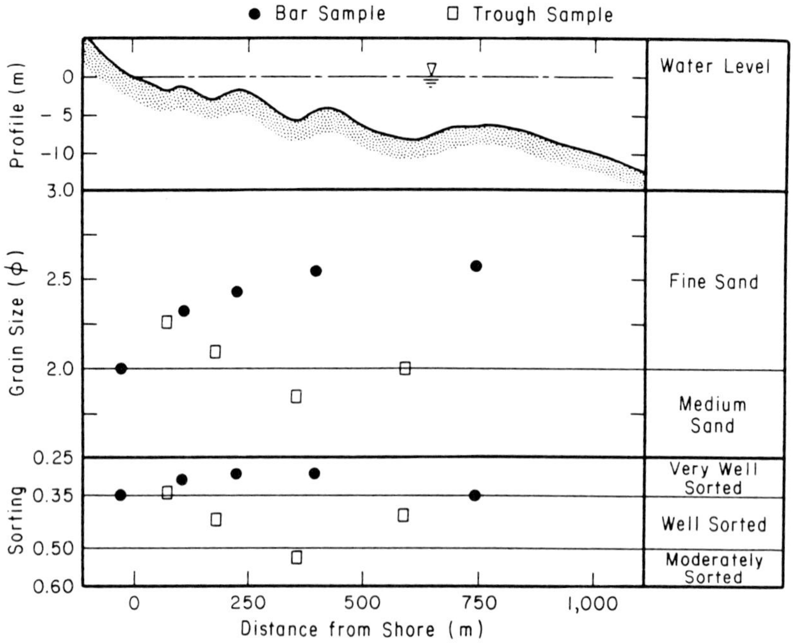

FIGURE 7. Nearshore profile and sediment distribution from a Lake Michigan beach. Well-sorted sands become progressively finer on the deeper longshore bars. Poorly sorted sand and gravel are coarser in deeper offshore troughs.[32]

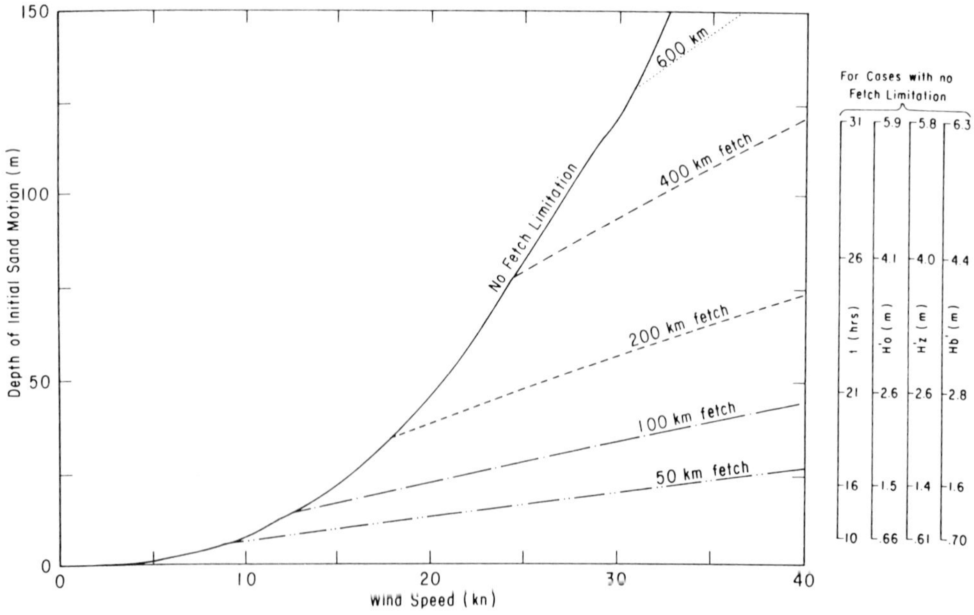

FIGURE 8. The effect of a restricted fetch in limiting the maximum depth of bed motion. Deep water wave conditions were forecast for various fetch distances and average persistent wind speeds (X axis). As waves move into shallower water, orbital velocities increase at the bed. The depth at which the bottom velocity reaches 15.24 cm/sec (0.5 ft/sec) is taken to be the depth of initial sand motion (Y axis). Four auxillary axes on the right indicate the duration of the winds necessary to develop the conditions depicted in the figure and also provide a feel for the resulting significant wave heights in deep water (H_0'), over the point of initial sand motion H_z') and at breaking (H_b'). The auxillary axes are nonlinear and assume no fetch limitations.

Most stretches of the Great Lakes shore are exposed to fetches greater than 100 km and some stretches of the southeastern shores of the lakes are exposed to 300 km fetches (Figure 1). As can be seen in Figure 8, as long as there is at least 100 km of fetch, the depth of bottom disturbance is not limited for a developed sea under persistent winds of up to 12.5 knots (6.4 m/sec). This means for example, a persistent wind with an average speed of 12 knots has the same potential for moving sediments to a depth of 13 m or 42 ft, regardless of whether it blows over the lake or an infinite ocean. At wind speeds above 12 knots, the available fetch may or may not limit the depth of sand motion. For example, assuming a fetch of 100 km, a persistent 31 knot wind would have the potential for moving sediment down to a respectable depth of 30 meters on the Great Lakes. Theoretically, this wind could cause sand motion much deeper than 100 m if the fetch were unrestricted; but as seen in Figure 8, this would require a wave generation area approximately 600 km wide in the direction of the average 31 knot winds. Thus, fetch may be limited not just by basin dimension, but also by the size of the storm.

Even though restricted basin dimensions do introduce a major difference in wave climates and, therefore, in the profile shape, Figure 8 nevertheless suggests that profile development in the Great Lakes and ocean is similar down to depths far below those expected for a small lake. And for winds of modest speeds, the size of the basin does not limit the depth of sand motion.

The three auxillary vertical axes in Figure 8 refer only to waves that were not fetch limited. These axes demonstrate the general feature that shoaled wave heights may be either greater or less than the deep water wave heights, depending on the point where they are measured (Chapter 1). The equations relating wind speeds to resultant wave length and height, as well as those described in the transformations of length, height and bottom orbital velocities as the wave shoals, are all well known, but the nearly constant proportionality between depth of disturbance and wind speed for a fixed fetch (evident in Figure 8) has not, to this author's knowledge, been previously noted. The nested double iterations necessary to obtain depth of disturbance from wind speed and fetch, obscures the nearly linear dependence between wind speed and depth of bed disturbance by fetch-limited waves.

In review of the Great Lakes as a model for oceanic shore erosion, the principal differences are: lack of tides, lower wave heights, and the lack of long-period swell on the Lakes. All three of these major differences do have the advantage, however, of facilitating field work. While maximum wave heights are less on the Lakes, they are nevertheless sufficient to move sediment out to considerable depths. With the exception of certain features of bar development, nearshore zones on the Lakes are generally similar in profile to ocean beaches. The lack of tides and the restricted range of wave heights and periods may explain the full development of a sequence of persistent longshore bars on enclosed seas.

IDEALIZED SEDIMENT BALANCE

This section presents an idealized concept for predicting profile adjustment to new water levels simply as a function of the magnitude of the water level change and two measurements of nearshore profile geometry. Corrections are then added to account for offshore losses of fine sediment and longshore imbalances in sediment flux.

As described by Bruun,[5] an increase in the mean elevation of the water level tends to shift the equilibrium sand profile landward. As water levels rise, erosion prevails on the upper beach causing the shoreline to recede. Conceptually, this shore erosion supplies material to build upward the outer part of the responding profile. If it is assumed that the initial profile shape will be reestablished farther inland at a distance above its initial position equal to the change in water level z, the ultimate retreat of the profile x can be calculated from the dimensions of the responding profile, X and Z, as depicted in Figure 9. This yields the simple relationship

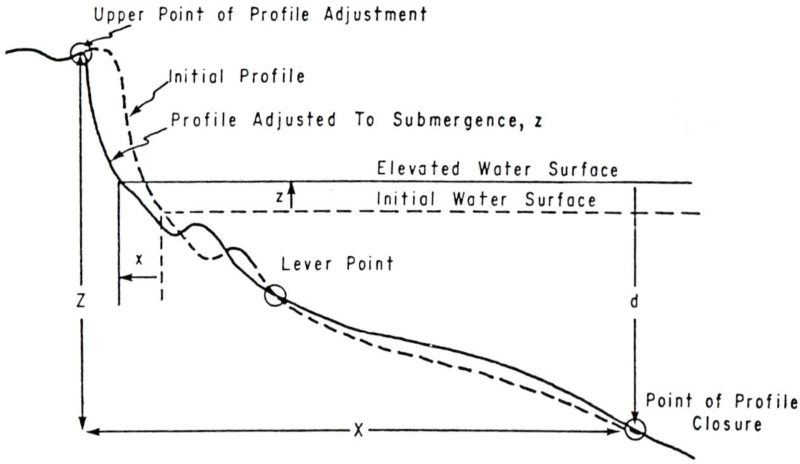

FIGURE 9. Sketch of profile measurements required to predict shore adjustment to an elevated water surface. Providing there is no net gain or loss outside the control volume, constancy of profile shape requires that the ultimate shore retreat x be approximately zX/Z.

$$x = \frac{zX}{Z} \tag{1}$$

If only P percent of the shore-eroded sediment remains in the active zone, greater shore recession will compensate for this loss. One method of estimating the proportion of shore-eroded material lost is to use the textural characteristics of the active beach as a guide. This method was originally proposed by Krumbein and James[25] to estimate the volume of material required to build a beach that will be of design dimensions after winnowing by wave action. Calculation of the overfill ratio, R_A, for beach fill has since been refined by Dean[8] and by James.[22] Hobson[21] explains procedures for estimating R_A. The same procedures will apply in the case of profile response due to high water, except that the "borrow material" characteristics must be based on a composite sample of the eroding section of the shore, i.e., the upper beach in the case of increases in lake level. If the water level declines, the lower part of the responding profile is eroded to supply material to prograde the upper profile. In this case the "native material" characteristics must be based on a composite sample of the lower profile (i.e., the zone of offshore erosion). In either case, the native material characteristics must be based on a composite sample of the entire responding profile, from the upper point of profile adjustment to the offshore point of profile closure (Figure 9). If there are no longshore imbalances, the increased shoreline retreat will be

$$x = \frac{zX\,(R_A)^{sg(z)}}{Z} \qquad \begin{aligned} &\text{where } sg(z) = 1 \text{ if } z > 0 \\ &\text{where } sg(z) = -1 \text{ if } z < 0 \end{aligned} \tag{2}$$

When eroding shore deposits are too fine the overfill ratio is positive, $R_A > 1$, and the predicted retreat is increased by a factor of R_A. When lake levels fall and eroding offshore sediments are too fine, R_A is again positive, but the predicted response (shore advance) must be decreased by a factor $1/R_A$. The exponent sg(z) serves to flip the overfill ratio from the numerator to denominator as appropriate.

Any longshore imbalances, or other uncompensated exchanges of sediment beyond the control volume, must also be accounted for to correctly predict the total shoreline change. Losses can occur offshore, onshore, or alongshore (Chapter 1). If Qt is the net exchange of sediment volume in time t across the boundaries of a control area with longshore length, Y, then the earlier estimate of retreat should be reduced by Qt/YZ. In its general form, the predicted shore response becomes:

$$x = \frac{zX\,(R_A)^{sg(z)}}{Z} - \frac{Qt}{YZ} \qquad (3)$$

as given by Hands.[18] Thus without introducing anything really new, Bruun's[5] concept can be applied to more complicated nonequilibrium conditions, and to the predictions of shore response as a consequence of falling as well as rising water levels.

DIFFICULTIES IN APPLYING THE SEDIMENT BALANCE APPROACH

There has been a continued interest in Bruun's concept since it was initially expounded in 1962. A special symposium on this subject was held in 1979.[33] Yet, in spite of widespread evidence of a sea level rise, the sediment balance approach has not been widely applied for predictive purposes. The following difficulties may explain a reluctance to routinely apply this approach:

1. Skepticism as to the *adequacy of an equilibrium model* for explaining short-term dynamic changes
2. Difficulties in determining *sediment lost* from the active zone
3. Problems of establishing a *realistic closure depth* below which water level changes have no effect on profile stability
4. Confusion arising from a typographic *error in one of the equations* defining profile retreat[5]
5. The perplexity caused by a *discontinuity in the profile* at the closure depth which appeared in the original and all subsequent diagrammatic sketches illustrating the concept

The first three difficulties (1, 2, and 3) warrant serious consideration before applying Equation 3; items 4 and 5, although perhaps confusing, should in no way discourage or limit use of Equation 3. The following paragraphs address each of these difficulties in reverse order.

Discontinuity in the Profile (Item 5) — Previous diagrams have illustrated the adjustment of a profile to higher water level by literally disconnecting the responding part of the bottom from a static region offshore. The apparent profile discontinuity at the juncture between the static and responding regions has some didactic value as it emphasizes congruency between initial and final profile shapes in the active region. Unfortunately, it also creates the impression that the model is inadequate for explaining the transition between the active and static parts of the profile. This discontinuity is not, however, an inherent part of the concept. It is only an artifice of the diagrams. Rigidly translating a profile upward and shoreward does not necessarily lead to a discontinuity nor even a change in slope as is demonstrated later in this report.

Error in Equation (Item 4) — Bruun's equation (Reference 5, Equation 1a, p. 124) is dimensionally incorrect as published. This apparent typographic error may have confused and discouraged some readers from giving Bruun's concept their full consideration. The problem equation is, however, unnecessary to the development of his concept which was correctly expressed in Equation 1b of that paper. The validity of the Bruun concept and of Equation 1 in this chapter is demonstrated geometrically in Figure 10.

Figure 10a depicts a nearshore profile in quasiequilibrium with wave-related forces. Note

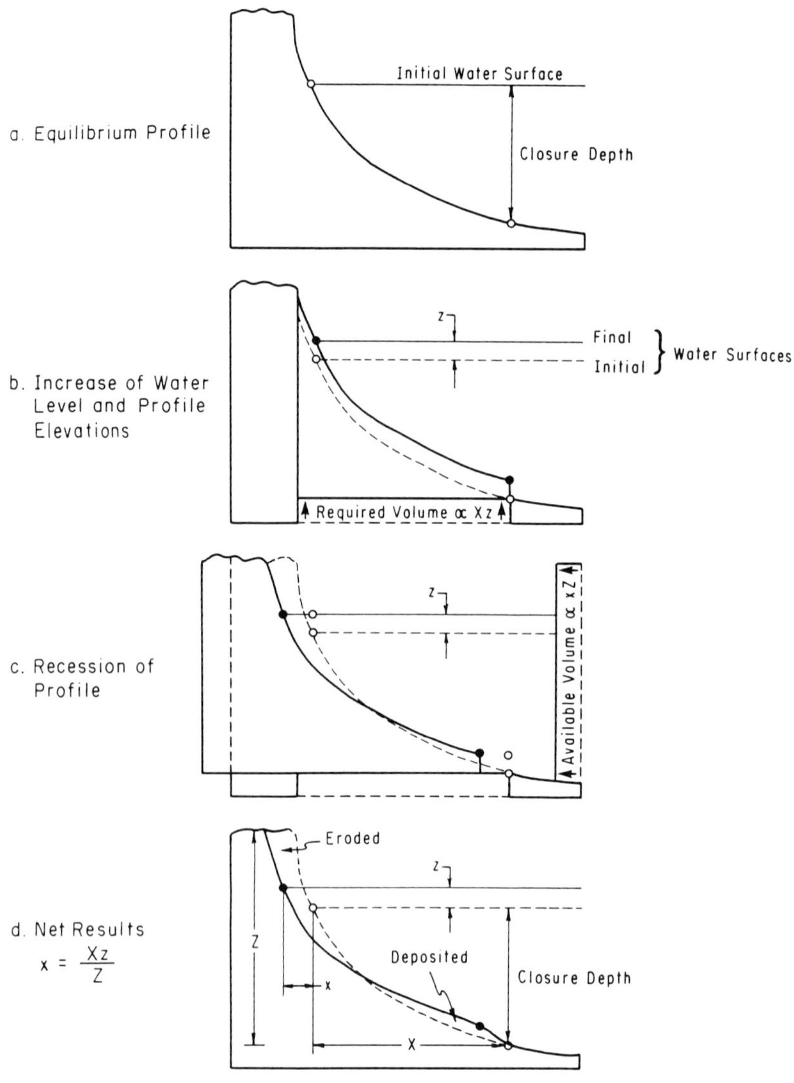

FIGURE 10. A schematized view of profile adjustment as two rigid translations, demonstrating the basis for the Bruun Rule (Equation 1).[18]

the closure depth below which the bottom presumably does not adjust to surface wave and current conditions. To estimate the ultimate shore retreat, the adjustment of the active profile is next depicted as two rigid profile translations. The first translation moves the active profile (i.e., the profile between the closure depth and the upper point of profile adjustment) up an amount z and reestablishes equilibrium depths below the elevated water surface (Figure 10b). This step requires a volume of sediment proportional to the product of X (the width of the active zone) times z (change in water level); the volume is made available by the second translation which is a recession of the profile. Figure 10c shows that x units of recession provide a volume of sediment proportional to the product of x times z (the vertical extent of the active profile from the closure depth up to the average elevation of the highest erosion on the backshore). Equating the volumes produced and required per unit length of shoreline by these two translations yields Equation 1.

In reality, both translations occur simultaneously and the closure point migrates upslope as the water level rises. A more formal development would integrate between profiles, allowing the closure point to move in infinitesimal steps with the water surface. This approach results in the more precise relationship:

$$x = X \ln \frac{Z}{Z - z} \qquad (4)$$

Equation 4 is generally unnecessary because the change in water level, z, is usually so small relative to the total height, Z, that equations 1 and 4 provide essentially the same results. For example, if $z < 0.1Z$ all results agree within less than 10%.

Thus, the simple expression $x \sim zX/Z$, is not only valuable as a close approximation, but is also most useful because it is easily (a) recalled by visualizing the adjustment of two rigid translations, (b) explained in the same manner, and (c) used as a quick mental check on the ultimate retreat expected for various values of the independent variables.

A Realistic Closure Depth (Item 3) — Determining a realistic closure depth is usually extremely difficult. The most direct approach is to compare historic bathymetric surveys of the site in question. Unfortunately, adequate survey data of this type are rare. Neither pier nor stadia surveys extend deep enough, and if a hydrographic survey does extend to deep water, allowances must be made for the fact that both sounding errors and boat-positioning errors usually increase significantly with depth and with distance from shore. The effect of long period swell may be impossible to distinguish from real bathymetric changes, and apparent bottom waves may or may not have been smoothed out during data reduction. Some surveyors may even "adjust" their profiles to obtain the best fit offshore, assuming there has been no change there. It is thus often impossible to substantiate apparent offshore changes unless you have taken repetitive profiles yourself.

Finally, even if you have such profiles, evaluation of the closure depth requires two surveys separated by an appropriate time period during which profile adjustment actually occurred in response to a known change in water levels. Hallermeier[11] demonstrates the dependence of profile closure depth on local wave conditions for essentially stable water levels. The difference between depths of closure at two sites with identical wave and sediment characteristics, one with a stable mean water level and the other with a recently displaced water level, has not been studied. It seems plausible that storm waves could cause a net profile change where equilibrium had been perturbed by the recent shift in the mean water level, and yet cause only sediment motion and (almost by definition) no net change in bottom elevation where the profile was in equilibrium with a constant water level. This being the case, real water level changes are essential if repetitive profiles are to reveal a closure depth suitable for testing the Bruun concept. Clearly, many problems plague the determination of the appropriate closure depth and therefore discourage application of Bruun's concept for predicting future shore retreat.

Sediment Losses (Item 2) — By adding terms, the Bruun concept can be extended to complicated, nonequilibrium situations.[18] This is useful, however, only if field data permit accurate evaluation of the added terms. Often field data are inadequate. The average rate of uncompensated sediment flux, Q, across the control boundaries must be known. Submarine canyons, hurricanes, and the long term effect of winds complicate this evaluation. Lastly, the composite textural characteristics of the active beach may be necessary to evaluate R_A.

Adequacy of an Equilibrium Model (Item 1) — Use of equilibrium assumptions to model dynamic coastal changes also deserves scrutiny. The idea of an "equilibrium beach profile" has had a long history (e.g., Reference 9, Fenneman, 1902); however, opinions still differ as to exactly what the concept actually entails. By one definition, the profile of equilibrium refers to the ultimate shape which coastal processes strive to impart to a beach. Of course nature seldom remains constant long enough for a strict equilibrium to develop. In the present context, the term *equilibrium profile* refers to a curve of fixed size and shape which "adequately" represents the "average" profile shape before perturbation by a shift in water level. By assumption, shore erosion eventually returns the profile to this same shape after it is displaced as a result of the water level change (Figure 9).

A willingness to accept equilibrium as a reasonable approximation is not inconsistent with recognition of seasonal, storm, or other temporary profile fluctuations. Careful judgment should be made on a case-by-case basis, if field profiles claim to represent quasiequilibrium conditions. Generally, the claim will be more reasonable the longer the time frame and the wider the spatial extent of the study. Greater sampling density in time and space produce more precise estimates of mean parameters.

DIFFICULTIES AVOIDED IN THE GREAT LAKES

The above discussion has shown that the Bruun concept is theoretically sound but is difficult to apply in the field. However, some of these difficulties can be avoided in the case of the Great Lakes.

Establishing a realistic closure depth (Item 3) depends on accurate repetitive profiling, the profile errors increasing with distance from shore. Fortunately, the bottom drops off to suitable depths relatively rapidly in the Great Lakes study area. Furthermore, the Lakes are free from tidal variations as well as from long-period swell that complicate profile comparison. Although the Great Lakes are notorious for their large storm surges and seiches, extensive water-level measurements verify that they are negligible in the present data set and so did not affect the determination of the closure depth.[18]

The difficulty of determining sediment losses (Item 2) on the Great Lakes is greatly simplified by the absence of submarine canyons, hurricanes, and overwash events. All rivers entering eastern Lake Michigan flow through deep inland sediment traps. Inlet losses have a negligible effect on the overall sediment budget for the broad study area. Thus, in the present application Q (in Equation 4) can be taken as essentially zero.

The section of shore is an isolated littoral cell, so the only process supplying new sediment to the active profile is shore recession. Furthermore, shore deposits and backshore bluffs within the study area contain less than 1 percent silt, making it unnecessary to correct for any unstable fine fraction (i.e., $R_A = 1$ in Equation 4). Thus, a number of site-specific attributes simplify the sediment balance for this particular study area.

SHORE RETREAT MEASUREMENTS

Encroachment refers to the immediate loss of beach width due to *submergence*. The total shore retreat will exceed encroachment as the shore *recedes* due to erosion (Figure 11). Between 1969 and 1975 (a period of persistent high water levels on Lake Michigan) the overall shore retreat exceeded encroachment by a factor of 5 (average retreat = 17.9 m, but only 3.4 m was due to encroachment). The final amount of encroachment at individual stations, while predictable, gave no clue as to the final amount of shore retreat.[17] Shore recession depends on the local sediment supply plus the exposure and resistance of the beach to erosive forces. Within the range of conditions observed on the lake, flatter foreshores showed no tendency to recede more or less than steeper foreshores. Moreover, shore recession continued for some time even after the water levels began to decline. Hence, encroachment, depending only on steepness of the foreshore and the change of water levels, though easy to predict, is a poor measure of total shore retreat.

FIELD CONFIRMATION OF THEORY

The extent of shore covered (25 profiles spread over 50 km), the length of time the entire

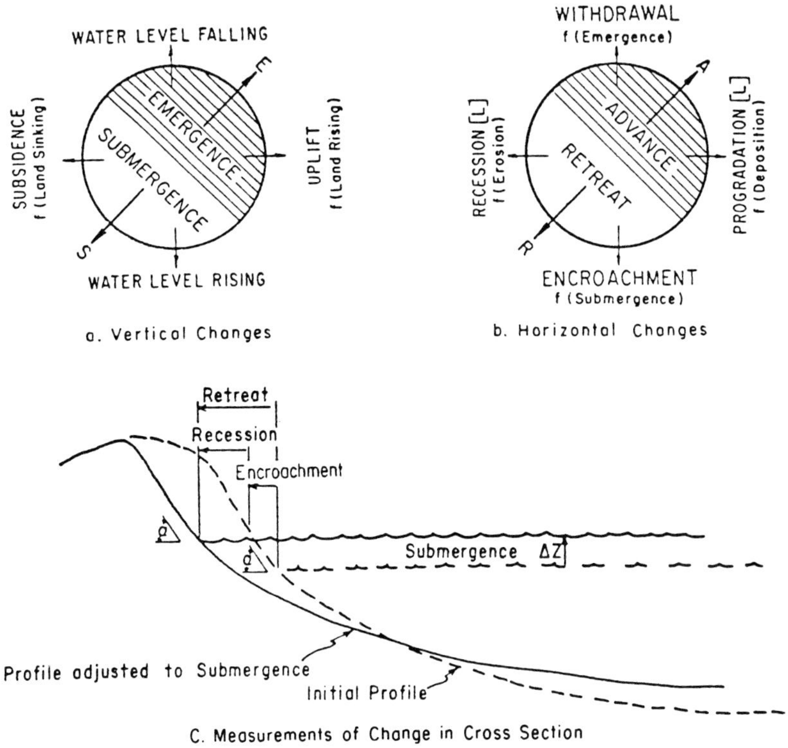

FIGURE 11. Terminology of shore retreat where retreat = encroachment + recession and encroachment = z cot α.[17]

stretch was monitored (7 years), the sizable increase in lake level experienced during the study, and the near-ideal site conditions (absence of silt and clay in eroding shore deposits, no overwash, hurricanes, seiches, submarine canyons, fluvial or longshore sediment input to contend with) makes this study the most realistic field test to date of an equilibrium profile response to water-level changes.

As points of a bar bifurcation migrated, the number of bars on some profiles varied, but overall the number and size of bars remained constant throughout the study. In keeping with the equilibrium assumption, the bars migrated landward and maintained a nearly constant depth beneath the gradually rising lake level (Figure 12). The landward migration of bars and troughs created shore parallel bands of alternating erosion and deposition rather than the simple lever point of Figure 9. Landward of the inner trough, erosion removed 10.3 m^3 per meter of beach per year averaged over the study period. Deposition measured offshore compensated for the inshore volume losses verifying conservation of sediment volume as required by the model. But how well would the sediment balance model have predicted the observed shore retreat?

Measurements of the width of each profile from the vegetated dune line to the closure depth for each station averaged X = 923 m. The heights of the scarps which waves had cut in the foredune, when averaged and added to the average closure depth, established the vertical dimension of the adjusting profile (Z = 13.6 m in Equation 1 and Figure 9). Equation 1 is applicable because Q = 0 and R_A = 1; thus the ratio X/Z times the measured water level change (z = 0.20 meters) yields a predicted ultimate shore retreat of 13.6 meters. The retreat actually measured between 1969 and 1976 also averaged −13.6 meters. Considering the measurement and sampling errors involved in determining each independent variable, a predictive capability of less than a tenth of a meter certainly is not claimed, but

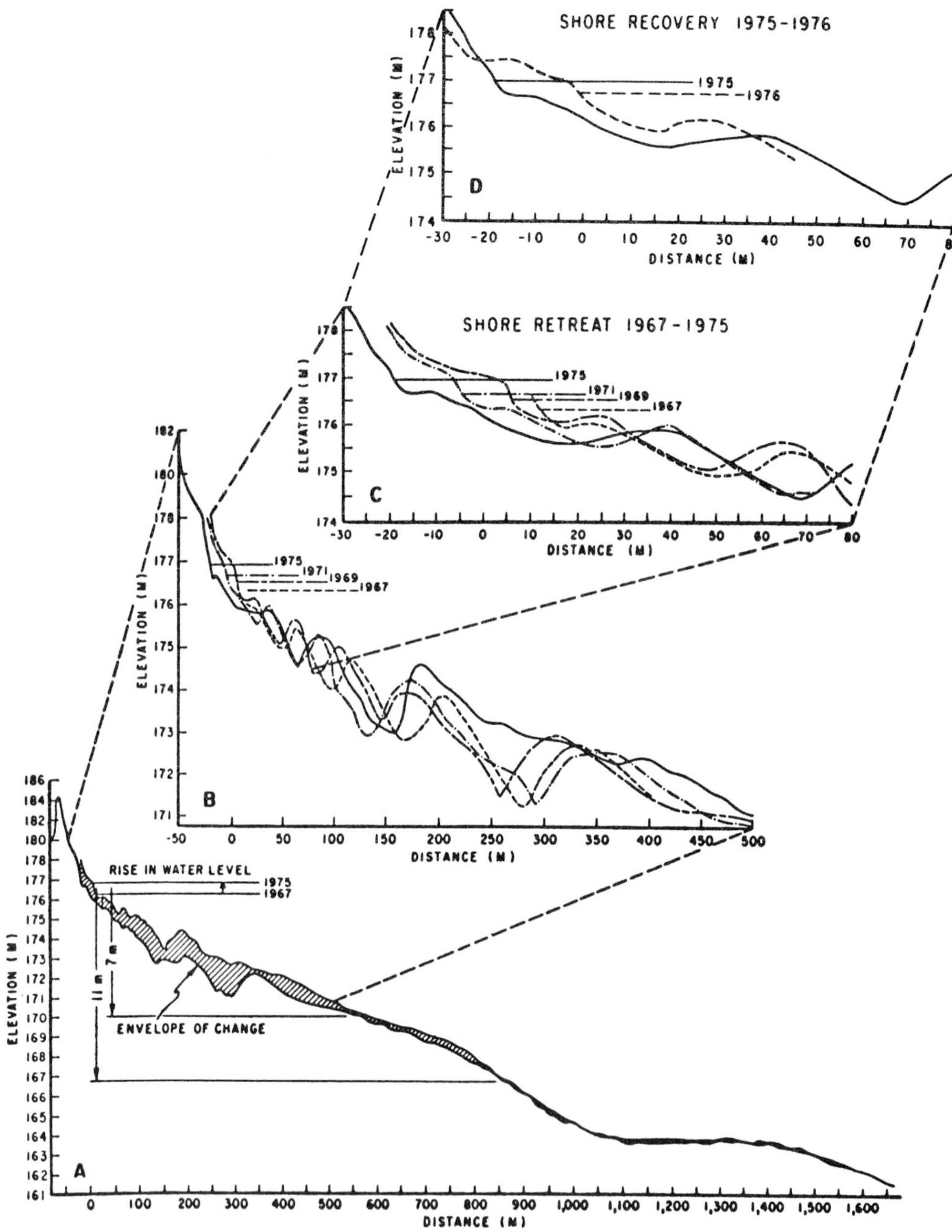

FIGURE 12. Measured profile adjustments over a 9-year period of rising and then stable water levels on Lake Michigan. Bars move upward to maintain constant depth beneath the rising water surface.

the results clearly confirm the appropriateness of the equilibrium-sediment balance approach when applied in the proper setting.

Closure depths were deeper south of Little Sable Point than to the north. The eroding dunes were also higher there, which even further enlarged the vertical dimension of profile adjustment. Consequently, the equilibrium prediction can be applied separately to the two regions. Likewise, because additional surveys were conducted in 1971 and 1975, separate predictions could be applied as well to these shorter time intervals (1969 to 1971 and 1969

Table 2
PREDICTED AND OBSERVED PROFILE RETREAT

Study Areas	Survey Periods 1969-1971 z = 0.12 m X = 870 m	1969-1975 z = 0.39 m X = 1020 m	1969-1976 z = 0.20 m X = 923 m
AVERAGE HEIGHT, Z (m)			
Northern section (stations 1 to 15)	10.84	12.15	12.50
Southern section (stations 16 to 29)	12.90	14.28	14.80
Whole area (stations 1 to 29)	11.86	13.21	13.60
PREDICTED RETREAT, Xz/Z (m)			
Northern section	9.63	32.74	14.77
Southern section	8.09	27.86	12.47
Whole area	8.80	30.11	13.57
OBSERVED RETREAT x (m)			
Northern section	4.6	20.0	12.6
Southern section	3.6	16.8	14.8
Whole area	4.3	18.8	13.6
OVERPREDICTION (pct)			
Northern section	109	40	10
Southern section	164	66	−7
Whole area	117	45	0

to 1975), thus partitioning the original data into nine individual, though not independent tests (Table 2). Considering prediction versus measurements, the predicted retreat from 1969 to 1971 was too high for all three areas (117% high for the areas as a whole). The prediction for 1969 to 1975 was also high, but not as far off as before (45% high for the whole area). These overestimations of retreat are attributed to the fact that profile retreat was actually lagging behind the lake level rise. Rising water levels establish a potential for erosion and realization of that potential requires sediment redistribution, i.e., work which depends on the energy being available. Eventual convergence of measured and predicted retreat in both regions, 3 years after annual lake levels had stabilized, suggest that several storm seasons may be required to readjust the profile to changes in mean water level of several tenths of a meter (Figure 13).

Between 1967 and 1975, persistent rapid shore recession occurred at almost all sites, as documented by five separate field surveys.[17] The last of these surveys indicated no decrease in recession rates even though the annual mean lake level had peaked 2 years earlier. Then recession rates dropped dramatically in 1976. The beach even prograded for the first time at 12 of 34 survey sites. This sudden interruption of the previous erosive trend indicated equilibrium was finally being approached. The continuation of high recession rates after lake levels peaked was consistent with the assumptions of the sediment balance model, as were the early overestimates of recession. The crucial proof of the model's usefulness, however, was that it correctly predicted shore retreat in 1976 just as the long-term period of high erosion was coming to a close.

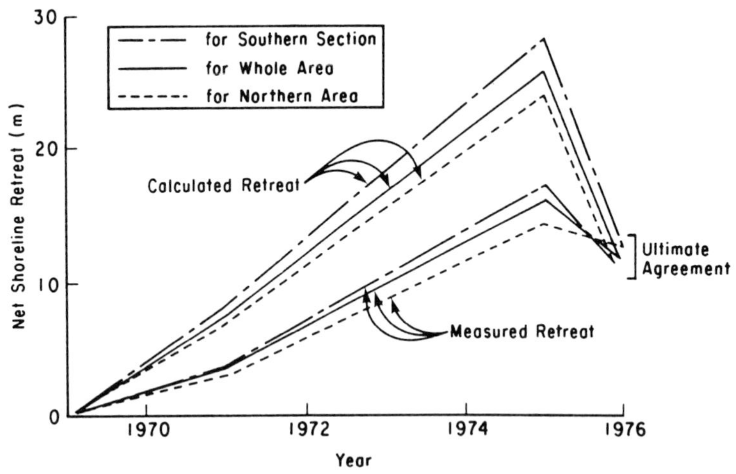

FIGURE 13. Calculated versus measured shoreline retreat. The predicted ultimate retreat, in response to post-1969 changes in mean lake level, exceeded the observed retreat by more than 100 percent in 1971 and by about 50 percent in 1975 presumably because the active profile had not had time to completely readjust to the higher water levels. Almost perfect agreement had developed by the time of the last survey, 3 years after the lake level maximum.[18]

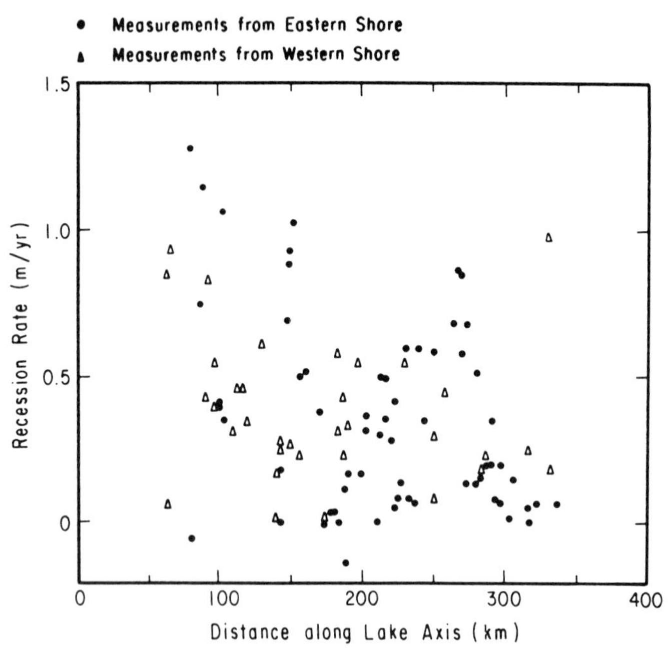

FIGURE 14. Historic rate of recession vs. measurement position projected on Lake Michigan mid-lake axis. Rates of recession are based on data from Powers (1958) and cover a period of about 125 years. There is a trend toward lower recession rates at the more rapidly rising, northern end of the basins.[5]

ADDITIONAL SUPPORT FROM HISTORIC DATA

In an earlier study, Hands[16] had reported an apparent north to south increase in the rate of historic shore retreat on Lake Michigan. The trend was suggested by plotting historic shore recession measurements[30] versus the position of the shore measurements projected on a mid-lake axis (Figure 14). Examination of longshore variations in resistance of the shore to erosion, in offshore bathymetry, wave power, and winter ice protection, all failed to provide any explanation for this apparent increase in erosion toward the south.[16] The basin is slowly tilting, however, as it continues to recover from Pleistocene glacial depression centered near Hudson Bay to the north. This is confirmed by geomorphic,[27] geophysical,[40] and geodetic[28] evidence of differential uplift, as well as by differences in the water level records up and down the shore.[23] Both the recession data and the water level data cover approximately the last 125 years. The least square regression of the 94 bluff measurements was 19 m per century per 100 km along the axis of tilt. The rate of tilt obtained by comparing 1929 and 1955 releveling data (0.087 m/100 km yr) was roughly comparable to that estimated from lake level differences (0.063 m/100 km yr). Thus, each cm of submergence would have to be responsible for between 1 to 4 m of shore recession if the regional trends in recession were to be attributed solely to submergence. For each unit that a high bluff recedes, more sediment is supplied to build the outer profile than would result from equal recession on a low shore. Interestingly, the increase in historic recession per unit of subsidence is greater on the relatively low western shore of Lake Michigan, in keeping with the concept of sediment balance.

LIMITS OF PREDICTION

Assuming constancy of profile shape, we have seen that the problem of predicting the ultimate shore erosion in response to a long term water level change becomes equivalent to simply identifying a profile closure depth. The confirmation of this concept with Lake Michigan data, encouraged a generalization to other regions of the Great Lakes where wave energies, and therefore closure depths, were different. Tables of closure depths for all sections of the U.S. shoreline of the Great Lakes were prepared based on the assumption that closure depth is proportional to the height of waves forecast with a 5-year return period.[19] The proportionality constant evaluated on Lake Michigan was about 2; i.e., the closure depth was equal to about twice the height of the 5-year return period wave height on the existing shore test site.

Theoretically, the depth of bottom motion should depend on the wave steepness, the shoaled and refracted wave height, and the grain size. The roles of these variables are however usually secondary to variations in offshore wave height. Given the restricted range of wave periods and sediment sizes off sandy shores of the Great Lakes, it was felt that a simple correlation of closure depth to wave height was justified. Over a period of varying wave conditions, it is reasonable that closure depth depends more strongly on the character of the higher waves. The 5-year return period wave was chosen because it was readily available and any consistent relative measure is probably satisfactory for extrapolation within the Great Lakes.

The greater response to submergence ratio, x/z, in the historic Lake Michigan study (1:100 and 1:400), as compared to the ratio (of 1:69) obtained during the shorter, more precise study, suggests that the dimensions of the responding profile expand and/or sand leakage occurs over a more extensive area as time scales increase. Thus, the results of monitoring profile responses on Lake Michigan did not indicate the ratio of wave height to profile relief (H_5/Z) that would be appropriate for periods of time greatly longer than 10 years. In as much as extreme storms during a 100 or 1000 year period would move sediment deeper, the responding profile would be expected to exceed $2H_5$ in relief. Indeed the historic study of erosion tends to support this suggestion. The responding profiles must have had a relief of about $4H_5$ for this less well established historic change.

Profile response at the other extreme (much less than 5 years) will probably not reach as deep as $2H_5$. The deeper part of the profile may lie unaffected while changes in the upper profile represent a scrambled mixture of responses to waves and water level changes that have occurred both during and prior to the shorter period of observation. Thus, application of the sediment budget prediction on short time scales involves great risk and should be discouraged.

The application of this approach on ocean beaches will be complicated not only by greater difficulties in establishing a sediment budget at most locations, but also (a) by unknown effects that tidal currents and other oceanographic processes have on the closure depth, (b) by the possible differences due to the presence of a broad continental shelf, and (c) by masking of long-term trends by tides and storm surges being superimposed on the much slower submergence resulting from sea level changes and crustal motion. With the longer periods of time necessary for substantial changes in sea level, additional uncertainties arise in evaluating points (a) and (b) above, and in choosing the appropriate wave height, rate of longshore sediment losses, and even the sea level change to best represent long-term net effects. Rates of relative sea level rise over the last century have been measured at many sites. However, rates of change earlier in this most recent epoch of deglaciation probably varied widely from these historic measurements. Future changes are even more problematic.

On ocean coasts, wave periods vary more widely than they do on the Great Lakes. Thus, the effect of wave period should be considered when estimating a closure depth for oceanic sites. Hallermeier[13] developed a Froude number that defines the effect of wave height, wave steepness, and grain size on the initiation of bed motion. Clearly, this parameter would correlate to closure depths better than wave height alone and should be considered in extensions of the sediment balance concept to ocean environments. Hallermeier[12] calculated the depths to which sand is moved by the annual median wave condition at numerous sites but expressed some uncertainty as to whether actual motion extended as deep as indicated for a few of these ocean beaches. An attempt to apply the same criterion to the Lake Michigan study site produces an offshore limit shallower than the observed profile changes. Further refinement therefore seems desirable to firmly establish the critical value of this Froude number associated with the depth of ocean profile response to changes in sea level.

As discussed, additional research in many areas could help clarify how to apply the sediment budget concept to ocean beaches. An unavoidable limitation of this approach will be that the sediment balance concept doesn't even address the question of when the predicted shore response will occur. It merely reveals the horizontal distance the shoreline must *ultimately* move to reestablish the presumed equilibrium profile at its new elevation. The magnitude of the change in water level controls the volume of sand that must be rearranged on the profile. Storms are necessary to provide the energy to accomplish this work and thus control the rate of adjustment. The quantity YZ is an upper limit on the cubic units of sand that must be moved per unit length of beach. For given wave conditions at least the *magnitude of potential* longshore transport rate can be estimated (Chapter 1); the rate of on or offshore transport is even less well established. So even if storminess were predictable, several additional steps would be necessary to calculate the lag time requited for profile equilibration. Meanwhile, empirical data provide some rough guidance: Under not unusual storminess on the Great Lakes it took several years for the rate of shore erosion to drop off following six years of water level increase amounting to 0.2 m. If the water level had risen appreciably slower, the lag time would have been shorter. If shore retreat undermines a steep backshore bluff, several more years of bluff adjustment and revegetation will follow after the beach response is complete.

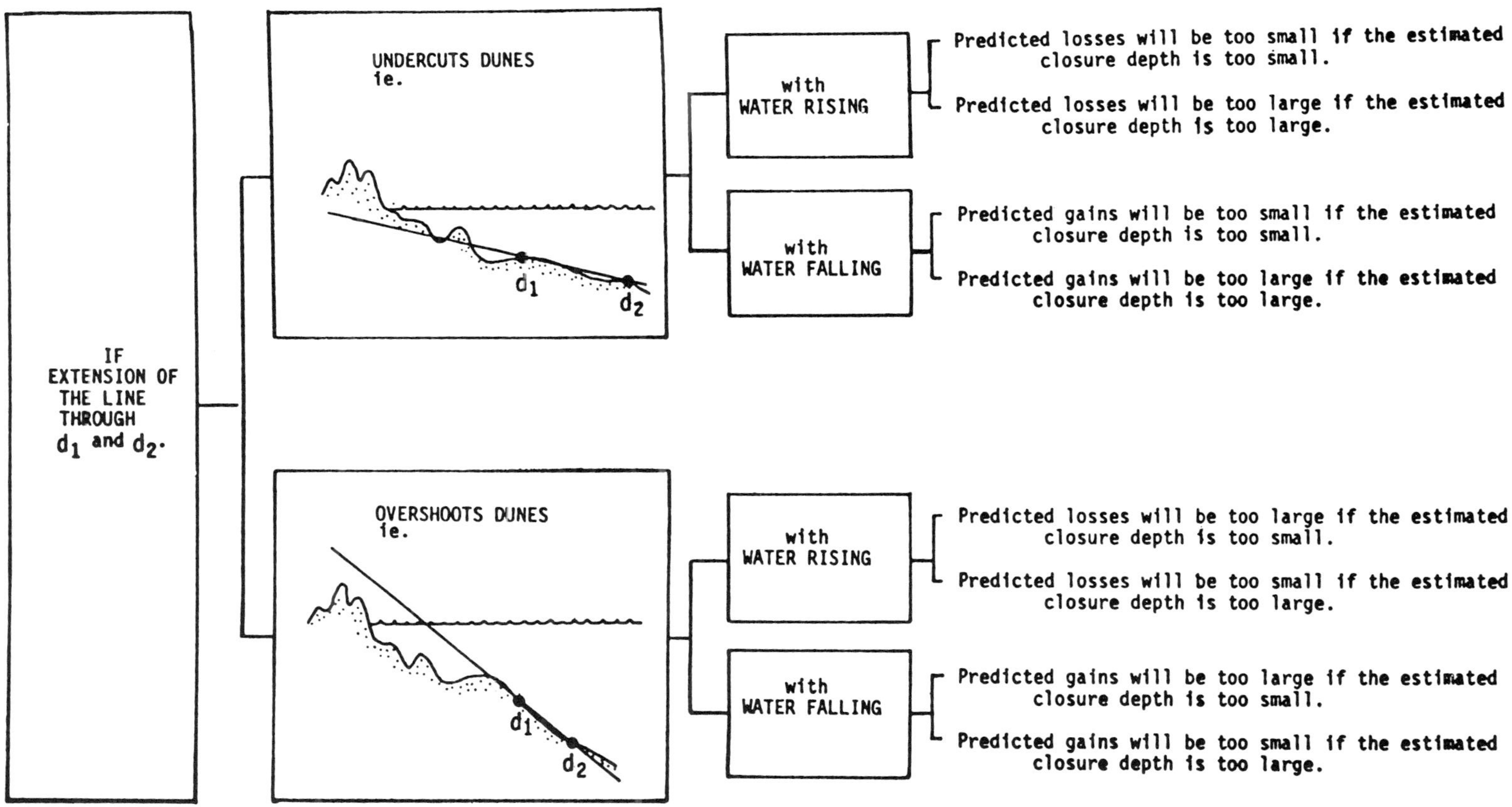

FIGURE 15. Relationships between closure depth and shore response.

CONCLUDING REMARKS

In the absence of other strong evidence as to the correct closure depth, this author suggests "twice the wave height" ($Z \simeq 2H_5$) as a rule of thumb to estimate the 5 to 10 year profile response under Great Lakes-like wave conditions; and possibly doubling this value for 100 year estimates. An empirically based critical value of Hallermeier's seaward Froude number may be a better guide on beaches exposed to a wider range of wave periods.

While it is true that confirmation of this approach in the marine environment is lacking (and not likely to be forthcoming due, among other things, to the problem of measuring small, slow changes), there is a saving grace in that the predicted shore response is not very sensitive to moderate errors in the estimation of Z. If Z is too large, there will be a compensatory error in X. In fact, if the bottom slope in the neighborhood of the suspected closure depth is tangent to a line between the closure depth and the highest point of response onshore, then any depth in that neighborhood will provide an identical prediction of shore response (other possibilities are depicted in Figure 15).

Lastly, progradation of the shore in response to falling levels was observed on the Lakes, but the model presented here has not been adequately tested under the conditions of lowered water levels. Such a test could be easily conducted during the next extended period of declining water levels on the Great Lakes.

ACKNOWLEDGMENTS

The author greatly appreciated suggestions from Robert Dean, Robert Hallermeier, Phil Vitale, and Todd Walton. The tests described and the resulting data presented herein, unless otherwise noted, were obtained from research conducted under the Coastal Engineering Research Program of the United States Army Corps of Engineers by the Coastal Engineering Research Center. Permission was granted by the Chief of Engineers to publish this information.

REFERENCES

1. **Armstrong, J. M., Seibel, E., and Alexander, Z. Q.,** Determination of Quantity and Quality of Great Lakes U.S. Shoreline Eroded Material, International Joint Commission, Chicago, September 1976.
2. **Berg, R. C. and Collinson, C.,** Bluff erosion, recession rates, and volumetric losses on the Lake Michigan shore in Illinois, *Environ. Geol. Notes Ill. State Geol. Surv.*, No. 76, July 1976.
3. **Birkemeier, W. A.,** The effect of structures and lake levels on bluff and shore erosion in Berrien County, Michigan, 1970-74, *U.S. Army Coastal Eng. Res. Cent.*, MR 80-2, 1980.
4. **Birkemeier, W. A.,** Coastal changes, eastern Lake Michigan, 1970-74, *U.S. Army Coastal Eng. Res. Cent.*, MR 81-2, Jan. 1981.
5. **Bruun, P.,** Sea-level rise as a cause of shore erosion, *J. Waterways, Harbors Coastal Eng. Div.*, Vol. 88 (No. WW1), 117, 1962.
6. **Davis, R. A., Jr., et al.,** Comparison of ridge and runnel systems in tidal and non-tidal environments, *J. Sediment. Petrol.*, 42(2), 413, 1972.
7. **Davis, R. A., Jr., et al.,** Beach profile changes: east coast of Lake Michigan, 1970-72, *U.S. Army Coastal Eng. Res. Cent.*, MP 10-75, 1975.
8. **Dean, R. J.,** Compatability of borrow materials for beachfills, *Proc. 14th Int. Coastal Engineering Conf.*, Vol. II, American Society of Civil Engineers, New York, 1974, 1319.
9. **Fenneman, N.M.,** Development of the profile of equilibrium of the subaqueous shore terrace, *J. Geol.*, 10, 22, 1902.
10. **Hails, J. R.,** A review of some current trends in nearshore research, *Earth-Sci. Rev.*, 10, 171, 1974.
11. **Hallermeier, R. J.,** Calculating a yearly limit depth to the active beach profile, *U.S. Army Coastal Eng. Res. Cent. Tech. Pap.*, TP 77-9, 1977.
12. **Hallermeier, R. J.,** Seaward limit of significant sand transport by waves: an annual zonation for seasonal profiles, *U.S. Army Coastal Eng. Res. Cent.*, CETA 81-2, 1981.

13. **Hallermeier, R. J.**, A profile zonation for seasonal sand beaches from wave climate, *Coastal Eng.*, 4, 253, 1981; also Reprint 81-3, U.S. Army Coastal Engineering Research Center, Fort Belvoir, Va., 1981.
14. **Hands, E. B.**, Observations of barred coastal profiles under the influence of rising water levels, Eastern Lake Michigan, 1967-71, *U.S. Army Coastal Eng. Res. Cent. Tech. Rep.*, TR 76-1, 1976.
15. **Hands, E. B.**, Comparative Bar Anatomy, unpublished data file, U.S. Army Corps of Engineers, Coastal Engineering Research Center, Fort Belvoir, Va., 1976.
16. **Hands, E. B.**, Some Data Points on Erosion and Flooding for Subsiding Coastal Regions, Proceedings of the Symposium on Anaheim, 1976 Land Subsidence, International Association of Hydrological Sciences, Reading, Berkshire, England, 1977, 629; also Reprint 78-11, U.S. Army Corps of Engineers, Coastal Engineering Research Center, Fort Belvoir, Va., NTIS AD AO51 796.
17. **Hands, E. B.**, Changes in rates of shore retreat, Lake Michigan, 1967-76, *U.S. Army Coastal Eng. Res. Cent. Tech. Pap.*, TP 79-4, 1979.
18. **Hands, E. B.**, Prediction of shore retreat and nearshore profile adjustments to rising water levels on the Great Lakes, *U.S. Army Coastal Eng. Res. Cent. Tech. Pap.*, TP 80-7, 1980.
19. **Hands, E. B.**, Predicting adjustments in shore and offshore sand profiles on the Great Lakes, *U.S. Army Coastal Eng. Res. Cent.*, CETA 81-4, 1981.
20. **Haras, W. S., et al.**, "Canada/Ontario Great Lakes Shore Damage Survey, Coastal Zone Atlas," Ontario Ministry on Natural Resources, Ottawa, 1975.
21. **Hobson, R. D.**, Review of design elements for beach-fill evaluation, *U.S. Army Coastal Eng. Res. Cent. Tech. Pap.*, TP 77-6, 1977.
22. **James, W. R.**, Techniques in evaluating suitability of borrow material for beach nourishment, *U.S. Army Coastal Eng. Res. Cent. Tech. Memo*, TM-60, Dec. 1975.
23. **Kite, G. W.**, An Engineering Study of Crustal Movement Around the Great Lakes, Tech. Bull. 63, Dept. of the Environment, Ottawa, Canada, 1972, 57 p.
24. **Komar, P. D. and Miller, M. C.**, The threshold of sediment movement under oscillatory water waves, *J. Sediment. Petrol.*, 43(4), 1101, 1973.
25. **Krumbein, W. C. and James, W. R.**, A lognormal size distribution model for estimating stability of beach fill material, *U.S. Army Coastal Eng. Res. Cent. Tech. Memo*, TM-16, 1965.
26. **Keulegan, G. H.**, An experimental study of submarine sand bars, *U.S. Army Beach Erosion Board, Tech. Rep.*, TR-3, 1948.
27. **Leverett, F. and Taylor, F. B.**, The Pleistocene of Indiana and Michigan and the History of the Great Lakes, Monograph Vol. LIII, U.S. Geological Survey, Washington, D.C., 1915, 523 p.
28. **Meade, B. K.**, Report of Sub-Commission on Recent Crustal Movements in North America, presented at the XV General Assembly of IVGG Moscow, USSR, unpublished, Aug. 1971.
29. **Munk, W. H.**, The solitary wave theory and its application to surf zone problems, *Ann. N.Y. Acad. Sci.*, 51, 376, 1949.
30. **Powers, W. E.**, Geomorphology of the Lake Michigan Shoreline, ONR Final Report on Project #NR387-015, Northwestern University, Evanston, Ill., 1958, 103 p.
31. **Resio, D. T. and Vincent, C. L.**, Lake Michigan, *U.S. Waterw. Exp. Stn., Vicksburg, Miss.*, Rep. 3, Tech. Rep. H-76-1, Nov. 1976.
32. **Saylor, J. H. and Hands, E. B.**, Properties of longshore bars in the Great Lakes, *Proc. 12th Conf. Coastal Engineering*, Vol. 2, American Society of Civil Engineers, New York, 1970, 839.
33. **Schwartz, M. L. and Fisher, J. J.**, Proceedings of the Per Bruun Symposium, International Geographical Union Commission on the Coastal Environment, Washington, D.C., April 1980, 83 p.
34. **Seibel, E.**, Shore Erosion at Selected Sites Along Lakes Michigan and Huron, Ph.D. Thesis, University of Michigan, Ann Arbor, 1972, 175.
35. **Thompson, E. F.**, Energy spectra in shallow U.S. coastal waters, *U.S. Army Coastal Eng. Res. Cent. Tech. Pap.*, TP 80-2, 1980.
36. U.S. Army Engineer District, Detroit, 1956, Beach Erosion Control Berrien County, Michigan, unpublished.
37. U.S. Army Corps of Engineers, Coastal Engineering Research Center, *Shore Protection Manual*, 3rd ed., Vol. 1, Stock No. 008-022-00113-1, U.S. Government Printing Office, Washington, D.C., 1977, 1, 262 pp.
38. U.S. Army Corps of Engineers, Coastal Engineering Research Center, *Shore Protection Manual*, 4th ed., Coastal Engineering Research Center, Fort Belvoir, Va., (in preparation 1982).
39. U.S. Congress, 1946, Beach Erosion Study, Lake Michigan Shoreline of Milwaukee County, Wisconsin: House Doc. 526, 79th Congress, 2d session.
40. **Walcott, R. I.**, Isostatic response to loading of the crust in Canada, *Can. J. Earth Sci.*, 7(2), 716, 1970.
41. Wisconsin Department of Natural Resources, "Survey Records of Milwaukee Co.: Exhibit #2", Public Works Hearing, H.R. 12712, Madison, Wis., 1969.

Chapter 9

COASTAL EROSION IN RESPONSE TO THE CONSTRUCTION OF JETTIES AND BREAKWATERS

Paul D. Komar

INTRODUCTION

Many occurrences of destructive coastal erosion have resulted directly from the construction of jetties, breakwaters, or other engineering structures. In the majority of these cases the analysis of the erosion is straightforward; the jetties block the natural littoral drift that had prevailed along the coast, depriving the beaches of sand in the downdrift direction and thus initiating erosion of the adjacent coastal properties. In other cases, however, it is not initially apparent why the engineering structure produced significant erosion. For example, jetties have induced erosion on coasts where no net littoral drift exists.

This chapter will examine the response of coastlines to the construction of engineering structures. This will be done by means of a series of case studies that exemplify various facets of the problem.

BLOCKAGE OF A NET LITTORAL DRIFT

In Chapter 1 it was pointed out that when waves break at an angle to the shoreline they produce a longshore movement of beach sediment known as the *littoral drift*. In many cases this sand movement under a given set of wave conditions can be evaluated with Equation 13 of Chapter 1.

Most areas of the coast are acted upon by numerous wave trains throughout the year, arriving from various offshore directions and producing variable amounts of littoral drift. There commonly is even a reversal of the drift direction under different storms. The *net littoral drift* is the summation of these many individual sediment transport episodes, summed over at least a 1-year time span. Clearly, with a reversing transport direction the net drift may in the end be quite small, smaller even than individual episodes of longshore sediment transport under a given storm. It is this net littoral drift that is of importance here — it being the volume of sand that will be blocked by the construction of engineering structures such as jetties or breakwaters.

Examples of jetties that have blocked this littoral drift, resulting in erosion, are so numerous that it was difficult to decide which to utilize here as illustrations. The harbors on the east coast of India are of special interest in that they provide some of the earliest examples, and the littoral drift in that region is one of the largest on the world coastlines.

There are four major ports on India's east coast, respectively, from south to north, Madras, Kakinada, Vishakhapatnam, and Paradip (Figure 1). This is also the direction of the net littoral drift, which progressively increases in the northward direction such that at Paradip it reaches nearly a million m^3/year. This northward sand transport is the product of waves generated by the monsoons blowing over the Bay of Bengal. During April to September the southwest monsoon produces waves approaching from the southwest, the average wave approach at Madras being 145°N after accounting for refraction. This, of course, produces a northward sand transport along the coast. The wave periods are between 8 and 9 sec and the wave heights variable, reaching a maximum of about 3 m. During the northeast monsoon of October to March the transport is in the opposite direction, the waves approaching Madras from about 65°N. But this return transport is much smaller, yielding a net annual littoral

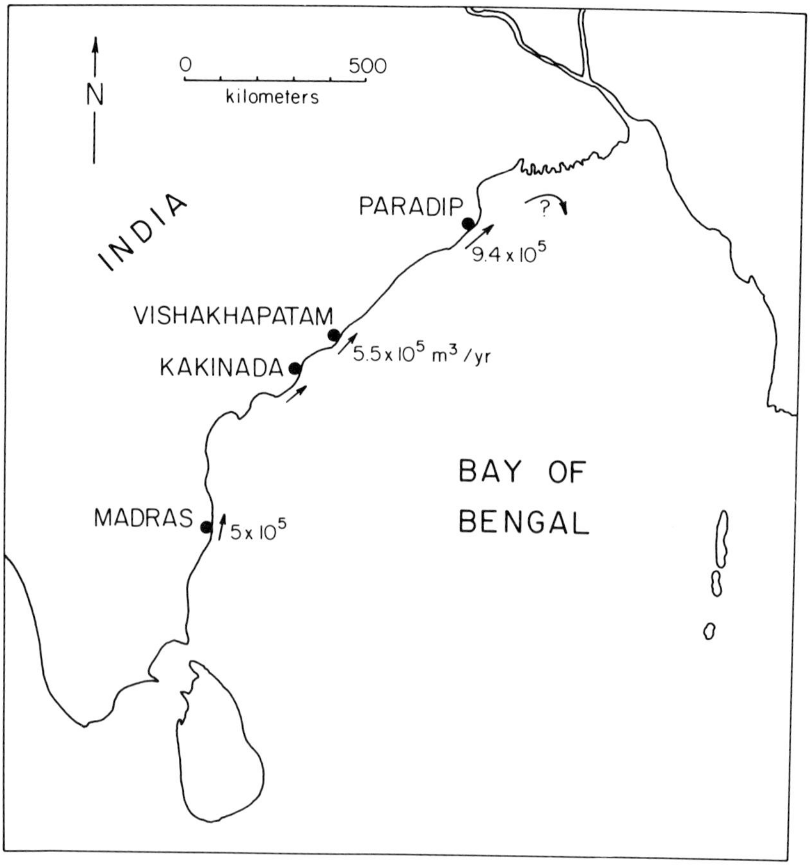

FIGURE 1. Locations of the principal harbors on the east coast of India.

drift to the north, evaluated at 5×10^5 m^3/year. This evaluation is based on both observed sand accumulation rates at the Madras harbor breakwater and by calculations using a formula such as Equation 13 of Chapter 1. The accumulations also demonstrate the reversal of drift directions, erosion occurring to the north of the harbor during the southwest monsoon and deposition in the same area during the northeast monsoon. But the erosion cycle is greater than the subsequent deposition due to the northward transport being greater than the southward return. Net erosion rates are large, having persistently been on the order of 20 m/year for nearly a century.

The Madras area lacked any natural protection for a harbor, and so the harbor had to be developed on the open coastline. Construction of the harbor was sanctioned by the British Government in 1875 and initial construction began soon thereafter (Figure 2). This initial construction quickly revealed the existence of a strong northward littoral drift, the beach to the south of the breakwater building out as rapidly as the construction, and erosion to the north requiring the placement of groins. Cyclones during this period of construction delayed the work, and the initial harbor was not ready until 1895. At that stage, Figure 2, the harbor consisted of an enclosed basin of about 0.75 km^2 with an entrance to the east.

Sand continued to accumulate on the south side of the breakwater until it had built out the shoreline to nearly the seaward length of the harbor, at which point the sand began to drift around the eastward end. This produced shoaling of the harbor entrance, so that the entrance had to be moved to the northward side of the harbor and an outer quay built, Figure 2, in an attempt to deflect the sand away from this new entrance. This is largely the

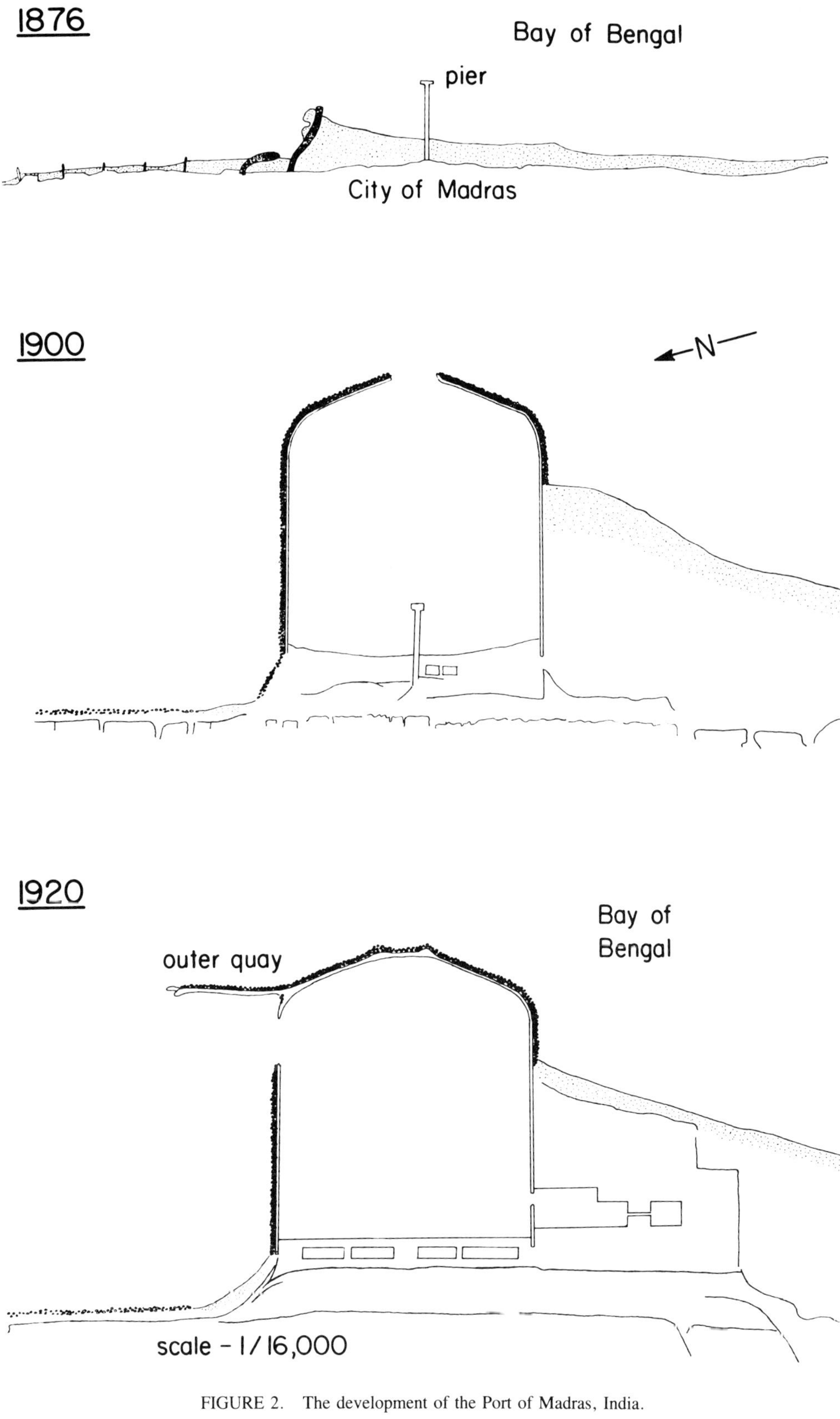

FIGURE 2. The development of the Port of Madras, India.

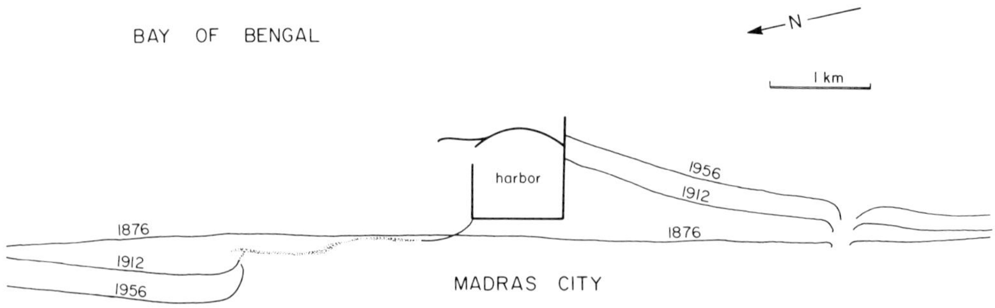

FIGURE 3. Shoreline changes at the Port of Madras, 1876 representing the pre-breakwater shoreline. Riprap placement immediately downdrift of the harbor limited the extent of erosion there.

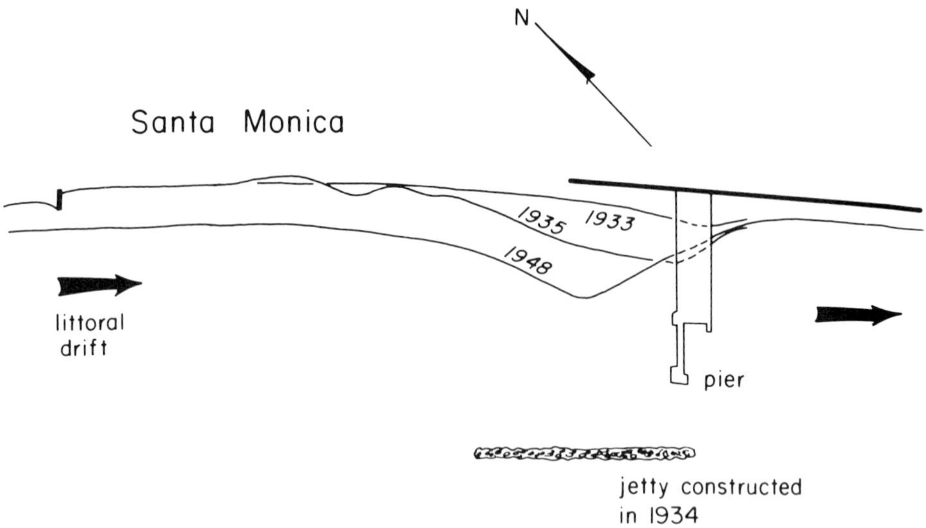

FIGURE 4. The build-out of the shoreline resulting from the construction of a detached breakwater at Santa Monica, Calif. (From Johnson, J. W., *J. Waterways Harbors*, 83, 1, 1957. With permission.)

configuration of the main harbor today, although a second docking area and fisheries harbor have been constructed to the north, the latter placed in a former land area that had eroded away.

The overall pattern of erosion and deposition that emerges is the one that is expected for any harbor installation which blocks a net littoral drift. As diagramed in Figure 3, there have been extensive shoreline changes produced by the breakwater construction, the 1876 survey showing the shoreline just prior to installation and the 1912 and 1956 surveys revealing the subsequent positions. As expected from the breakwater having acted as a dam to the littoral drift, there was an extensive shoreline advance on the updrift side of the harbor, the south side, and considerable erosion on the downdrift side. It is seen that both the deposition and shoreline erosion extend for many kilometers from the breakwater itself, so that the construction affects a considerable portion of coastline. The changes were most rapid during the initial stages of construction. From the date of commencement in 1876 until 1898, the south shoreline advanced at an annual rate of 17 m; from 1898 to 1950 the rate was reduced to about 9.4 m/year during the period 1898 to 1921 and to about 7.3 m/year during 1921 to 1950. The erosion was similarly at a maximum during the early construction, maximum losses of 90 m having occurred during the first southwest monsoon.

The next major port to the north is Kakinada, but this port is in the sheltered zone of a

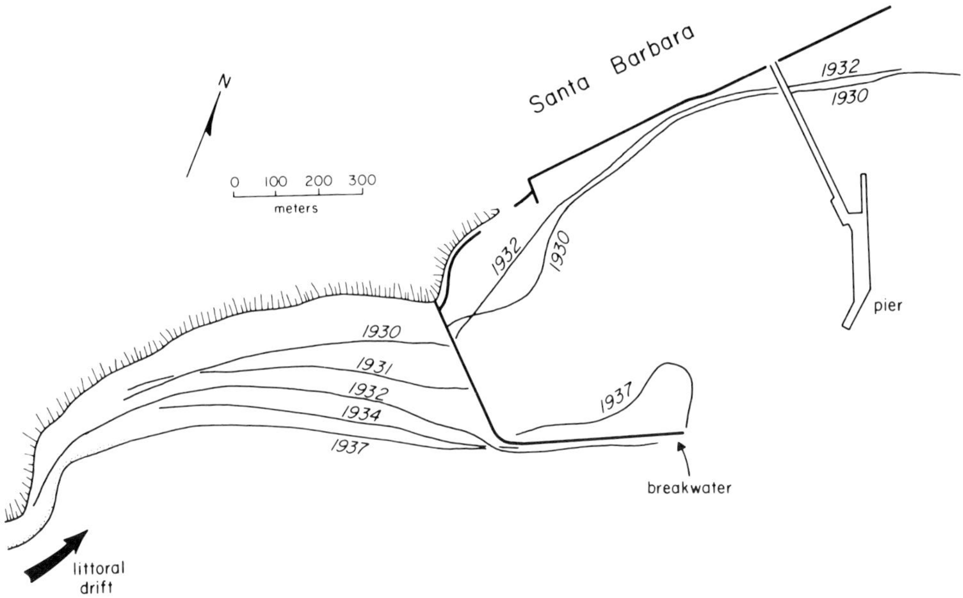

FIGURE 5. Deposition-erosion pattern around the Santa Barbara breakwater, California. (From Johnson, J. W., *J. Waterways Harbors*, 83, 1, 1957. With permission.)

large northward-growing sand spit. Although the harbor construction produced large shoreline changes, these resulted from the altered currents and paths of fine-grained sediment movement in this protected zone, not from a blockage of the littoral drift. The port of Vishakhapatnam consists of an inner harbor cut into the land and an outer harbor protected by a jetty-breakwater system which slants obliquely toward the seaward-facing entrance. The littoral drift is estimated to be 5.5×10^5 m^3/year, approximately half of which is naturally bypassed due to this harbor design. The other half does find its way into the dredged channel leading into the harbor, requiring maintenance dredging and bypassing. The last harbor is Paradip, located at the far north end of the Bay of Bengal (Figure 1). Due to its location, Paradip receives waves predominantly from the southwest monsoon, producing a large northward littoral drift estimated to be 9.4×10^5 m^3/year. The breakwater there was built between July 1963 and September 1965, producing shoreline changes much like those at Madras. During the actual construction, the shoreline to the south advanced at a rate of 120 m/year with erosion to the north being as much as 30 to 40 m during individual storms. The erosion averaged about 21 m/year between 1963 and 1971. A system of groins and sea walls has been constructed in the immediate downdrift zone. A bypassing scheme of dredging on the updrift side and trucking to the downdrift side has been in operation since 1969, but has been able to achieve only about 50% of the estimated littoral drift.

A comparable example to the Madras harbor, but in the U.S., is the breakwater constructed at Santa Barbara, Calif. As originally built in 1927 to 1928, this breakwater was detached, that is, constructed parallel to the shoreline a short distance offshore. The notion was that this design would provide a sheltered zone from the waves, but at the same time would somehow let the littoral drift "slip through". This same design can still be seen in the nearby Santa Monica breakwater, Figure 4. At Santa Monica a breakwater 600 m long was constructed about 600 m offshore. Immediately following its construction sand began to deposit in its protected lee.[2] On the updrift side of the breakwater the shoreline advanced, while in the downdrift direction the shoreline eroded. Only by repeated dredging has the entire closure of the harbor been prevented, the shoreline attaching to the breakwater. It is apparent then that such a detached breakwater will not permit the littoral drift to "slip through"; the

FIGURE 6. The blockage of the littoral drift by the jetties on the north inlet of Lake Worth, Fla., sand accumulating on the updrift side of the jetties and erosion occurring on the downdrift side. (*Photograph from* University of Florida archives, *courtesy* of R. G. Dean and T. Walton.)

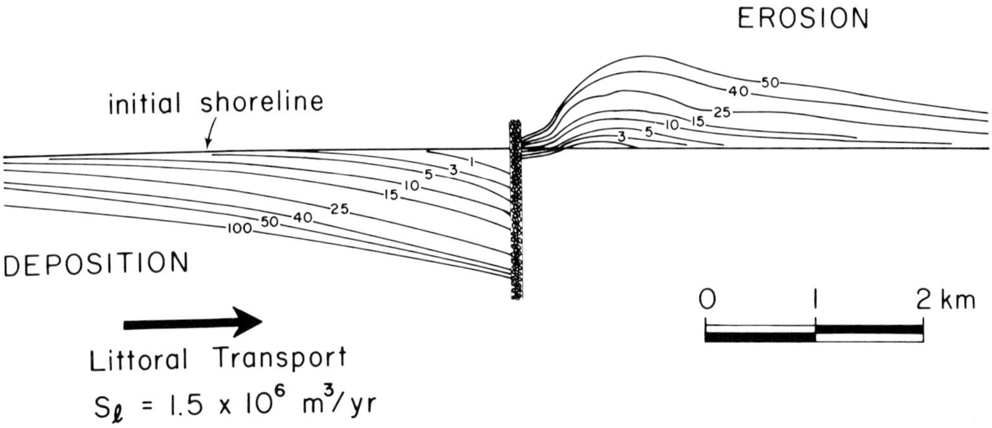

FIGURE 7. The blockage of the littoral drift by a jetty or groin; this example being from a physical laboratory model of a proposed jetty at Cotonou, Dohomey. Included are the shoreline changes, given in years, as well as the drift rate pertaining to the prototype, 20 min of model operation being equivalent to a year of prototype conditions according to the model scaling. (From Sireyjol, P., *Proc. 9th Conf. Coastal Eng.*, American Society of Civil Engineers, New York, 1965. With permission.)

breakwater decreases or nearly eliminates the wave energy at the shoreline, and according to littoral drift formulas (such as Chapter 1, Equation 13) the longshore sand movement will come to a halt even though there is not a physical barrier across the nearshore.

The original detached breakwater at Santa Barbara had the same effect, so that in order to prevent harbor shoaling, in 1930 the detached breakwater was connected to the shoreline on its updrift side (Figure 5). The littoral drift averages about 2.2×10^5 m^3/year from the west so that sand quickly accumulated at its western side with erosion in the downdrift eastern direction. After only 7 years of accumulation on the updrift side the shoreline had built out to the end of the breakwater, Figure 5, and in 1937 began to spill around its outer end, forming a sand spit which began to grow across the harbor opening. Left unattended, this spit eventually would have grown and attached itself to the downdrift shoreline, and once this had been accomplished the entire littoral drift would then pass around the breakwater and a new equilibrium shoreline configuration would have been achieved. However, this would of course close the harbor, and to prevent that a dredging operation was initiated and presently operates continuously. The dredge is located at the tip of the spit, dumping the sand on the beach in the downdrift direction so as to reduce the erosion. A more complete case history of the problems at Santa Barbara can be found in Wiegel.[12,13]

Jetties built at the mouth of a river or tidal inlet to a bay, lagoon, or estuary have much the same effect when they block a littoral drift. As seen in the example of Figure 6, the shoreline again builds out on the updrift side and retreats due to erosion in the downdrift direction. The jetty acts as an effective dam to the littoral drift, halting its advance at the updrift jetty. At the same time, on the downdrift side the sand transport processes continue to operate and so cause sand to drift away from the jetties with a resulting shoreline erosion. This same effect is shown in Figure 7, based on a laboratory model where a single "jetty" blocks the longshore sand movement. Here it is seen that the impact on the shoreline, either as accretion or erosion, is felt earliest and with maximum effect close to the littoral drift obstacle. This same pattern is observed in the prototype jetty systems.

Jetties have been constructed at many inlets on the east coast of the U.S. For example, jetties were built in 1935 to stabilize the inlet south of Ocean City, Md., a seaside resort. These blocked the north-to-south littoral drift. Fortunately, the main community of Ocean City lay to the north of the jetties so experienced a shoreline advance. However, due to the loss of its supply of sand, the barrier Assateague Island to the south eroded and the shoreline

FIGURE 8. A groin field south of Tampa Bay, Fla., the groins trapping the littoral drift and building out the beach locally. (Photograph from University of Florida archives, courtesy of R. G. Dean and T. Walton.)

retreated about 450 m in the 20 years following jetty construction.[8] By 1961 the south beach had eroded to the point where it actually separated from the inner end of the south jetty, leaving a gap of almost 240 m of open water. In 1962 a severe storm opened a breach over a kilometer wide 2.5 km south of the jetties. Subsequent dredging and fill has partially restored the beach, but it can be expected that such problems will continue.

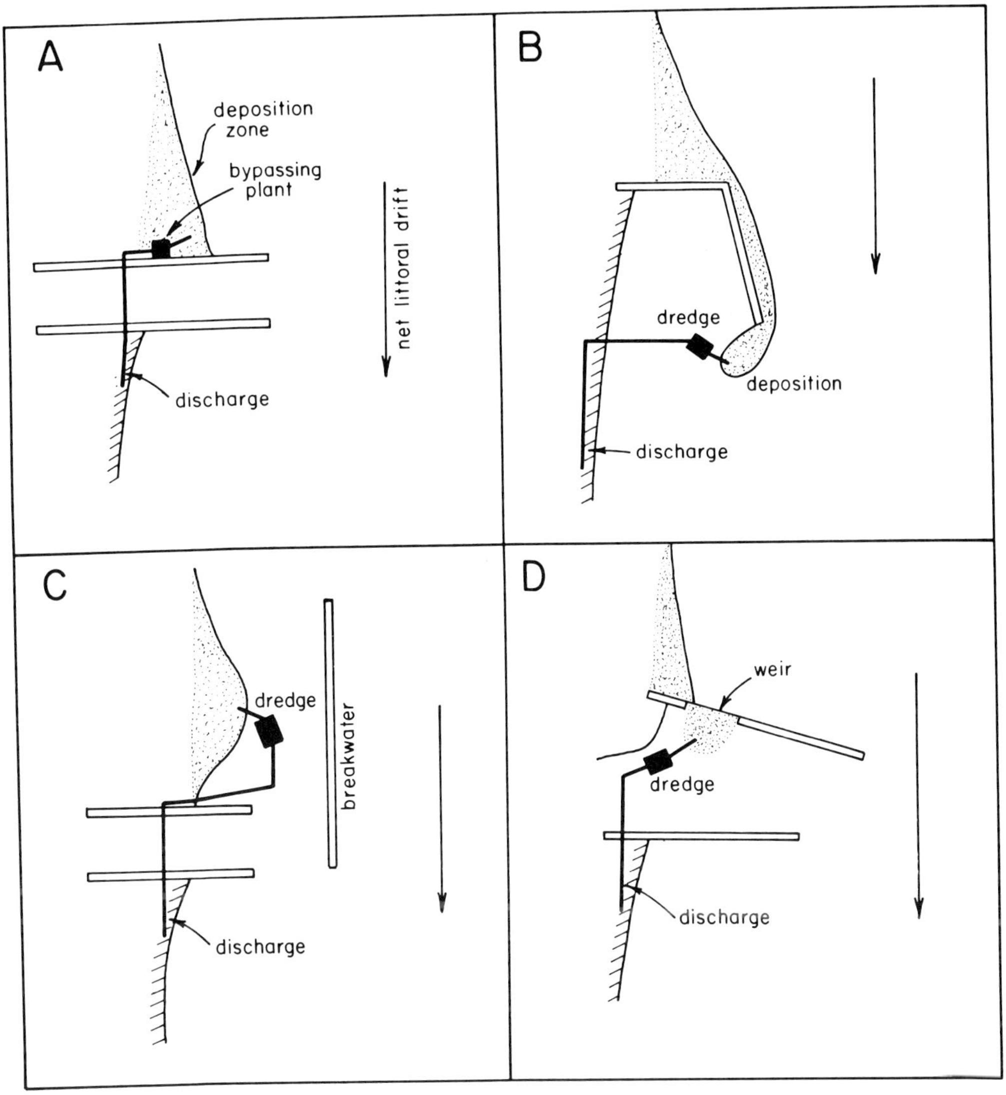

FIGURE 9. Schematic of the different types of bypassing systems that have been used to transfer the littoral drift past jetties or breakwaters.

A groin is a rib built approximately perpendicular to the shoreline to trap the littoral drift and thereby locally build out the shoreline (Figure 8). Its purpose is to widen the beach and thereby protect the landward property. Groins are relatively narrow in width and may vary in length from less than 10 m to over 200 m. A single groin has much the same effect as a jetty, like that of Figure 7, the trapped littoral drift widening the beach. The impact is smaller, however, due to the shorter length of the groin, the updrift side becoming filled with sand so that the remaining littoral drift eventually spills around the groin and contributes to the downdrift beach. Some sand is permanently trapped and thereby lost to the downdrift beaches. Thus, the construction of a groin will tend to enhance the erosion of the downdrift beaches and property. In order to shift the erosion further downdrift, it is common practice that a series of groins (a groin field) are built (Figure 8). This enables an extended stretch of beach to be built out and shifts the zone of erosion out of your immediate area to the down-coast neighbor. And of course the greater the number of groins built the larger the total volume of sand trapped, and hence the greater the erosion that does result to your neighbor's property. One way to partially alleviate this impact of groin construction is to nourish the beach at the same time (Chapter 11), by obtaining an outside source of suitable sand which can be placed on the beach to fill the volumes that would be trapped by the groins. This reduces the quantity of sand that is extracted from the natural littoral drift, which would then be relatively undiminished when it reaches the downdrift beaches.

Such beach nourishment schemes also have been utilized to alleviate the downdrift erosion resulting from jetty or breakwater construction. However, here the sand is placed directly onto the downdrift beach rather than attempting to fill the updrift side which would cause sand to spill over and into the harbor, much as occurred naturally at Santa Barbara. Although in some cases a foreign sand source is obtained for the eroding beach, the most common practice is to obtain the sand from the updrift beach where there is the simultaneous problem of its accumulation, a problem in the sense that it will eventually cause harbor shoaling. This removal from the updrift beach and placement on the downdrift side of the jetties constitutes a bypassing scheme. In its simplest form sand is carried past the jetties in trucks. More elaborate schemes involve dredging of the sand on the updrift side and pumping it as a fluidized sand-water mixture to the downdrift beach where the sand can continue to move as part of the natural longshore transport regime. Such bypassing schemes are shown schematically in Figure 9. Figure 9A illustrates the oldest approach used in the U.S., where a fixed dredging plant on the updrift side consists of a boom supporting an intake pipe that can be swung in an arc over the deposition zone. This intake pipe is connected to a pump and discharge line that crosses the inlet. South Lake Worth Inlet, Fla. is an example of such a system.[11]

Most floating dredges are sensitive to wave attack and so so must be protected in some way. The system used at Santa Barbara, Calif. is diagramed in Figure 9B, the harbor itself providing shelter for the dredge which removes sand from the spit as it spills into the harbor. Figure 9C shows the system employed at the Channel Islands Harbor, Calif. where a detached breakwater shelters the dredge.[3] Construction of such a breakwater for the sole use of the dredge is expensive, and so a scheme such as that of Figure 9D is sometimes used. The weir has an elevation at or near mean sea level at the shoreward end of the updrift jetty. This permits sand to wash over the weir into a deposition basin where the dredge can operate in safety, the sand again being pumped to the downdrift side. Such a system is employed at Masonboro Inlet, N.C.[7] It is apparent that in all of these schemes considerable effort and expense is required to provide a sand-moving operation to replace the natural sand transport processes which have been interfered with by jetty or breakwater construction.

JETTIES ON COASTS WITH A ZERO NET LITTORAL DRIFT

It might be expected that if a jetty or breakwater is constructed in an area of coast where there is essentially a zero net littoral drift, then no sand would be dammed and no shoreline erosion would result. This is in large part true, and there are many examples of jetties that have been built with little coastline changes. However, under certain circumstances jetties constructed on coasts with a zero net drift have produced major shoreline alterations with considerable loss of property. One example of this is the beach erosion and sand spit destruction that occurred at the entrance to Tillamook Bay, Ore.[6,10] A north jetty was constructed at the mouth of the bay in 1914 to 1917 and was rebuilt and extended in 1933 (Figure 10). Bayocean Spit to the south of the entrance, separating Tillamook Bay from the ocean, suffered progressive erosion following construction of this jetty, culminating in its breaching in 1952. The breach was subsequently closed by the construction of a dike in 1956. The resort community of Bayocean Park was entirely destroyed by the erosion.

While erosion was occurring along much of the spit length, the shoreline advanced to the

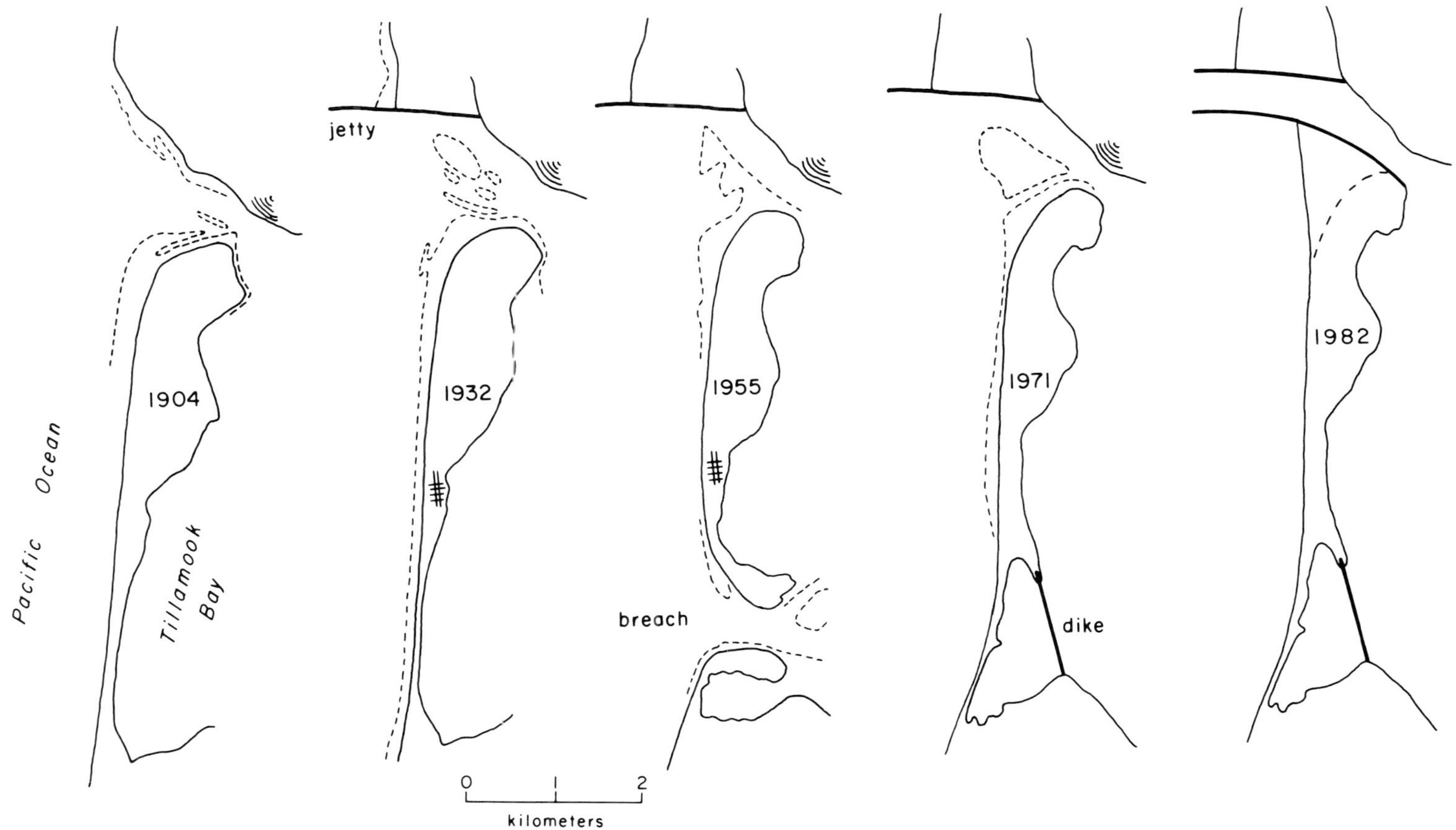

FIGURE 10. Alterations of the shoreline, including erosion of Bayocean Spit, due to the construction of a north jetty in 1914 to 1917, and a south jetty in the mid-1970s on the entrance to Tillamook Bay on the northern Oregon coast. Extensive erosion occurred on the spit resulting in the complete loss of the town of Bayocean, this occurring in spite of the existence of a zero net littoral drift on this portion of coast. (From Terich, T. A. and Komar, P. D., *Shore Beach*, 42, 3, 1974. With permission.)

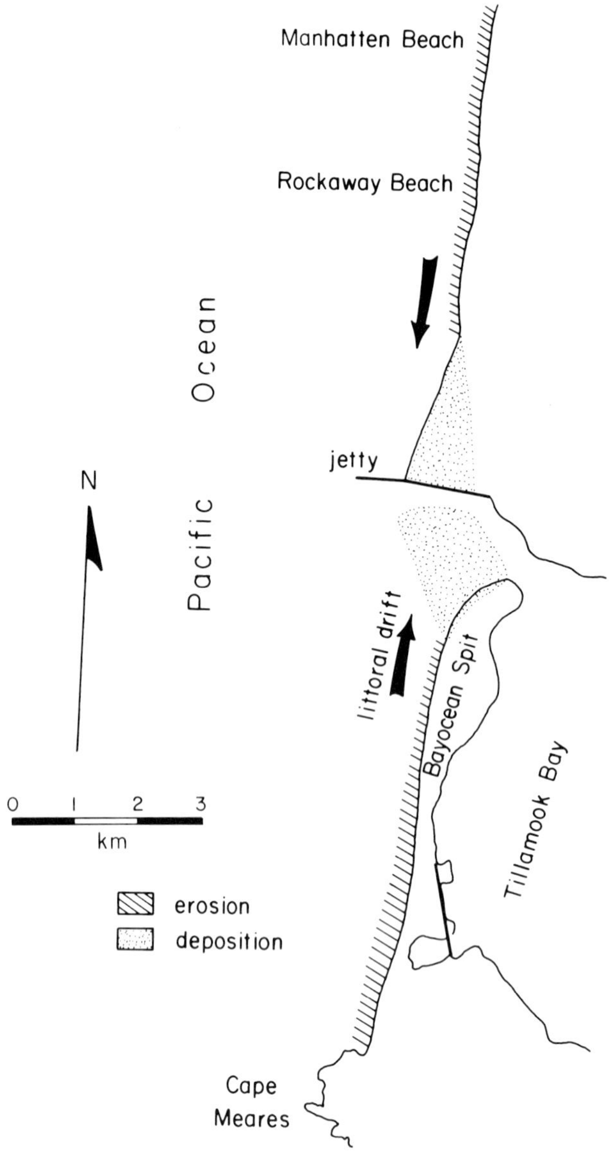

FIGURE 11. A schematic illustration of the patterns of deposition and erosion due to jetty construction on the entrance of Tillamook Bay, Ore. On coasts with a zero net littoral drift jetty construction tends to result in deposition and shoreline advancement adjacent to the jetties with erosion occurring at greater distances from the jetties to supply that sand. (From Terich, T. A. and Komar, P. D., *Shore Beach*, 42, 3, 1974. With permission.)

immediate north of the north jetty. This would at first imply the existence of a north to south littoral drift, and the U.S. Army Corps of Engineers, Portland District, actually estimated a mean annual southward drift of approximately 6×10^5 m^3/year. However, closer study by Terich and Komar[10] and Komar et al.[5] demonstrated that there is in fact a zero net littoral drift along most of the Oregon coast including the Bayocean Spit region. If a strong littoral drift did exist in the region, the large rocky headlands such as Cape Meares to the immediate south of Bayocean Spit, Figure 11, would act much like a jetty, blocking the littoral drift. But there is no observed deposition on one side of the headland and erosion on the opposite side. A study of the effects of jetties on the shoreline also indicates that a zero net drift prevails, deposition occurring to both the immediate north and south of the jetties with erosion existing at still greater distances from the jetties.[5] This pattern is diagramed schematically in Figure 11 for Bayocean Spit and another example is shown in Chapter 10 Figure 5 for the case of the jetties on the Siuslaw River, Ore. The pattern is more complex at the Tillamook Bay entrance because only one jetty was constructed rather than a pair but the study by Terich and Komar[10] demonstrated that while the shoreline was advancing to the north of the north jetty, a large shoal also developed within the inlet itself, nearly closing it off. The effects on the inlet morphology are further documented in Komar et al.[5] Therefore, a symmetrical pattern of erosion and deposition was still evident at the Tillamook Bay entrance even though only one jetty was initially constructed (Figure 11). In the mid-1970s the second jetty was finally built, reaffirming the conclusion of a zero net drift in the region (Figure 10).

Therefore, major readjustments of the shoreline following construction of jetties can even occur on a coast with a zero net drift. The amount of resulting erosion depends on two main factors: (1) the size of the embayment formed between the pre-jetty shoreline and the newly constructed jetty, this embayment having to become filled with sand in order to achieve a new equilibrium shoreline, and (2) the length of shoreline experiencing erosion to supply sand to this depositional area adjacent to the jetty. The embayment factor is illustrated by the Siuslaw River jetties (Chapter 10, Figure 5), the jetty construction having formed a large embayment to the north which required a considerable volume of sand to fill it. In contrast, the embayment to the south of the Siuslaw jetties was small and produced only minor shoreline advances. The second factor, shoreline length, is illustrated by Bayocean Spit. As shown in Figure 11, Cape Meares lies to the immediate south, acting as a barrier to this stretch of coast. The sand needed for deposition near the inlet therefore had to be gained from erosion of a relatively short stretch of coast so that the amount of erosion per unit shoreline length was large. In contrast, the sand volume deposited to the north of the jetty came from a longer stretch of coast so that the erosion per unit shoreline length was smaller with a negligible retreat of the coastline.

Another important difference between effects of jetties on coasts with a net littoral drift and those with a zero net drift is the final equilibrium. In the case of jetties blocking a net drift, the problem continues indefinitely unless the harbor is allowed to fill (as Santa Barbara nearly did), usually requiring a bypassing system to prevent such a harbor filling. In contrast, jetties built on a coast with zero net drift may cause an appreciable readjustment and reorientation of the shoreline, but the shifts are usually toward a new equilibrium where a zero net drift is again achieved. This was the case for all of the jetty systems built on the Oregon coast, the shoreline changes being limited to some 10 to 20 years after jetty construction with little subsequent erosion attributable to the jetties.

SUMMARY

Whenver jetties or a breakwater block a net littoral drift along the coastline, the damming effect produces accretion and a shoreline advance on the updrift side of the obstacle, and erosion in the downdrift direction. The impact is noted first and with the greatest shoreline changes closest to the engineering structure, but the effects may eventually extend for many kilometers along the coast. The problems associated with blocking a net littoral drift are persistent and usually require the implementation of a bypassing scheme which dredges the accumulating sand from the updrift side and transfers it to the downdrift side where it is returned to the littoral drift regime.

Although jetties may often be constructed in coastal regions having a zero net drift with minimal adverse impact, in certain circumstances there can be considerable alterations to the shoreline. These usually involve shoreline advances adjacent to the jetties to fill "embayments" formed between the jetties and the pre-jetty shoreline, with erosion at greater distances from the jetties to supply that sand. The changes are transient, however, usually being completed within 10 to 20 years, and no permanent bypassing measures are required.

ACKNOWLEDGMENTS

The material in this chapter dealing with the east coast of India is based on studies undertaken at the Central Water and Power Research Station in Poona, India. I became acquainted with this material in 1979 during a visit sponsored by the United Nations. I would especially like to thank P. C. Saxena, J. G. Dixit, and R. Srinivasan for familiarizing me with their work on the coast of India and for my pleasurable visit.

REFERENCES

1. CERC, *Shore Protection Manual,* U.S. Army Corps Eng., Coastal Eng. Res. Cent., U.S. Government Printing Office, Washington, D.C., 1973.
2. **Handin, J. W. and Ludwick, J. C.,** Accretion of beach sand behind a detached breakwater, *U.S. Army Beach Erosion Board Tech. Memo,* TM-16, 1950.
3. **Herron, W. J. and Harris, R. L.,** Littoral bypassing and beach restoration in the vicinity of Port Hueneme, Calif., *Proc. 10th Conf. Coastal Eng.,* American Society of Civil Engineers, New York, 1966, 651.
4. **Johnson, J. W.,** The littoral drift problem at shoreline harbors, Pap. 1211, *J. Waterways Harbors,* 83, 1, 1957.
5. **Komar, P. D., Lizarraga-Arciniega, J. R., and Terich, T. A.,** Oregon coast shoreline changes due to jetties, *J. Waterways, Harbors Coastal Eng.,* 102(WW1), 13, 1976.
6. **Komar, P. D. and Terich, T. A.,** Changes due to jetties at Tillamook Bay, Oregon, *Proc. 15th Conf. Coastal Eng.,* American Society of Civil Engineers, New York, 1977.
7. **Magnuson, N. C.,** Planning and design of a low-weir section jetty at Masonboro Inlet, N.C. Proc. Santa Barbara Coastal Eng. Spec. Conf., Santa Barbara, Calif., 1966, 807.
8. **Shepard, F. P. and Wanless, H. R.,** *Our Changing Coastline,* McGraw-Hill, New York, 1971.
9. **Sireyjol, P.,** Communication sur la construction du port de Contonou (Dahomey), *Proc. 9th Conf. Coastal Eng.,* American Society of Civil Engineers, New York, 1965.
10. **Terich, T. A., and Komar, P. D.,** Bayocean Spit, Oregon: History of development and erosional destruction, *Shore Beach,* 42, 3, 1974.
11. **Watts, G. M.,** A study of sand movement at South Lake Worth Inlet, Florida, *U.S. Army Beach Erosion Board Tech. Memo.,* TM-42, 1953.
12. **Wiegel, R. L.,** Sand bypassing at Santa Barbara, California, *J. Waterways Harbors,* 85(WW2), 1, 1959.
13. **Wiegel, R. L.,** *Oceanographic Engineering,* Prentice-Hall, Englewood Cliffs, N. J., 1964.

Chapter 10

COMPUTER MODELS OF SHORELINE CHANGES

Paul D. Komar

Although they have not seen widespread use to date, computer-simulation models of nearshore processes and shoreline changes have a high potential for aiding in the analysis of certain erosion problems. The types of problems that have been examined relate to predicting equilibrium beach shapes in groin fields,[14] predicting changes in the shoreline due to the effects of offshore dredging,[12] and in the examination of shoreline changes in response to jetty construction.[9] Other models have been developed whose primary objectives have been the analysis of coastal geomorphic features such as river delta shapes,[6] and the shape of "hooked" or "logarithmic spiral" beaches.[10,11,15]

This chapter will review the applicability of computer modeling to shoreline changes, discussing the general techniques of developing such a model, and then examining those published models that have particular relevance to coastal erosion problems. Additional reviews of computer modeling of shorelines can be found in Dean[2] and Komar.[7,8] Harbaugh and Bonham-Carter[5] give a thorough presentation on the general philosophy of simulation modeling and the various techniques employed.

TECHNIQUES OF MODELING

The general approach to developing computer models of shoreline changes is closely analogous to the computer models of fluid flow that have been used for many years by engineers and scientists. The continuity equation for the fluid is replaced by a comparable continuity relationship for the beach sand, one that keeps track of the total volume or mass of sand and makes sure that none is miraculously created or destroyed. In place of the Navier-Stokes equations which govern the flow of fluids, the shoreline models employ a relationship such as Equation 13 of Chapter 1 which relates the sand movement along the beach to the nearshore wave conditions.

Also analogous to the fluid-flow models, the environment is divided into a number of cells, and the shoreline models consist of following the shift of sand from one cell to another. The models that have been employed to date in the nearshore are principally two-dimensional, a stretch of shoreline being divided into a series of cells (Figure 1A), the end of each cell terminating in a schematic beach profile (Figure 1B). As shown, all the cells have a uniform width, Δx, though this need not necessarily be the case, and have individual lengths $y_1 \ldots y_i \ldots y_n$ beyond some baseline. The series of cells is seen to approximate the smoothly curving shoreline, the greater the number of cells and the smaller their Δx widths, the more nearly they represent the smooth shoreline.

The basis of any shoreline model is the evaluation of the quantities of sand entering and leaving each cell, and the resulting changes in the shoreline position due to the balance of input vs. exit. The littoral drift is usually the main cause of sand moving from one cell to another. If Q_i is the rate of littoral drift from cell i into cell i + 1 and Q_{i-1} is the drift into cell i from cell i − 1 (Figure 1A), then the net volume of sand, ΔV_i, gained or lost from cell i is

$$\Delta V_i = (Q_{i-1} - Q_i)\Delta t \qquad (1)$$

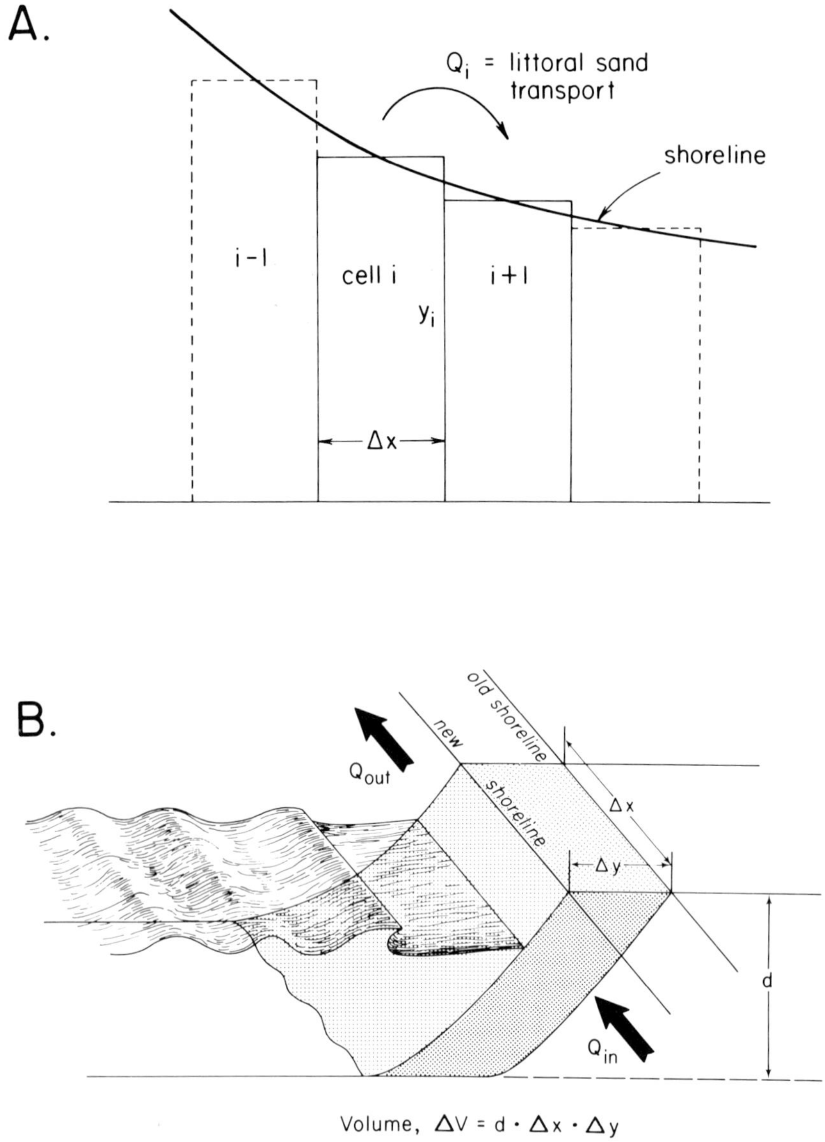

FIGURE 1. (A) The smooth shoreline divided up into a series of cells of width Δx and variable y_i lengths beyond some base line. (B) One shoreline cell demonstrating how a change in sand volume ΔV is produced by the littoral drift into and out of the cell and how this results in a shoreline change, Δy. (From Komar, P. D., *Geol. Soc. Am. Bull.*, 84, 2217, 1973. With permission.)

where Δt is the elapsed time, the increment of time over which the model is run. To further generalize, there may be other sand contributions to cell i from an adjacent river, onshore sand movement, or from some other source, and there may be possible losses including being blown inshore to form dunes or lost offshore. Grouping these various other sources or sinks of beach sand into a term, Q_r, Equation 1 is modified to

$$\Delta V_i = (Q_{i-1} - Q_i \pm Q_r)\Delta t \qquad (2)$$

This development of Equation 2 constitutes a minibudget of sediments, applied to cell i.

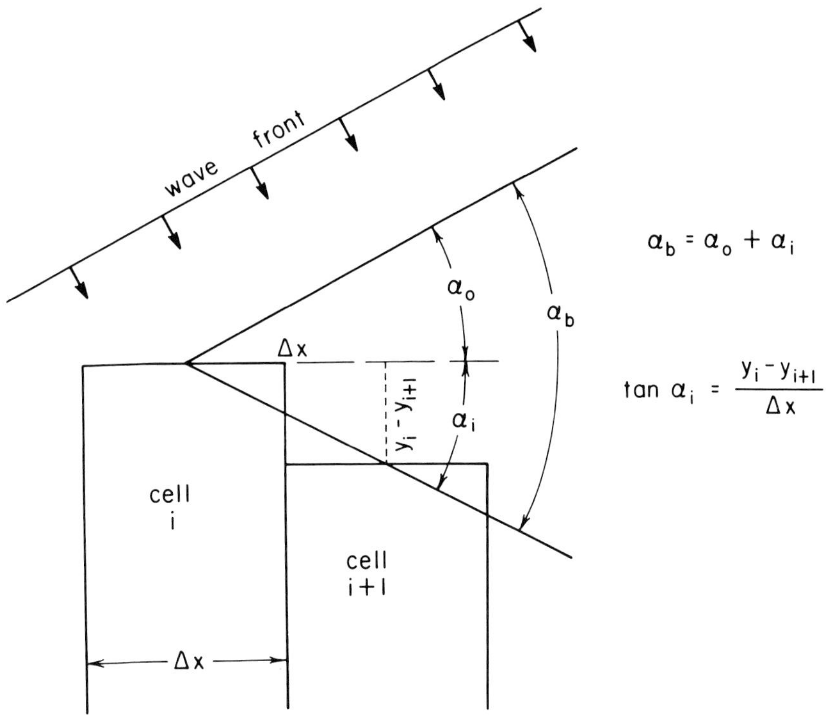

FIGURE 2. Geometric relationships between the wave approach angle, α_o, the beach orientation, α_i, and the resulting total breaker angle, α_b.

Note that ΔV_i can be positive, negative, or the various inputs and outputs can balance to yield $\Delta V_i = 0$, in which case the shoreline position does not change in that particular cell. $\Delta V_i = -$ signifies erosion in that more sand has left the cell than has entered, and $\Delta V_i = +$ represents a net deposition.

It is desirable to express ΔV_i as an actual change in the shoreline position, that is, as a change in the length y_i of the cell (Figure 1A). If Δy_i is the corresponding change in y_i in the time Δt, then from the geometry of the cell shown in Figure 1B,

$$\Delta V_i = d\, \Delta y_i\, \Delta x \tag{3}$$

The linear dimension d must be chosen to yield the correct correspondence between ΔV_i and Δy_i, and will depend on the nature of the beach profile. From the geometry of Figure 1B it is seen that $d\Delta y_i$ equals the cross-sectional area of the wedge of sand deposited or eroded — multiplication by Δx then yielding the volume. The linear dimension d tends to be on the order of the water depth at the breaker position.

Combining Equations 2 and 3 gives

$$\Delta y_i = (Q_{i-1} - Q_i \pm Q_r)\, \frac{\Delta t}{d\Delta x} \tag{4}$$

for the change in shoreline position of cell i as a function of the sand inputs and exits. Since in the models the parameters Δt, d, and Δx are generally fixed, the Δy_i shoreline changes are produced simply by the balance of the Q-terms. Again, $\Delta y_i = -$ represents erosion and a shoreline retreat, and $\Delta y_i = +$ is a shoreline advance.

Equation 4 can perhaps more easily be recognized as a sand continuity equation if the finite elements are decreased to their limits, yielding

$$\frac{dy}{dt} = -\frac{1}{d}\frac{dQ_s}{dx} \tag{5}$$

Q_r has also been dropped, leaving only the littoral drift. This form of the relationship is also informative in showing the dependence of the time-rate of shoreline change, dy/dt, as depending on the longshore gradient of the littoral drift, dQ_s/dx. If $dQ_s/dx = +$ then Q_s is increasing in the longshore direction, and this requires more sand which comes from erosion of the beach, i.e., $dy/dt = -$ (assuming, of course, no other sources). And the greater the magnitude of dQ_s/dx the greater the rate of shoreline recession. In the opposite direction, a decreasing Q_s gives $dQ_s/dx = -$ and $dy/dt = +$, and advancing shoreline. Greenwood and McGillivray[4] and Davidson-Arnott and Pollard[1] provide examples of where this approach has been used to explain locations of shoreline erosion.

The littoral drift terms of Equation 4, Q_i and Q_{i-1}, are evaluated with a littoral drift formula such as Equation 13 in Chapter 1. This evaluation in turn requires a knowlege of the wave power, $(ECn)_b$, and breaker angle, α_b. The breaker angle will generally change from cell to cell due to the changing orientation of the shoreline. As diagramed in Figure 2, the angle α_i which the shoreline makes with a parallel to the x-axis, between cells i and i + 1, is given by

$$\tan\alpha_i = \frac{y_i - y_{i+1}}{\Delta x} \tag{6}$$

If the incoming waves make an angle α_o with the x-axis direction, Figure 2, then the breaker angle at the shoreline is $\alpha_b = \alpha_i \pm \alpha_o$, the subtraction occurring when α_o opens in the negative x-direction. This breaker angle is most easily obtained with the trigonometric identity

$$\tan\alpha_b = \tan(\alpha_i \pm \alpha_o) = \frac{\tan\alpha_i \pm \tan\alpha_o}{1 + \tan\alpha_i \tan\alpha_o} \tag{7}$$

Table 1 presents a routine written in Fortran for the evaluation of breaker angles and the littoral drift from cell i to cell i + 1. TANO is $\tan\alpha_o$ from the above relationships and is assumed to already be known, having been fixed as the overall wave approach angle. TANI represents $\tan\alpha_i$ according to Figure 2 and has to be determined for each pair of cells in the model. If $y_i = y_{i+1}$ then $\alpha_i = 0$ and one simply has $\alpha_b = \alpha_o$; this case is covered by Statements 7 through 10, TANB representing $\tan\alpha_b$. The parameter DIR indicates the direction of the resulting littoral drift, DIR = 1.0 corresponding to sand transport in the positive x-direction and DIR = −1.0 indicating that the drift will be in the negative x-direction. Statements 12 through 15 cover the condition when $y_i > y_{i+1}$ and Statements 16 through 22 are for $y_i < y_{i+1}$. Note that in the latter case the direction of the resulting transport depends on the comparative magnitudes of α_o and α_i. Statements 13, 18, and 21 are all Fortran expressions of Equation 7.

This completes the evaluation of $\tan\alpha_b$ and the determination of the littoral drift direction, and Statements 24 through 26 calculate the corresponding values of $\sin\alpha_b$ and $\cos\alpha_b$ required in the littoral drift evaluation. Statement 28 calculates Q_i, the drift between cells i and i + 1, the relationship corresponding to Equation 13 of Chapter 1, where P is the Fortran equivalent to the wave energy flux, $(ECn)_b$. Note that DIR is included in the Fortran statement so that Q_i can be either positive or negative according to its direction. A positive Q_i value signifies that sand is leaving cell i and is entering cell i + 1, whereas a negative Q_i value is just the opposite.

The above calculations are performed for each cell in the model, the program of Table 1 having a do-loop to repeat the calculations for 100 cells. Once this is completed we have evaluated the littoral drift rates between all adjacent cell pairs at this particular shoreline

Table 1
COMPUTER ROUTINE FOR MODELING SHORELINE CHANGES

```
(1)              DO 100 K = 1,50
(2)              TIME = TIME + DELT
(3)   C          CALCULATE BREAKER ANGLES AND SAND TRANSPORT
(4)              I = 1
(5)              DO 12 J = 1, 100
(6)   C          CALCULATE WAVE APPROACH ANGLES
(7)              IF(Y(I).NE.Y(I + 1)) GO TO 6
(8)              TANB = TANO
(9)              DIR = 1.0
(10)             GO TO 10
(11)          6  IF(Y(I).LT.Y(I + 1)) GO TO 7
(12)             TANI = (Y(I) - Y(I + 1))/DELX
(13)             TANB = (TANI + TANO)/(1.0 - TANI*TANO)
(14)             DIR = 1.0
(15)             GO TO 10
(16)          7  TANI = (Y(I + 1) - Y(I))/DELX
(17)             IF(TANI.LT.TANO) GO TO 8
(18)             TANB = (TANI - TANO)/(1.0 + TANI*TANO)
(19)             DIR = -1.0
(20)             GO TO 10
(21)          8  TANB = (TANO - TANI)/(1.0 + TANI*TANO)
(22)             DIR = 1.0
(23)  C          CALCULATE SINES AND COSINES OF BREAKER ANGLES
(24)         10  A = TANB**2
(25)             SINB = SQRT(A/(A + 1.0))
(26)             COSB = SQRT(1.0 - SINB**2)
(27)  C          CALCULATE LITTORAL DRIFT RATES
(28)             Q(I) = DIR*6.8*P*SINB*COSB
(29)             I = I + 1
(30)         12  CONTINUE
(31)  C          CALCULATE RESULTING SHORELINE CHANGES
(32)             I = 1
(33)             DELY(1) = (QE - Q(1))*DELT/(DELX*D)
(34)             Y(1) = Y(1) + DELY(1)
(35)             I = 2
(36)             DO 20 J = 1,98
(37)             DELY(I) = (Q(I - 1) - Q(I))*DELT/(DELX*D)
(38)             Y(I) = Y(I) + DELY(I)
(39)             I = I + 1
(40)         20  CONTINUE
(41)             I = 100
(42)             DELY(100) = (Q(99)*DELT)/(DELX*D)
(43)             Y(100) = Y(100) + DELY (100)
(44)        100  CONTINUE
(45)  C          PRINT OUT RESULTING SHORELINE
```

stage. Statements 32 through 43 calculate the shoreline changes resulting from this littoral drift over a time increment DELT = Δt. Statement 37 corresponds to the continuity expression of Equation 4. Statement 38 alters the previous value of y_i by the calculated Δy_i = DELY(I), assigning the new length to y_i. Note that cells i = 1 and 100 have to be dealt with separately due to the absence of cells i = 0 and 101. For i = 1 there is a QE term which could have a fixed value due to a river adjacent to cell i = 1 or it could represent a littoral drift into the stretch of shoreline under consideration, accounting for an end effect in the finite-size model. For cell i = 100, Statement 42, there is only a Q(99) drift. The absence of a Q(100), which would be a drift out of the model, in effect places an obstacle to the right of cell 100, an obstacle such as a jetty or headland which blocks any possible drift beyond cell 100.

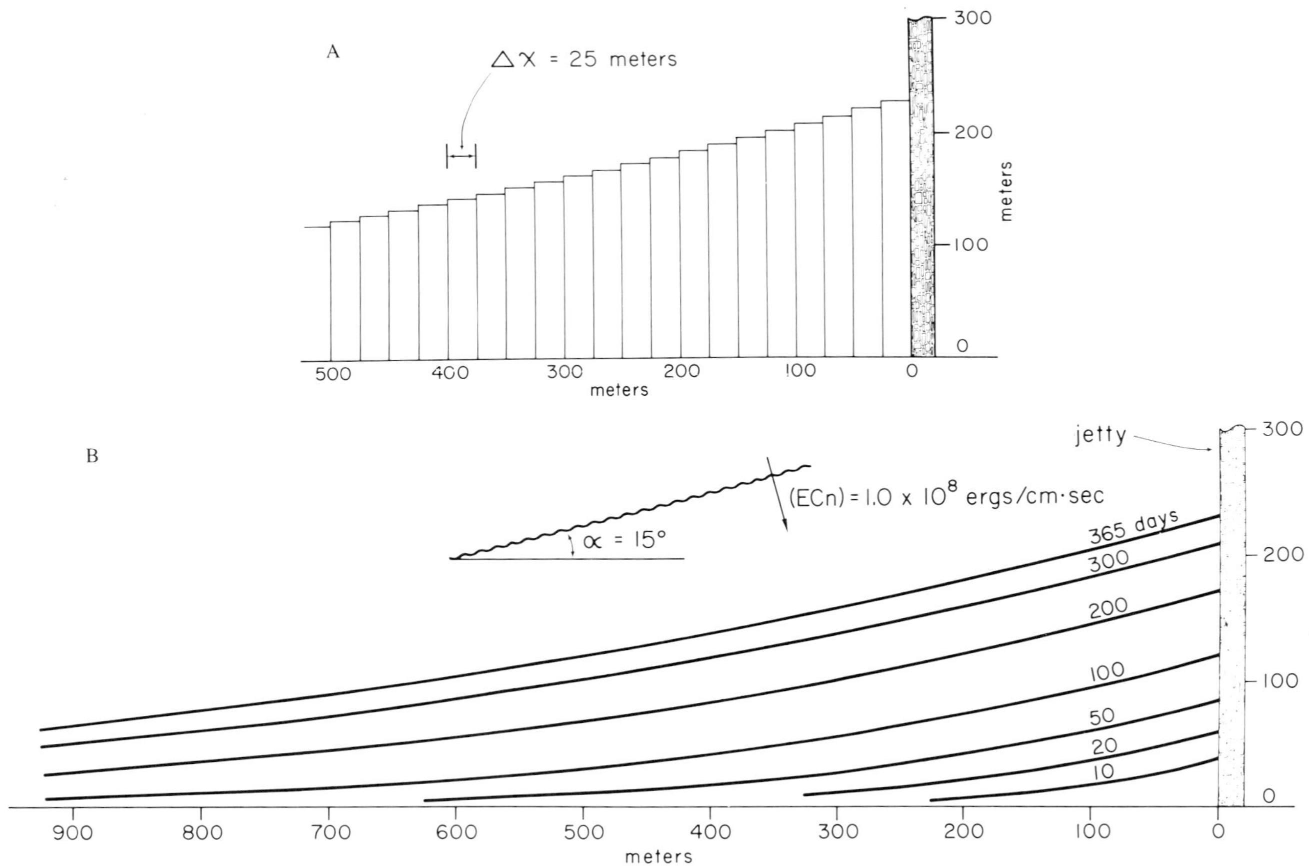

FIGURE 3. (A) Model of a jetty blocking a littoral drift from left to right in the figure. Only the last 20 cells are shown (i = 80, . . . 100), the model actually extending updrift of the jetty for 2.5 km. (B) The progressive advance of the shoreline as obtained in the model, smooth shorelines having been drawn through the 25-m-wide cells. (From Komar, P. D., in *The Sea*, Goldberg, E., McCare, I., O'Brien, J., and Steele, J., Eds., John Wiley & Sons, New York, 1977, 499. With permission.)

The routine of Table 1 is the heart of many models of shoreline changes but is presented here in its simplest terms. A complete model must also establish the initial shoreline configuration at time t = 0, and may involve sand sources and losses in addition to the littoral drift, changes in wave conditions through time, and contain statements for printing the results or producing a graphical representation of the model.

In summary, simulation models of shoreline changes usually involve the following sequence of operations:

1. Define an initial shoreline configuration
2. Establish the various sources and losses of sand to the beach, the Q_r terms of Equation 4
3. Provide the model with the offshore wave conditions (heights, periods, and approach angles)
4. Indicate how the littoral transport of sand will be governed by the wave parameters, usually employing a relationship such as Equation 13 of Chapter 1
5. Run the model through increments of time, Δt, for some total span of time to determine how the shoreline evolves from its initial configuration

As in other numerical models, it is important that the Δy_i changes in cell lengths remain relatively small so that there are no sudden "jumps" in its position. This usually requires that the time increment, Δt, be kept small, commonly a fraction of one day. From Equation 4 it is apparent that the smaller the cell width Δx selected for the model, the smaller the time increment required to avoid such shoreline jumps.

Figure 3 provides a simple example of a simulation model where a jetty is blocking a littoral drift. This model is a direct application of the routine in Table 1, the jetty being present to the immediate right of cell 100 with a littoral drift arriving from the far left at a rate appropriate for the set wave conditions, $P = ECn = 10^8$ erg/cm·sec $= 10^3$ W/m and $\alpha_o = 15°$. Initially the shoreline is straight, corresponding to the x-axis ($y_i = 0$), and the model employs the increments $\Delta x = 25$ m and $\Delta t = 0.1$ day. Figure 3A shows cells 80 through 100 after 1 year of the littoral drift being blocked by the jetty (the program itself took just under 10 min to run). As expected, the cells have built outward to represent the accumulating sand, cell 100 adjacent to the jetty building out the most. Figure 3B shows the progressive accretion through the year, smooth curves having been fitted to the finite cells. The shorelines obtained in the model are seen to be very similar to those observed in the field where a new jetty blocks a littoral drift. Even with this simple model one could predict the rates of shoreline advance at various positions updrift from a proposed jetty. Perhaps of greater interest, a similar model could be developed for the down-drift direction to predict the resulting rates of erosion.

APPLICATIONS

One of the very first studies that involved the development of computer models of shoreline changes is that of Price et al.[14] Their investigation is of particular interest in that the model is on the scale of a wave basin rather than on a prototype scale, and the results from the computer model are compared with actual tests within a wave basin. The purpose of the study is to examine changes in a beach brought about by the construction of a groin or jetty blocking a littoral drift, similar to the model presented in the preceding section.

Figure 4 shows an example of the results obtained by Price et al.,[14] the waves approaching the initially straight beach at a 4° angle with a 4-cm offshore height. It is seen that the shoreline prediction from the computer model compares favorably with the actual tests in the wave basin. Being only a short stretch of beach, as expected the shoreline rotates around

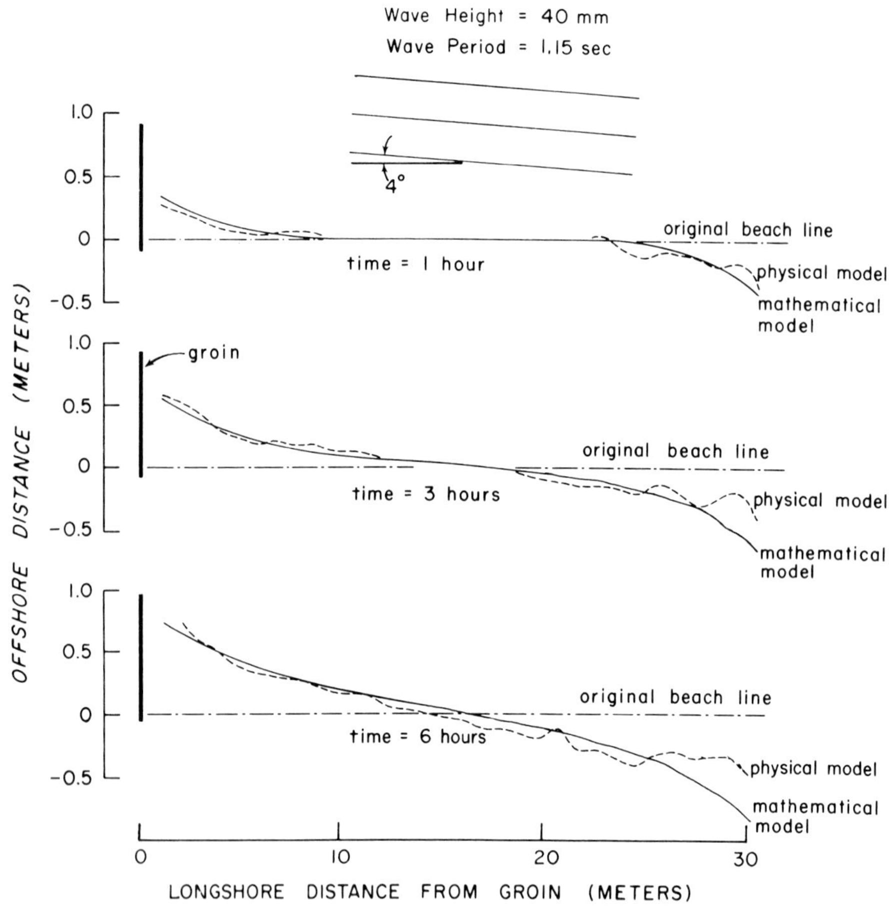

FIGURE 4. Computer simulation model of shoreline reorientation between two groins compared with the observed changes of a physical model in a laboratory wave basin. (From Price, W. A., Tomlinson, K. W., and Willis, D. H., *Proc. 13th Conf. Coastal Eng.*, American Society of Engineers, New York, 1973, 1321. With permission.)

so as to parallel the wave crests such that $\alpha_b \simeq 0$ and $Q_s \simeq 0$ at which time little subsequent change is experienced. This is shown by both the computer model and physical model.

Komar et al.[9] developed a model to examine the shoreline changes brought about by jetty construction at the mouth of the Siuslaw River on the Oregon coast. There is no net drift along this coast, so the shoreline changes did not result from a simple blockage of the drift as examined in the above models. Instead, the jetty construction produced embayments between the jetties and the pre-jetty shorelines which curved inward toward the river mouth (the 1889 shoreline of Figure 5A). Prior to jetty construction this curved shoreline was apparently in equilibrium with the ocean waves and river and tidal flow through the inlet. But jetty construction removed the contribution from the river and tidal flows, leaving only the waves which broke obliquely to the inward-curving shoreline and so tended to carry sand into the embayments. As seen in Figure 5A, the embayments progressively filled both to the north and south of the jetties, the process taking some 20 years for complete filling of the larger embayment to the north. The shoreline advancement continued only until it became parallel to the prevailing wave conditions at which point a new equilibrium had been achieved, one with a zero net drift where no further rearrangements occur. This was achieved by the mid-1930s, the present-day shoreline being basically the same.

Komar et al.[9] developed a computer model to simulate this filling of the large north

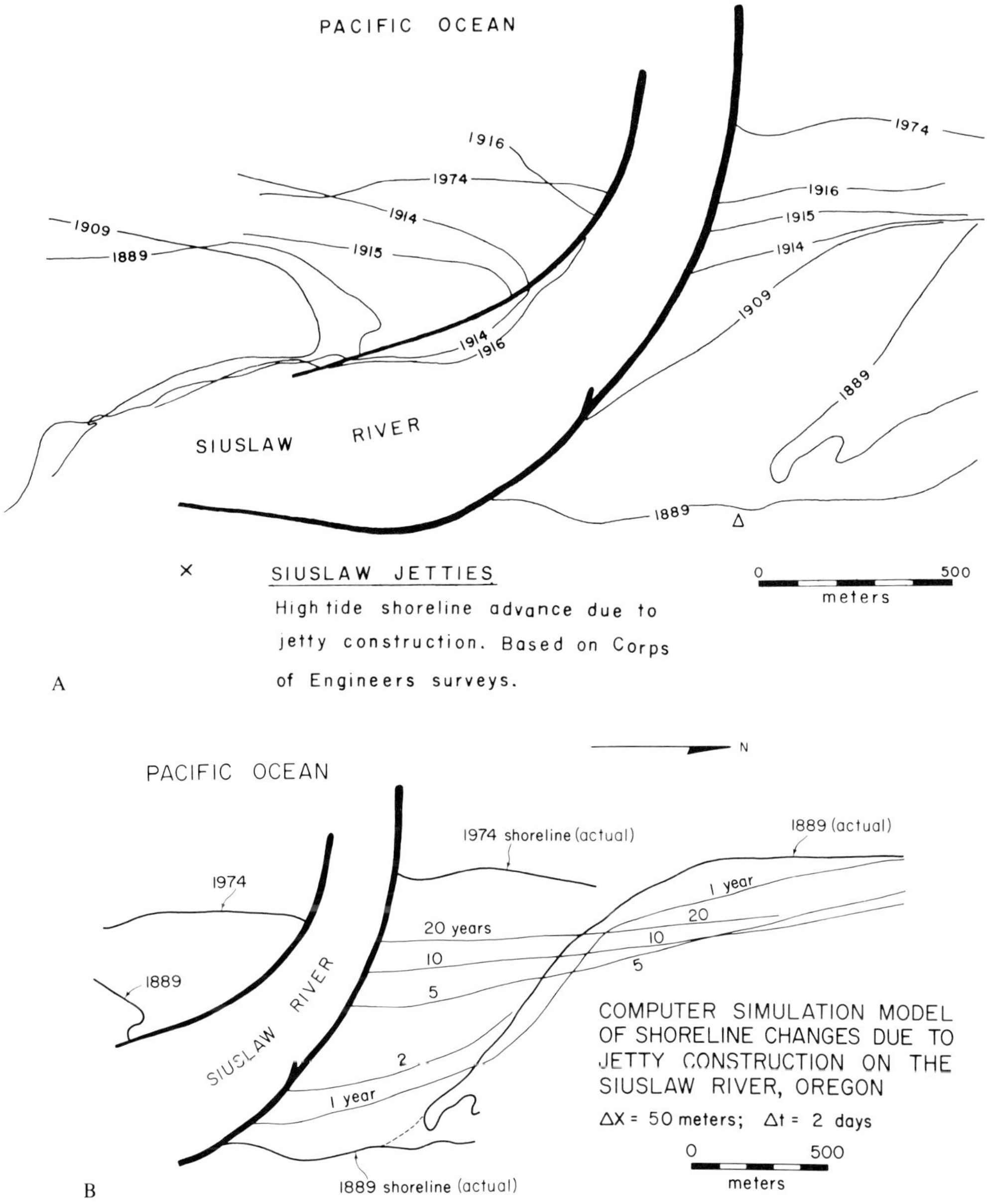

FIGURE 5. (A) Shoreline changes brought about by jetty construction at the entrance to the Siuslaw River, Ore. based on U.S. Army Corps of Engineers surveys obtained during jetty development. (B) A computer model of the shoreline changes, the actual 1889 shoreline being used as the starting configuration in the calculations. (From Komar, P. D., Lizarraga-Arciniega, J. R., and Terich, T. A., *J. Waterways Harbors Coastal Eng.*, 102(WW1), 13, 1976. With permission.)

embayment, and the results are shown in Figure 5B. The actual 1889 shoreline is used in the model as the initial shoreline prior to jetty construction. This shoreline was divided into a series of cells of uniform width, $\Delta x = 50$ m, and the model was run with the time increment of $\Delta t = 2$ days for a total period of 20 years. The waves were held constant with an energy flux that is representative of average wave heights and periods on the Oregon coast, and were made to approach parallel to the shoreline to the far right of the jetties beyond the effects of the river mouth, this yielding the desired zero net drift along that

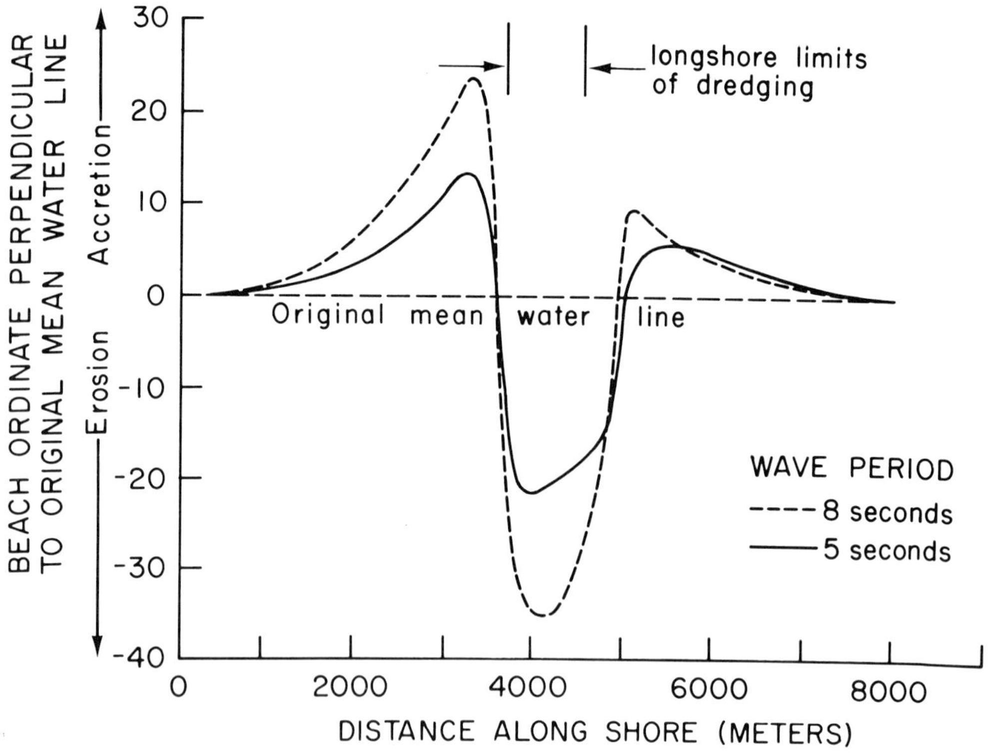

FIGURE 6. Patterns of shoreline erosion and deposition obtained with a computer model designed to simulate the effects of offshore dredging on the wave refraction which in turn cause the beach changes. (From Motyka, J. M. and Willis, D. H., *Proc. 14th Conf. Coastal Eng.*, American Society of Civil Engineers, New York, 1975, 615. With permission.)

portion of coast. The severest shortcoming of the model is the absence of considerations of wave refraction within the embayments so that breaker angles there would be larger than had actually occurred, resulting in higher sand transport rates.

It is seen in Figure 5B that the shoreline changes obtained with the computer model agree reasonably well with those that had actually occurred (Figure 5A). The main disagreement results from the actual jetty construction not having kept pace with the shoreline advance so that some sand bypassed the jetty and entered the estuary. Note that the model not only simulates deposition within the embayment next to the jetty, but also evaluates the beach erosion further from the jetty that is required to supply sand to the accreting area. There is even an intermediate zone which first erodes and then later accretes. Such predictions would have been useful at the time the jetties were actually constructed, especially if homes were located in this erosion area.

This model makes it possible to predict actual quantities of beach erosion and deposition at various positions along the coast resulting from jetty construction. This particular example, being a "hindcast" model, enabled a quick verification of the model by comparison with the actual shoreline changes that occurred; but it is apparent that such a modeling approach could be useful in predicting shoreline changes and rates of erosion for proposed jetties.

Quite another type of problem was examined by Motyka and Willis,[12] a study which also provides a conceptual advance in shoreline modeling in that they coupled their model with a computer scheme which computes wave refraction patterns. This considerably increases the complexity of the models in that it requires an offshore array to account for the bottom topography that affects the wave refraction. This approach was necessary in the study of

Motyka and Willis in that they were investigating the effects of offshore dredging on the wave refraction, which in turn alters the shoreline configuration, so wave refraction is a necessary component to their analysis. The results have application to offshore mining of sand and shingle, and answers such questions as how deep must the water be and how much material can be removed without causing shoreline erosion problems.

One example of the resulting models of Motyka and Willis[12] is shown in Figure 6. This model considers the effects of dredging a 4-m deep hole in a 7-m depth of water, 500 m offshore. The model used realistic profiles of the beach and offshore with a dredged hole inserted. The root-mean-square wave height was 0.4 m and periods of 5 and 8 sec were used. Wave directions were selected so as to yield a net longshore sediment transport of 30,000 m^3/year. The model performed the following sequence of operations:

1. Calculated breaking wave conditions by refracting the deep-water waves over the seabed
2. Calculated the longshore sand transport rates at each cell position along the shoreline
3. Calculated the resulting change in beach plan shape
4. Distributed accretion and erosion over the inshore bed
5. Recalculated the wave refraction and returned to step 2

The results shown in Figure 6 demonstrate that the dredge hole causes erosion of the shoreline in its lee and an advance to either side. The pattern is asymmetrical owing to the superimposed littoral drift and overall oblique wave approach. The effects are greater under the 8-sec waves than with the 5-sec waves in that longer-period waves undergo more refraction. The models are seen to predict major erosion resulting from the offshore dredging, a shoreline retreat of 20 to 35 m existing over a kilometer of shoreline length.

Additional investigations that involve the application of numerical models of shoreline changes are those of Perlin and Dean[13] further examining jetty effects, and those of Rea and Komar,[15] Komar,[7] and Dean[3] which investigate the shoreline shape within an embayment or the "logarithmic shape" downdrift of a headland (or jetty). The Perlin and Dean models[13] were developed to examine permeable vs. impermeable jetties, shoreline perturbations caused by jetties which are aligned with the incoming waves vs. jetties perpendicular to shore, as well as the general effects of a littoral barrier. Simplified refraction and diffraction schemes are incorporated into the models. By way of providing examples of simple numerical shoreline models, Komar[7] shows results for the shoreline shape within an embayment where the shoreline takes on the shape of the refracted-diffracted waves until an equilibrium of zero longshore drift is everywhere achieved. Dean[3] provides a more complete analysis including considerations of the wave diffraction patterns emanating from narrow openings between a series of closely spaced offshore breakwaters (such systems are sometimes used for coastal defense). Another interesting problem is the shoreline shape downdrift of a headland or jetty which is commonly described by a logarithmic spiral or ζ-curve.[16,17] Rea and Komar[15] developed a computer model to simulate the formation of such hooked shorelines, employing a combination of two sets of one-dimensional cell systems at right angles to each other so that the beach erosion can proceed in two directions. The resulting model shorelines were found to closely follow logarithmic spiral curves adjacent to the "headlands". Dean[3] also developed models for "headland beaches", and although not strictly a computer model, the study of LeBlond[11] also analyzed these shoreline shapes.

The above models involve shoreline changes brought about by a littoral drift alone, there being no extra sand sources or losses. To the author's knowledge, the only computer models that have included additional sources are those of Komar,[6,8] derived to examine the morphology of river deltas as a function of the supply of sand from the rivers and wave-energy level which acts to distribute that supply along the coastline. Models relevant to coastal erosion have not as yet incorporated the many potential sources and losses of sand, partly due to the common difficulties of evaluating these and because the models devised to date have been comparatively simple. In principal one could develop a computer model that incorporates all aspects of the budget of littoral sediments (Chapter 1), evaluating all the sand sources and sinks, any littoral drift, and then predicting how the shoreline will evolve through time. In such a model one could, for example, easily eliminate a river source due to dam construction in order to assess the resulting impact on the coastline.

REFERENCES

1. **Davidson-Arnott, R. G. D. and Pollard, W. H.,** Wave climate and potential longshore sediment transport patterns, Nottawasaga Bay, Ontario, *J. Great Lakes Res.*, 6(1), 54, 1980.
2. **Dean, R. G.,** Beach erosion: causes, processes, and remedial measures, *CRC Crit. Rev. Environ. Control*, 6(3), 259, 1976.
3. **Dean, R. G.,** Diffraction calculation of shoreline planforms, *Proc. 16th Coastal Eng. Conf.*, American Society of Civil Engineers, New York, 1979, 1903.
4. **Greenwood, B. and McGillivray, D. G.,** Modelling the impact of large structures upon littoral transport in the cnetral Toronto waterfront, Lake Ontario, Canada, *Z. Geomorph. N.F.*, 34, 97, 1980.
5. **Harbaugh, J. W. and Bonham-Carter, G.,** *Computer Stimulation in Geology*, John Wiley & Sons, New York, 1970.
6. **Komar, P. D.,** Computer models of delta growth due to sediment input from rivers and longshore transport, *Geol. Soc. Am. Bull.*, 84, 2217, 1973.
7. **Komar, P. D.,** *Beach Processes and Sedimentation*, Prentice-Hall, Englewood Cliffs, N.J., 1976.
8. **Komar, P. D.,** Modeling of sand transport on beaches and the resulting shoreline evolution, in *The Sea*, Goldberg, E., McCave, I., O'Brien, J., and Steele, J., John Wiley & Sons, New York, 1977, 499.
9. **Komar, P. D., Lizarraga-Arciniega, J. R., and Terich, T. A.,** Oregon coast shoreline changes due to jetties, *J. Waterways, Harbors Coastal Eng.*, 102(WW1), 13, 1976.
10. **LeBlond, P. H.,** On the formation of spiral beaches, *Proc. 13th Conf. Coastal Eng.*, American Society of Civil Engineers, New York, 1973, 1331.
11. **LeBlond, P. H.,** An explanation of the logarithmic spiral plan shape of headland-bay beaches, *J. Sediment. Petrol.*, 49, 1093, 1979.
12. **Motyka, J. M. and Willis, D. H.,** The effect of refraction over dredged holes, *Proc. 14th Conf. Coastal Eng.*, American Society of Civil Engineers, New York, 1975, 615.
13. **Perlin, M. and Dean, R. G.,** Prediction of beach planforms with littoral controls, *Proc. 16th Coastal Eng. Conf.*, American Society of Civil Engineers, New York, 1979, 1818.
14. **Price, W. A., Tomlinson, K. W., and Willis, D. H.,** Predicting the changes in the plan shape of beaches, *Proc. 13th Conf. Coastal Eng.*, American Society of Civil Engineers, New York, 1973, 1321.
15. **Rea, C. C. and Komar, P. D.,** Computer simulation models of a hooked beach shoreline configuration, *J. Sediment. Petrol.*, 45, 866, 1975.
16. **Silvester, R.,** Growth of crenulate shaped bays to equilibrium, *J. Waterways Harbors*, 275, 1970.
17. **Yasso, W. E.,** Plan geometry of headland-bay beaches, *J. Geol.*, 73, 702, 1965.

Chapter 11

PRINCIPLES OF BEACH NOURISHMENT

R. G. Dean

INTRODUCTION

In many cases, beach nourishment, the placement of substantial quantities of compatible sand to advance the shoreline seaward is a viable approach to mitigation of shoreline recession. Although guidelines are available for selecting some design parameters, improved procedures are required to enhance the capability to: (1) realistically estimate the design life of the nourishment project, (2) make quantitative trade-offs between alternate borrow source materials, (3) establish optimal allocation of available funding between sand and stabilization structures, and (4) predict the slope adjustment that will occur due to placement of nourishment material on a slope steeper than that of the equilibrium profile.

This chapter reviews available principles and introduces new principles for the design of beach nourishment projects.

NEEDS FOR AND GOALS OF BEACH NOURISHMENT

Many shorelines of the world are experiencing a trend of beach recession. This shoreline retreat is due to both natural causes and human interaction and can vary from stable shorelines to those retreating at 15 m/year or more. The primary cause of natural shoreline retreat is the relative rise in sea level which is on the order of 15 to 30 cm/century (Chapter 1). Human interactions with the shoreline can also cause beach erosion and include: (1) the modification of inlets to improve navigation, (2) the reduction in supply of sand to the shoreline through the construction of dams, (3) the construction of shoreline structures updrift of an area, and (4) the removal (mining) of sediment from the shoreline.

Beach nourishment provides two distinctly different benefits to an area. The enhancement of the recreational features of the beach are obvious and may contribute significantly to the income generated by tourism. Additionally, a wider beach provides a useful (but generally unquantified) degree of protection to upland structures against large waves occurring as a result of extreme storms.

The rationale for beach nourishment can be viewed as follows. Suppose Figure 1 presents the long-term variation of the shoreline position at a particular beach of interest. The recession trend is due to the aforementioned sea-level rise or to some additional human interaction erosion stresses. The comparatively short-term erosional events followed by a slower recovery are due to steep storm waves and the mild wave conditions which follow. The idealized effects of a well-designed beach nourishment project are shown as the dotted line. The nourishment event displaces the beach seaward by a particular amount, in effect setting the system "back in time", assuming that the nourished beach behaves as the natural beach did at the earlier time. It will be shown later that nourished beaches may erode more rapidly than the natural beach due to losses at the ends of the fill segment or to losses of the fine fraction of the fill material.

SEDIMENT TRANSPORT PROCESSES IN BEACH NOURISHMENT PROJECTS

Sediment is lost from the original location of a beach nourishment project by several mechanisms and pathways. An understanding of these mechanisms is important to the design and maintenance of a beach nourishment project.

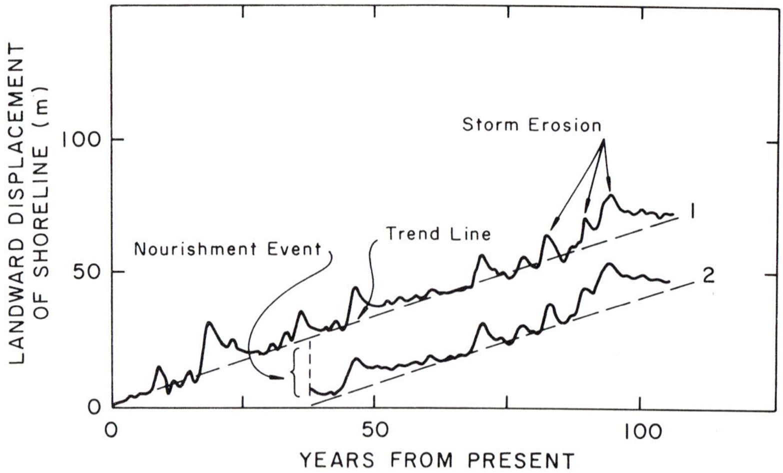

FIGURE 1. Erosion characteristics with and without nourishment event. Curve 1 — without nourishment. Curve 2 — with nourishment.

Consider the beach shown in Figure 2. A length, ℓ, of this beach is nourished resulting in a seaward extension of the shoreline and an increase in the sand elevation relative to the original contours. Sand can be transported from the nourished area by several pathways.

If the placed sand contains a significant fraction of fine material, then this material will be placed into suspension and eventually be carried seaward to a location where the wave agitation level is sufficiently low that the material is stable. Many early beach nourishment projects obtained material from inside bays which contained a substantial percentage of very fine material and the performance of the projects was understandably poor. As a rule of thumb, if the fall time of suspended material is greater than the wave period, T, then that material will probably be carried seaward beyond the surf zone. Considering the material to be suspended to some fraction, β, of the depth, h, this criterion would yield

$$w < \beta \frac{h}{T} \tag{1}$$

in which w is the fall velocity. For a suspension height βh of 10 cm and a wave period of 10 sec, a sediment with a fall velocity of less than 1 cm/sec would be lost from the surf zone. This corresponds to a diameter of 0.1 mm which is usually the approximate lower size limit of material found in the surf zone.

A second pathway whereby sand can be transported seaward from the nourished area is simply by a profile adjustment which occurs following the placement of the nourished material on a slope which is steeper than the equilibrium. Although the profile adjustment results in a shoreline recession, the material is still present offshore of the placed zone.

Sand is also lost from the ends of the project area. Basically the nourishment represents a perturbation to the shoreline which is thus out of equilibrium and the forces tend to restore equilibrium by spreading the sand in the alongshore direction to a condition characterized by straight and parallel contours. The simplified equation governing this process is the diffusion equation and thus the spreading of sediment in the alongshore direction is directly analogous to the diffusion of heat from a warmer segment of a bar which is insulated such that heat flux can occur only along the axis of the bar. Since this effect is due to sand transfer from the ends of the nourished area, it follows that short beach nourishment projects would lose a greater percentage of placed sand much more rapidly than long projects. This effect will be explored quantitatively later.

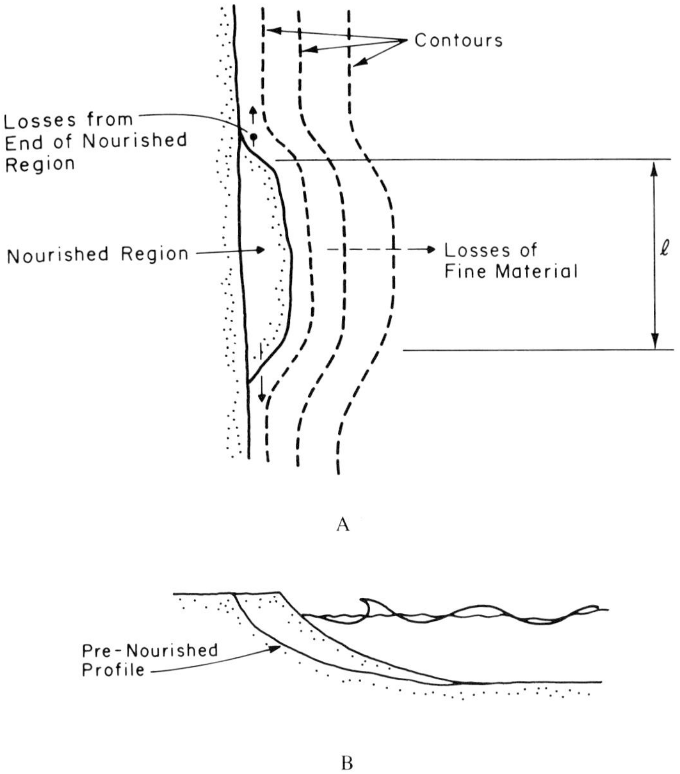

FIGURE 2. Geometric characteristics of nourished region and sediment loss pathways: (A) plan view of nourished area; (B) elevation view.

The Equilibrium Beach Profile

The equilibrium beach profile is an idealization of conditions which occur in nature for particular sediment characteristics and steady wave conditions. Thus, although beaches in nature may never achieve equilibrium and are continually responding to changes in wave conditions, the concept of an equilibrium beach profile is generally useful in problems and interpretations involving beaches and, in particular, in the design of beach nourishment problems.

Dean[1] analyzed over 500 beach profiles ranging from Long Island, N.Y. to the Texas-Mexico border and found that the profiles could be described by

$$h(x) = Ax^m \tag{2}$$

in which x is the distance offshore to a water depth h, A is a scale parameter which depends primarily on sediment size, and m is a shape parameter. The best-fit value of m was found to be 2/3 and this was shown to be compatible with and was interpreted as due to uniform wave energy dissipation by spilling breakers across the surf zone. The rationale for this interpretation is that the turbulence due to wave breaking is a mobilizing (destabilizing) agent and that a sediment of particular characteristics will be stable in the presence of a certain level of mobilizing forces.

Moore[4] analyzed a number of published beach profiles comprising sediments over a wide range of sizes and found the variation of A vs. D as presented in Figure 3. It is noted that the dimension of the scale parameter, A, is $(\text{length})^{1/3}$ and that Figure 3 presents A in metric dimensions.

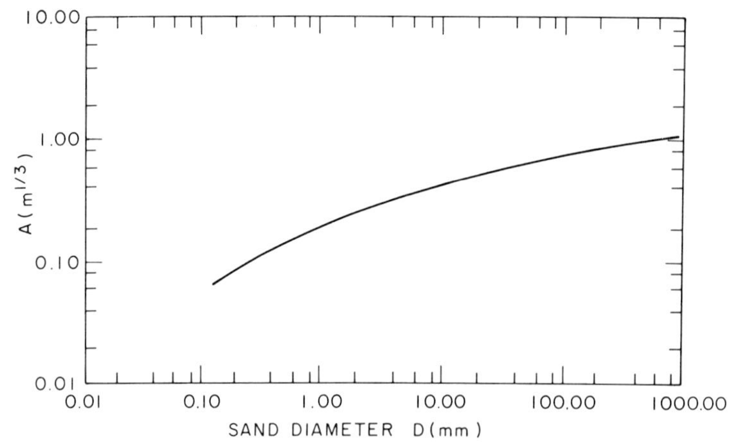

FIGURE 3. Variation of scale parameter, A, with sand diameter, D. (From Moore, B., M.Sc. thesis, University of Delaware, Newark, 1982. With permission.)

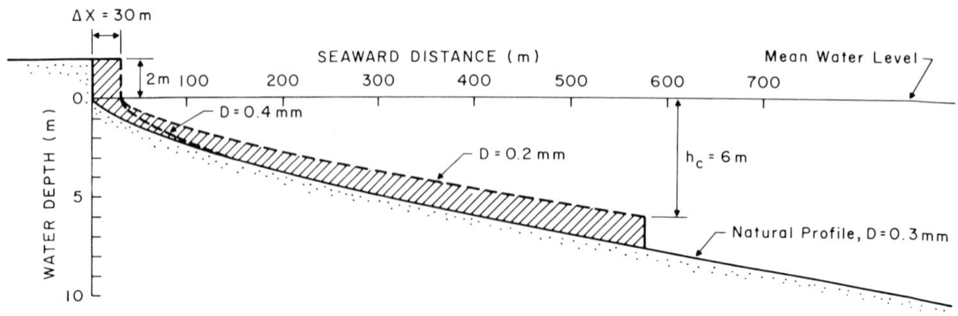

FIGURE 4. Characteristics of natural and two filled profiles.

Example — The concept of equilibrium beach profiles will be utilized and/or referenced a number of times in the present chapter. A simple example will serve to illustrate the relevance of equilibrium beach profiles to beach nourishment projects. Suppose that the sizes and associated scale factors of the native (N) and two alternate placed (P) sands are

Native sand:	$D_N = 0.3$ mm,	$A_N = 0.11\ m^{1/3}$
Placed sand:	$D_{P_1} = 0.2$ mm,	$A_{P_1} = 0.09\ m^{1/3}$
	$D_{P_2} = 0.4$ mm,	$A_{P_2} = 0.13\ m^{1/3}$
	Depth of closure (limiting depth of no motion):	$h_c = 6$ m
	Berm height:	$B = 2$ m

Suppose also that it is desired to advance the shoreline seaward a distance $\Delta x = 30$ m. The profiles as calculated from Equation 2 are presented in Figure 4.

The distances x_c to the depth of closure, h_c, for profiles composed of the three sand characteristics are given by

$$x_c = \left(\frac{h_c}{A}\right)^{3/2} \tag{3}$$

Table 1
EXAMPLE OF CHARACTERISTICS OF BEACH FILLS FOR TWO DIFFERENT FILL DIAMETERS AND SAME REQUIRED SHORELINE ADVANCEMENT

Sand	Diameter (mm)	Distance (m) to 6 m contour	Required volume per unit length of beach (m^3/m)
Native	0.3	403	—
Nourishment	0.2	544	719
Nourishment	0.4	314	112

which are presented in column 3 of Table 1. Recalling that the x-origins for the profiles are at their respective waterlines, then

$$x_{c_2} < x_{c_N} + 30 \text{ m} \tag{4}$$

which indicates that the profile composed of the 0.4-mm sand intersects the natural profile at a shallower depth than 6 m. The distance, x_*, to this depth, h_*, is found by equating the depths for the two profiles

$$h_* = A_N x_*^{2/3} = A_{P_2}(x_* - 30)^{2/3} \tag{5}$$

or

$$x_* = 135.4 \text{ m}$$

corresponding to

$$h_* = 2.9 \text{ m}$$

The volume of fill, $\forall_{p1}$, for the 0.2-mm sand is determined from

$$\forall_{P1} = B\,\Delta x + \int_0^{x_{P1}+\Delta x} A_N x^{2/3} dx - \int_0^{x_{P1}} A_{P_1} x^{2/3} dx \tag{6}$$

$$= B\,\Delta x + \frac{3}{5}\left[A_N(x_{P_1} + \Delta x)^{5/3} - A_{P_1} x_{P_1}^{5/3}\right] \tag{7}$$

$$= 719 \text{ m}^3/\text{m}$$

The corresponding required fill, $\forall_{p2}$, for the 0.4-mm sand is

$$\forall_{p2} = B\,\Delta x + \int_0^{x_*} A_N x^{2/3} dx - \int_0^{x_* - \Delta x} A_{P_2} x^{2/3} dx \tag{8}$$

$$= 112 \text{ m}^3/\text{m}$$

The differences in required volumes for the same shoreline advancement are quite significant. The results of this example are summarized in Table 1.

DESIGN PRINCIPLES

Effect of Fill Length

As noted previously, an unstabilized beach nourishment project will experience losses from the ends of the project that are due to alongshore sediment transport to the neighboring beaches (Figure 2). In effect, the nourished area acts as a "feeder beach" to the adjacent shoreline segments. It is possible to evaluate the effect of both representative wave height, H, and beach fill length, ℓ, on the time, t_p for p percent of the original volume to be transported out of the nourished region. Pelnard-Considére[5] showed that the following equation could be used to approximate the planform response of a beach system

$$\frac{\partial y}{\partial t} = K \frac{\partial^2 y}{\partial x^2} \tag{9}$$

in which t is time, x and y are presented in Figure 2, and K may be considered as a rate or activity constant, expressed in terms of the breaking wave height, H_b, as

$$K = G\, H_b^{5/2} \tag{10}$$

where G is a factor incorporating beach and sediment characteristics. Le Mehaute and Soldate[2] have presented the solution to Equation 9, attributed to Noda, for the evolution of a rectangular beach fill of length ℓ, and seaward displacement, Y, as

$$y(x,t) = Y \left\{ \text{erf}\left[\frac{\ell}{4\sqrt{Kt}}\left(\frac{2x}{\ell} + 1\right)\right] - \text{erf}\left[\frac{\ell}{4\sqrt{Kt}}\left(\frac{2x}{\ell} - 1\right)\right] \right\} \tag{11}$$

where "erf" represents the error function.

From the above, it can be seen that the time required, t_p, for a percentage, p, of the original volume to be lost is

$$(t_p)_2 = (t_p)_1 \frac{\ell_2^2\, K_1}{\ell_1^2\, K_2} \tag{12}$$

in which the subscripts (1) and (2) refer to conditions for two particular projects.

The effect of fill length on fill life is obtained as a square law relationship from Equation 12. Thus, for example, if a beach fill of a particular length loses one half of the volume from the nourished area in 10 years, a fill of twice the length and the same width under the same wave action would lose one half the volume from the filled region in 40 years. The explanation for the square law relationship is as follows. Examination of the derivation of Equation 9 will demonstrate that the net transport rate of sand from the nourished region is due to the slope (in planview), $\partial y/\partial x$, of the planform. Doubling the fill length reduces the slope by a factor of one half. Secondly, the double-length fill simply has twice the volume which accounts for the other factor of one half.

The relationship of effective wave height to time required for a percentage, p, of the fill volume to be transported from the nourished area is

$$\frac{(t_p)_2}{(t_p)_1} = \frac{H_1^{5/2}}{H_2^{5/2}} \tag{13}$$

Thus, the effect of doubling the wave height is a reduction in the life of the nourishment to 17.7%.

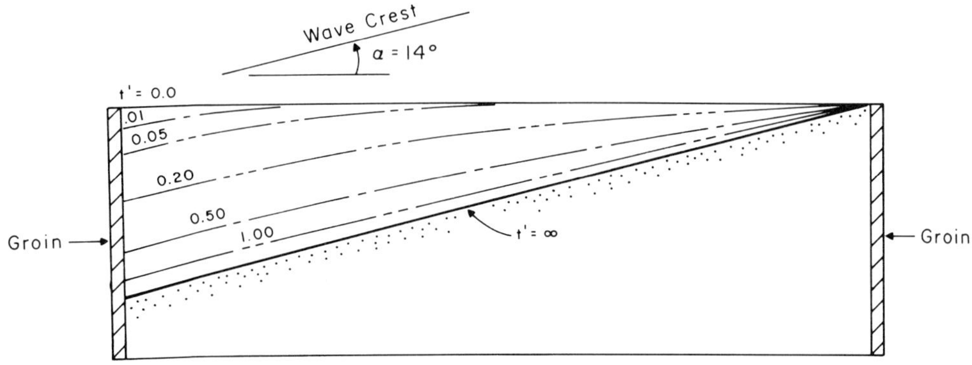

FIGURE 5. Evolution of an initially filled compartment between two groins.

In general, Equations 12 and 13 should be useful in transferring experience from an existing beach nourishment project to a contemplated project of different fill length and wave climate.

Effects of Structures

Terminal Structures and Groins — Structures placed near the ends of a beach nourishment project can be especially effective in reducing losses of nourished material. The simplest case, and that which will be examined here, is for an initially filled groin compartment of length, ℓ, and groin length, w, subjected to the action of unidirectional waves arriving at an angle α to the shoreline (Figure 5). It can be shown that, neglecting wave diffraction, the solution of Equation 9 with the appropriate boundary conditions, is

$$y = w - \ell\left(1 - \frac{x}{\ell}\right)\tan\alpha + \frac{2\tan\alpha}{\ell}\sum_{n=0}^{\infty} \left[\frac{2\ell}{(2n+1)\pi}\right]^2 \exp\left\{-K\frac{(2n+1)^2\pi^2 t}{4\ell^2}\right\}\cos\left[\frac{(2n+1)\pi x}{2\ell}\right] \quad (14)$$

Examination of Equation 14 will show that the shoreline in the vicinity of the structures approaches a straight line parallel to the incoming wave crests with the downdrift end of the shoreline in the groin compartment coincident with the downdrift groin. Figure 5 presents the evolution of the shoreline within a nourished compartment bounded by two terminal structures from an initially full condition to an equilibrium condition. The specific conditions of the example are $\ell/w = 3$ and $\tan\alpha = 1/4$ (or $\alpha = 14°$). It is seen that, ultimately, the approximate percentage loss of sand from the groin compartment is

$$\%\ \text{loss} = \frac{1}{2}\frac{\ell}{w}\tan\alpha \quad (15)$$

where the above does not account for losses due to onshore/offshore transport. From Equation 15 it is seen that the expected percentage losses would be reduced substantially if a groin field were installed. A practical matter to be considered near the ends of a stabilized fill is the possible perturbation caused to adjacent shoreline segments. To alleviate any such effect, it is desirable to either taper a stabilized fill including, of course, the length of the stabilizing structures and/or to place sand outside the limits of the nourished region.

The Perched Beach — The effects of shore perpendicular structures (groins and terminal

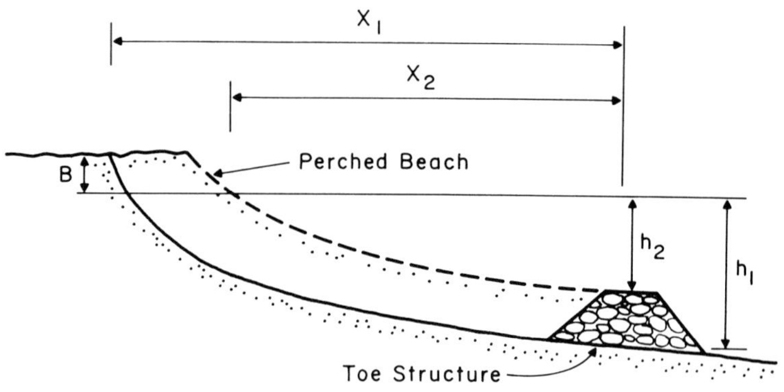

FIGURE 6. Definition sketch for perched beach.

structures) in retaining sediment have been noted. Designs have been developed in the past to utilize submerged shore-parallel structures to limit the seaward transport of sediment; this "perched beach" concept is illustrated in Figure 6. Although no examples are available of a perched beach which has been installed, the approach does appear to offer substantial advantages, especially for cases in which the placed material is finer than the native beach sand.

Example — To illustrate the reduction in sand volume that could be obtained through the use of a perched beach, consider the following:

Native beach: $D_N = 0.3$ mm, $A_N = 0.11\ m^{1/3}$
Nourished beach: $D_P = 0.2$ mm, $A_p = 0.09\ m^{1/3}$
Desired seaward advancement, Δx, = 30 m
Depth to which the nourished profile would be configured in the absence of a structure = 6 m
B = berm elevation = 2 m
$h_1 = 4$ m

It is desired to determine the height of the structure, $h_1 - h_2$. It will be assumed that the beach profiles can be represented by the equilibrium form

$$h = Ax^{2/3}$$

The volume of sand required to advance the shoreline seaward by 30 m without the structure present has been computed in a previous example (Equation 7)

$$\forall \text{ w/o} = 719 \text{ m}^3/\text{m}$$

With the structure present in a water depth of 4 m, the distance x_1 is

$$x_1 = \left(\frac{4}{0.11}\right)^{3/2} = 219 \text{ m}$$

and the depth, h_2, for the filled material and a shoreline advancement of 30 m is

$$h_2 = 0.09(219 - 30)^{3/2} = 2.96 \text{ m}$$

Thus the height of the structure must be

$$h_1 - h_2 = 4 - 2.96 = 1.04 \simeq 1 \text{ m}$$

The required fill volume with the structure present, $\forall_w$, is

$$\begin{aligned} \forall_w &= B\Delta x + \int_0^{x_1} A_N x^{2/3} dx - \int_0^{x_2} A_p x^{2/3} dx \\ &= B\Delta x + \frac{3}{5} [A_N x_1^{5/3} - A_p x_2^{5/3}] \\ &= 249 \text{ m}^3/\text{m} \end{aligned}$$

or approximately 35% of the volume required if the structure were not present. Factors which have not been discussed here but which would be important in design would be scour about the base of the structure and the possibility of fine material being lost over the crest of the structure as a result of storm waves which tend to cause a milder slope.

Slope Adjustment

It is important to anticipate and to convey to the public the slope adjustments and associated retreat in mean water-line position that will occur following the placement of a beach fill. As noted previously, the degree of slope adjustment depends on the native and placed material characteristics and the configuring wave height. Procedures for establishing the shoreline recession due to slope adjustment are presented in Maurmeyer and Dean.[3]

Proximity to Inlets

The location of a nourishment project in proximity to an inlet induces additional erosional stresses, primarily due to the flood tidal currents which, combined with the mobilization by waves, transport sediment into the inlet where it is deposited on the flood tidal shoals or from which it is jetted to the ebb tidal shoals. In every case where erosion adjacent to an inlet has occurred to the extent that nourishment is being considered, or perhaps where establishment of an artificial beach is desired, stabilization structures will undoubtedly be cost-effective. The optimal type of structure is not clear. In some cases, a single terminal structure may provide adequate stabilization, whereas in others a field of groins may be more effective.

Maintenance Nourishment Requirements and Sea Level Rise

A comprehensive attempt at estimating loss rates from a nourished area should include the winnowing of very fine material, slope adjustments, and losses from the ends of the fill including the effects of structures if present. As discussed earlier, one approach to estimating loss rates is the transferral of performance data from an existing beach fill in accordance with Equation 12. Probably the most effective approach would combine such information with numerical modeling in which the sediment transport and continuity equations are solved including the effects of structures, if present (Chapter 10). Measured wave characteristics would be employed, if available, to drive the sediment transport equations. If wave measurements are not available, hindcast waves or wave characteristics which are judged to be representative may be suitable.

As noted earlier, if a nourishment project is very long, the "end losses" will be relatively small. In view of this and the many qualitative conjectures of geologists and engineers as to the maintenance requirements due to sea level rise, it may be worthwhile to demonstrate a method for calculating this effect and to explore the effect quantitatively by a realistic example. Consider a nourished location with closure depth, h_c, and sediment characterized

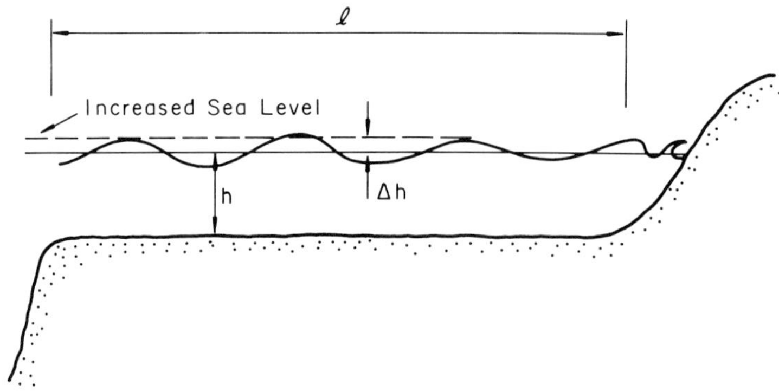

FIGURE 7. Idealized shelf characteristics.

by a scale parameter, A (Equation 2). If the shoreline is to be maintained in the same location, the volume of sediment, ΔV, per unit length of beach, to compensate for a sea level rise, ΔS, is

$$\Delta V = W\,\Delta S \tag{16}$$

in which W is the width of the active profile.

Example — An example will serve to quantify ΔV for a reasonable set of conditions. Consider a sediment size, D, of 0.3 mm and a closure depth, h_c, of 10 m. The rate of sea-level rise will be taken as 0.15 m per century. The value of the scale parameter, A, is 0.11 $m^{1/3}$ associated with the sediment size, D = 0.3 mm. The width W associated with a depth of 10 m is determined from Equation 2.

$$W = \left(\frac{h_c}{A}\right)^{3/2} = \left(\frac{10}{0.11}\right)^{1.5} = 867\ \text{m} \tag{17}$$

The *annual* beach nourishment requirement for an annual sea level rise of 0.0015 m is

$$\Delta V = (867)\,(0.0015\ \text{m}) = 1.3\ \text{m}^3/\text{m}$$

At present, the cost per cubic meter of sand in place on the beach is on the order of $6.00. Thus the annual costs per kilometer of beach would be approximately $7800.00. This cost is probably small compared to the protection provided if valuable upland structures are present.

Effect of Sea-Level Rise in Causing Accelerating Nourishment Maintenance Requirements

It has been suggested that the effect of sea-level rise will be to cause a continual deepening of the continental shelf beyond the depth of closure and a concomitant increase in effective wave height due to the reduction in bottom friction due to the greater depths. Although it is more probable that some sedimentation in the offshore region will occur as a result of deposition of fine sediments, it may be useful to estimate the rate of increase of beach nourishment maintenance due to sea-level rise assuming that no offshore deposition has occurred.

The idealized shelf considered is shown in Figure 7. The details are somewhat involved and the method will be presented here without derivation.

It can be shown that an increase in the nearshore wave height, H_2, due to an increase in water depth, Δh, is given by

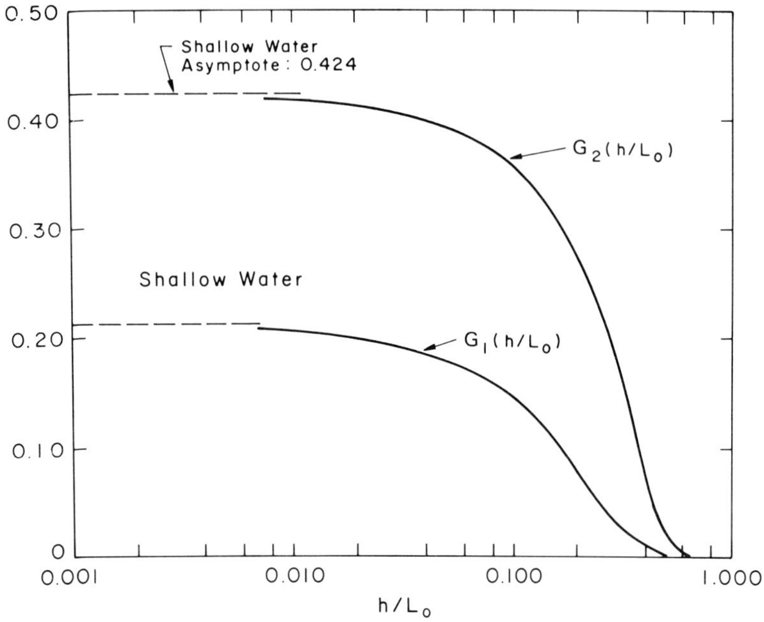

FIGURE 8. Variation of functions G_1 and G_2 with h/L_o.

$$\frac{\Delta H_2}{H_2} = \frac{H_1/h\ W/h}{\left[1 + C_f \frac{H_1}{h}\frac{W}{h} G_1(h/L_o)\right]} C_f G_2(h/L_o) \frac{\Delta h}{h} \tag{18}$$

in which C_f is the bottom shear stress coefficient and G_1 and G_2 are functions presented in Figure 8. Based on Equation 13, we can argue that the relative increase in erosion potential, Γ, (at least due to longshore sediment transport) is

$$\frac{\Delta\Gamma}{\Gamma} = \frac{5}{2}\frac{\Delta H}{H} \tag{19}$$

Example — To illustrate the application of Equations 18 and 19 and Figure 8, consider the following example:

$H_1 = 2$ m　　　$h = 10$ m

$T = 8$ sec　　　$C_f = 0.01$

$W = 60$ km $= 60{,}000$ m

$\Delta t =$ time span $= 10$ years　　　$\Delta h = 0.0015$ m/year $\times$ 10 $= 0.015$ m

$\frac{h}{L_o} = \frac{10}{(1.56)\ (8)^2} = 0.100$　　　$G_1(0.10) = 0.25$

$G_2(0.10) = 0.36$

$$\frac{\Delta H_2}{H_2} = \frac{(2/10)\,(60{,}000/10)}{\left[1 + 0.01\left(\frac{2}{10}\right)\left(\frac{60{,}000}{10}\right)(0.25)\right]}(0.01)\,(0.36)\left(\frac{0.015}{10}\right)$$

$$= \frac{1200}{4}(0.01)\,(0.36)\left(\frac{0.015}{10}\right) = 0.00162$$

Therefore, from Equation 19, the increase in erosion potential, Γ is

$$\frac{\Delta\Gamma}{\Gamma} = \frac{5}{2}(0.00162) = 0.00405$$

i.e., in a 10-year period, the rise in sea level would result in a 0.4% *increase* in the *erosion potential*. Thus, if a loss rate of 10,000 m^3/year were occurring without sea level rise, over a period of 10 years this loss rate would increase to 10,041 m^3/year, a comparatively small amount.

Effects of Compressibility of Marsh Sediments Underlying Barrier Islands on the Viability of Stabilization

It is well known that former marsh deposits underlie many barrier islands and that these barrier islands retreat reasonably rapidly. Dr. Rudy Slingerland of Pennsylvania State University has found that these marsh deposits consolidate greatly under loading. In fact, spoil piles of sand placed on present marshes to an original height of 2 m have compressed the underlying marsh to such an extent that, after a decade, the upper surface of the spoil pile is below that of the adjacent marsh. The possibility of wind ablation of the spoil sand exists, but in Slingerland's studies, this possibility has been negated by cores taken through the original spoil piles which confirmed negligible change in vertical dimension of sand.

Translating the scenario above to a barrier island retreating over a marsh surface, it is seen that the consolidation of the marsh and the associated decrease in elevation of the barrier island results in the increased vulnerability of the barrier island to overwash and an increased retreat rate due to the compressibility of the marsh deposits. Considering now the effect of stabilization by beach nourishment, it is seen, on a qualitative basis, that if a barrier island were stabilized in one location, the consolidation would approach its asymptotic limit and the nourishment required for future stabilization of the barrier island would diminish. Thus the retreat rate of a barrier island under natural conditions would provide an unrealistically pessimistic view of the maintenance requirements to stabilize the island. The quantitative effect of the compressibility of the marsh deposits over which a barrier island retreats is not known; however, it is believed that the relationship to beach nourishment considerations is quite significant.

EXAMPLES OF BEACH NOURISHMENT PROJECTS

With the complexities of transport of a placed volume of sand that may not be compatible with the native material and the possibility of a highly variable wave and current climate, the difficulties of developing accurate predictions of maintenance nourishment requirements are not surprising. In some cases, the best basis may be in transferring performance data from a project of similar size, sediment characteristics, and wave climate to the project of interest. Some of the principles presented in this chapter may aid in the transferral. A fairly large number of beach nourishment projects have been installed over a sufficiently long time to provide quantitative information that may be sufficiently similar to projects under consideration or may simply provide qualitative information relative to maintenance require-

Table 2
SUMMARY OF CHARACTERISTICS OF BEACH NOURISHMENT PROJECTS INCLUDING LOSS RATES WHERE AVAILABLE

Beach nourishment project	Date	Quantity $\times 10^{-6}$ yd^3	ℓ (miles)	Size (mm) Native	Size (mm) Borrow	Losses quantity (yd^3)	Duration losses measured	Cost ($)
1. Cape Canaveral Beach[a]	6/74—3/75	2.3	2.1	0.7—1.0	0.06—1.0	269,000 (11.7%)	12 mo	?
2. Fort Pierce-St. Lucie[b]	5/70—5/71	0.72	1.3	0.10—0.25	?	?	?	580,000
3. Jupiter Island	6/73—8/73 6/74—8/74	2.4 1.0	5.0	0.39—0.81	0.11—0.12	Negligible	?	3,000,000
4. Delray Beach	6/73—7/73	1.6	2.8	0.32—0.50	0.21—0.31	Negligible	?	2,074,000
5. Pompano Beach	5/70—10/70	1.03	3.2	0.35—1.50	?	?	?	1,873,437
6. Hallandale Beach	7/71—9/71	0.36	0.8	?	0.2—0.46	?	?	524,270
7. Virginia Key[c]	1969	0.18	1.3	≈0.35	0.27—0.47	61,000 (35%)	14 mo	544,000
	1973	0.10	1.3					1,450,000
8. Key Biscayne[d]	1969	0.20	1.2	≈0.35	≈0.27—0.47	≈0	14 mo	?
9. Key West[e]	3/59—4/60	0.067	0.6	0.07—1.0	0.1—1.0	25,000 (37%)	6 yr	285,000
10. Captiva Island	7/81—10/81	0.75	1.9	0.2	0.3	69,000		2,400,000
11. Lido Key Beach		0.35	0.7	0.47	0.14		?	333,000
12. Treasure Island	July 1969	0.76	1.7	0.20	0.17	0.20	14 mo 23% 21 mo 32%	700,000

[a] Downdrift of entrance to Cape Canaveral harbor.
[b] Downdrift of entrance to Fort Pierce Inlet.
[c] Next to inlet comparing profile only 4 to 5 ft.
[d] Next to inlet comparing profile only 4 to 5 ft.
[e] 4 Groins.

ments. Walton[6] has summarized (Table 2) the characteristics of a number of beach nourishment projects. These projects are discussed below.

Loss rate information is provided for only 7 of the 12 projects presented in Table 2. The annual losses range from a descriptive low of "negligible" to a high of 29% for Virginia Key, which is in close proximity to an inlet which induces rather strong alongshore currents. Moreover, in this case, the loss rates were based on profile comparisons only out to the 4- to 5-ft depth. Very preliminary efforts to correlate annual loss rates with beach nourishment parameters did not demonstrate any strong relations with beach fill parameters. Information which can be derived from Table 2 is that most projects on the order of 1 mi in length in a location where the wave climate is fairly mild, as in Florida, will experience annual loss rates on the order of 10% or less. Qualitatively, we know that the coarser the material and the greater the length of beach fill, the smaller will be the annual beach nourishment requirements. Secondly, it appears that much more and better information is required to evaluate the present methodologies for designing the characteristics of beach nourishment projects and the predictions of maintenance requirements. In fact, at present the methods for predicting maintenance nourishment requirements may be better than the general data base for evaluating those methods. This is unfortunate, since there are generally many critics of a particular project, thereby adding to its costs and delays.

At present, beach nourishment experience has been proven to be a technically viable approach to mitigating beach erosion. The question which must be answered on a case-by-case basis is whether the project costs are justified by the benefits to the public and private sponsors. Much more effective measurement programs at completed projects will be required to answer this question adequately. One consequence of beach nourishment is certain. If the sediment is coarse and durable, this material will remain within and thus benefit the general nearshore system. The difficult question lies in predicting the duration that the sediment will remain in the particular beach system providing recreational benefits and storm protection to the upland property of those individuals, municipalities, etc. that sponsored the project.

SUMMARY

In striving for an appropriate design of a beach nourishment project, it is important to recognize that since each shoreline section is unique in terms of the beach materials present and the affecting erosional force, the design that evolves will also be unique. In some cases, placement of sand alone may represent the best design; in others stabilization by structures may be necessary. As general principles:

1. Longer beach nourishment projects will experience substantially lower percentage volumetric loss rates due to alongshore transport than shorter projects. Thus it is effective to develop projects which extend over at least several miles, perhaps involving the unified actions of several communities. The formation of "erosion control districts" extending between significant littoral controls would be ideal.
2. Especially in cases where long beach nourishment projects are not feasible, the allocation of available funds to material placed on the beach and to stabilizing structures should be carefully evaluated. Stabilization structures could comprise either terminal structures or a groin field. Terminal structures would be most effective in cases where the dominant waves approach the shoreline with small obliquity. If the waves approach the beach at a high angle, a field of low-profile groins may be more effective. A groin field may also be desirable if the project is located downdrift of a littoral barrier such as an inlet which completely blocks the sediment transport.

REFERENCES

1. **Dean, R. G.,** *Equilibrium Beach Profiles; U.S. Atlantic and Gulf Coasts,* Ocean Eng. Tech. Rep. No. 12, University of Delaware, Newark, 1977.
2. **Le Mehaute, B. and Soldate, M.,** Mathematical modeling of shoreline evolution, *U.S. Army Coastal Eng. Res. Cent. Misc. Rep.,* No. 77-10, 1977.
3. **Maurmeyer, E. M. and Dean, R. G.,** Sediment and overwash considerations in the design of beach nourishment projects, *Proc. Coastal Zone '80,* American Society of Civil Engineers, New York, 1980, 2978.
4. **Moore, B.,** Beach Profile Evolution in Response to Changes in Water Level and Wave Height, Master of Science thesis, Department of Civil Engineering, University of Delaware, Newark, 1982.
5. **Pelnard-Considére, R.,** Essai de Theorie de l'Evolution des Formes de Rivage en Plages de Sable et de Galets, 4th Journees de l'Hydraulique, Les Energies de la Mer, Question III, Rapport No. 2, 1956.
6. **Walton, T. L.,** *Beach Nourishments in Florida and on The Lower Atlantic and Gulf Coasts,* Tech. Rep. No. 2, Coastal Oceanographic Engineering Department, University of Florida, Gainesville, 1977.

Chapter 12

PROCESSES OF SEA CLIFF AND PLATFORM EROSION

Tsuguo Sunamura

INTRODUCTION

Coastal cliff recession is one of the significant geomorphic events occurring at the interface of the lithosphere, hydrosphere, and atmosphere. In an attempt to understand cliff recession processes, geologists, geomorphologists, and physical geographers have studied this phenomenon from purely scientific interest. Recently, due to increased use of the coast, considerable attention has been focused on the erosion of sea cliffs in connection with coastal zone management and engineering problems. From the viewpoints of not only pure but also applied sciences, the elucidation of erosional mechanisms is of basic importance. Such studies also give a promising clue for understanding the formative processes of shore platforms which are primarily created by the retreat of sea cliffs. Applications of cliff erosion research are possible to the problems of (1) development of continental shelves with erosional origin,[139] and (2) formation of uplifted marine terraces, if wider spatial and longer temporal scales are considered.

When waves erode a cliff base, the cliff becomes unstable due to increases in the slope angle and in the slope stress caused by erosion at the toe. This instability induces mass movement of various types: (1) slumps including rotational and planar slidings (e.g., McLean and Davidson,[98] Hutchinson,[68] Clark et al.,[24] McGreal,[94] Brunsden and Jones,[17] Quigley et al.,[113] Quigley and Di Nardo[111]), (2) rock or debris slides (e.g., McLean and Davidson[98]), (3) mudslides (e.g., Prior and Renwick[109]) (4) mudflows (e.g., Prior et al.,[110] Hutchinson,[69] Bryan and Price[18]), and (5) rock or debris falls (e.g., Williams and Davies[170]). The occurrence of mass movement is clearly influenced by the lithology, geologic structure, and geotechnical properties of the cliff-forming material. Other contributing factors are precipitation (e.g., McGreal,[94] Kuhn and Shepard[85]), ground water level (e.g., McGreal and Craig,[97] McGreal,[94] Bryan and Price[18]), and wind (e.g., Cambers[21]). Most of the debris at a cliff base is supplied from mass movement. No toe erosion occurs while the debris obstructs wave attack at the cliff base. Continual wave action transports the debris alongshore and/or offshore. The cliff base is again exposed to waves to suffer erosion. Figure 1 shows a descriptive diagram of the cliff recession system. Toe erosion by waves is necessary for continued cliff retreat. Therefore, a thorough understanding of wave-induced erosion is essential for cliff recession studies.

Focusing on this point, the present chapter will review the state of the art of studies of cliff erosion processes and a related problem — shore platform formation. A book edited by Barnes[5] presents environmental management and ecological aspects, and includes interesting and useful discussions of earth cliffs[93] and rocky cliffs.[55]

CLIFF EROSION RATES AND MEASURING TECHNIQUES

Erosion rates have been documented or recorded in many countries in the world (Table 1). The erosion rates listed in this table are average values over the period of 10^0 to 10^2 years. Orders of erosion rates summarized on a lithological basis are generally: 10^{-3} m/year for granitic rocks; 10^{-3} to 10^{-2} m/year for limestone; 10^{-2} m/year for flysh and shale; 10^{-1} to 10^0 m/year for chalk and Tertiary sedimentary rocks; 10^0 to 10^1 m/year for Quaternary deposits; and 10^1 m/year for unconsolidated volcanic ejecta. These results show that lithology

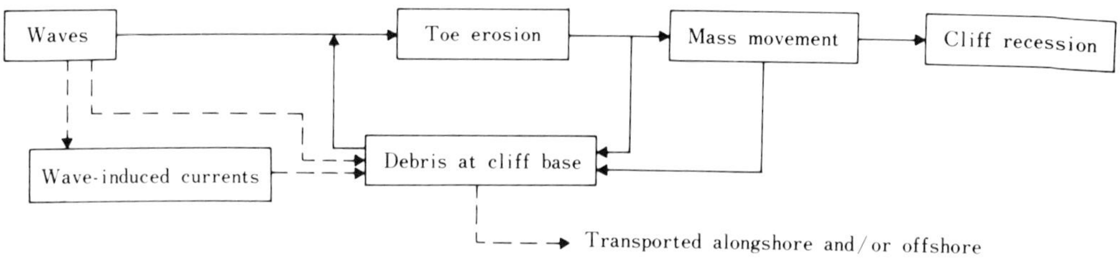

FIGURE 1. Sea cliff recession system. Toe erosion by waves is essential for continued cliff retreat.

greatly controls the erosion rate. It should be noted, however, that wave action originally induces cliff recession.

Extremely short-term erosional events caused by storm waves make such average rates as shown in Table 1 very unreliable, as pointed out by Steers.[133] Cliff recession is essentially an intermittent[21] and localized process.[15] A recession distance vs. time relationship at a given site along a stretch of coast is schematically shown in Figure 2. Here, τ is the length of interval under consideration, X/τ is the long-term average erosion rate, and $\Delta X/\Delta t$ is the erosion rate in a very short period of time during which erosion actually occurred. The two erosion rates are quite different, with the latter being much greater than the former. Table 2 shows some examples of recession distances in the short interval of 10^{-3} to 10^{-1} years. On the north shore of Long Island, New York, a hurricane of September 14, 1944 cut back the bluff at Shareham a horizontal distance of over 12 m in a single day,[29] producing an average erosion rate of 12 m/day, which is equivalent to a mean annual rate of 4380 m/year. This provides a striking contrast to an 80-year average of 0.5 m/year for this coast.[15]

A well-established technique for measuring recession distances is to compare the position of the cliffline on different editions of maps or plans. This technique has been widely used in many countries (Table 1). Other techniques similar to this are to use (1) old and recent terrestrial photographs (e.g., Byrne,[19] Zenkovich et al.[183]), and (2) old and new aerial photographs (e.g., Carter,[22] Gatto,[48]). Surveying has also been commonly employed. Rather unique techniques are to use dated inscriptions or graffiti carved on cliffs,[36,40] and to calculate the volume of talus deposits supplied from cliffs.[122] A microerosion meter technique originally developed for the measurement of the change on bare rock surfaces[60] was modified for coastal studies,[118] and the technique has been applied to studies of morphological changes of rocky coasts (e.g., Trudgill,[163] Kirk,[81] Robinson,[120,121]). Direct measurements using pegs or nails driven into cliffs have also been conducted (e.g., Hodgkin,[63] Lee et al.,[87] Bird et al.,[11]). Air-photogrammetric maps have been used for the accurate measurement of cliff profile changes (e.g., Horikawa and Sunamura,[65] Sunamura,[135] Norrman,[106]). This technique is especially powerful for coasts hindering human access due to high cliff angle and absence of a beach. When cliff faces are very steep, some expedients may be employed in the air-photographing technique.[146] In order to relate the cliff recession to wave action, it is absolutely necessary to obtain reliable erosion data of the cliff base.

FACTORS AFFECTING WAVE-INDUCED EROSION

Many factors are involved in wave erosion at a cliff base, as illustrated in Figure 3 showing possible interrelationships among them. Ultimately, the factors governing toe erosion are (1) the assailing force of waves at the cliff foot, and (2) the resisting force of the cliff-forming material. The relative intensity of the two determines whether erosion occurs or not.

Assailing Force of Waves

The wave energy in deep water has a direct relation to the assailing force exerted by

Table 1
WORLDWIDE COASTAL CLIFF EROSION RATES

Location	Lithology	Erosion rate (m/year)	Interval	Method	Ref.
Australia					
Point Peron near Perth	Limestone	0.0002—0.001	1953—1962	Steel pegs	63
Warrnambool, Victoria	Aeolianite	0.014	(130 yr)	Surveys	53
Barbados					
Mullins Bay	Coral rock	0.002	(4.5 yr)	Pegs	11
Paynes Bay	Coral rock	0.0005	—	Pegs	11
Canada					
Port Bruce, Ontario	Glacial deposits	2.2	1966—1976	Maps	111
Central part, Lake Erie, Ontario	Glacial deposits	0.25—2.75	1810—1964	Surveys, air photos	49
Scarborough, Ontario	Glacial deposits	0.76	1952—1976	Surveys, air photos	18
France					
Ault, Somme	Chalk	0.08—0.37	(100 yr)	—	108
Dieppe, Seine Maritime	Chalk	0.4	—	—	108
St. Jouin, Seine Maritime	Chalk	0	1834—1894	Maps	108
Germany					
North of Kiel, Baltic Sea	Glacial clay	0.6 0.8	1873—1934 1876—1938	— —	126
Helgoland, North Sea	Mesozoic sandstone	1	—	—	184
Cape Arkona, Rugen	Chalk	3—4	(100 yr)	—	184
Iceland					
Surtsey Island	Lava	25—37	1967—1975	Air-photogrammetric surveys	106
Indian Ocean					
Aldabra Atoll	Limestone	0.0010—0.0013	1969—1971	Micro-erosion meter	163
Indonesia					
Krakatoa Island	Volcanic ash	33	1883—1928	—	57
Japan					
Haranomachi, Fukushima	Pliocene sandstone, mudstone	0.3—0.7	1912—1959	Maps	177
Iwaki, Fukushima	Pliocene mudstone	0.13—2.05	1974—1978	Surveys	3
Okuma, Fukushima	Pliocene mudstone, sandstone	0.62 0.31 1.08	1947—1961 1961—1963 1963—1965	Air-photogrammetric maps, surveys	65

Table 1 (continued)
WORLDWIDE COASTAL CLIFF EROSION RATES

Location	Lithology	Erosion rate (m/year)	Interval	Method	Ref.
Suetsugi to Hirono, Fukushima	Pliocene siltstone	0.80—1.4	1947—1973	Air-photogrammetric maps	107
	Pliocene sandstone	0.74	1947—1973		
Byobugaura, Chiba	Pliocene mudstone	0.4—1.1	1888—1950	Maps, surveys	76
Byobugaura, Chiba	Pliocene mudstone	0.79	1946—1960	Air-photogrammetric maps	144
		0.91	1960—1965		
		1.47	1965—1967		
Byobugaura, Chiba	Pliocene mudstone	0.73	1884—1969	Maps	67
Taito-misaki, Chiba	Pliocene mudstone	1.11	1960—1966	Air-photogrammetric maps	135
		0.70	1966—1970		
Toban, Hyogo	Pleistocene deposits	1.0—1.5	1893—1955	Maps	103
Southern coast, Atsumi peninsula	Pleistocene deposits	0.03—0.6	1888—1959	Maps	178
Toyama coast	Alluvial fan deposits	0.5—1.0	—	—	179
West coast, Kanazawa city	Alluvial clay	1—2	—	Surveys	73
Nozuka, Tokachi, Hokkaido	Alluvium	2.3	1964—1974	Surveys	156
Habuseura, Niijima Island	Volcani sand	5.5	1961—1965	Maps, surveys	176
Nishinoshima Island	Volcanic ejecta	80	1974—1977	Air-photogrammetric surveys	100
New Zealand					
Kai-iwi Beach, Wellington	Pliocene siltstone	1.5	1876—1893	Surveys	44
Point Kean, Kaikoura	Tertiary mudstone	0.24	1942—1974	Air photos	80
Sharks Tooth Point, Kaikoura	Teritary mudstone	0.33	1942—1974	Air photos	80
Fifth Bay, Kaikoura	Tertiary mudstone	0.08—0.10	1942—1974	Air photos	80
Mudstone Bay, Kaikoura	Tertiary mudstone	0.06—0.13	1942—1974	Air photos	80
Poland					
Baltic Sea coast	Quaternary deposits	1	—	—	182
Sweden					
Hallshuk, Gotland	Silurian limestone	0.018	1899—1955	Photos	122
North of Ygne, Gotland	Silurian marl	0.025	1950—1962	Photos	122
U.K.					
Fourth Bight, Whitby, Yorkshire	Upper Lias shale	0.023	1971—1972	Micro-erosion meter	120
Tees estuary to Ravenscar, Yorkshire	Upper Lias shale, sandstone	0.036	1891— ?	Maps, surveys	1

Flamborough Head, Humberside	Chalk	0.3	—	—	91
Bridlington to Sewerby, Humberside	Glacial deposits	1.8	—	—	91
Holderness, Humberside	Glacial deposits	3.3	—	—	91
Holderness, Humberside	Glacial deposits	0.29—1.75	1852—1952	Surveys	164
Norfolk	Glacial deposits	0.9	1880—1967	Maps, air photos	21
West Runton to East Runton, Norfolk	Glacial deposits	0.2	1971—1973	Cliff-top stakes	21
Cromer to Mudesley, Norfolk	Glacial deposits	4.2	1838—1861	—	91
		5.7	1861—1905	—	
Suffolk	Glacial deposits	0.8	1880—1950	Maps	21
Hopton, Suffolk	Glacial deposits	0.8—0.9+	1926—1950	Maps	132
Pakefield, Suffolk	Glacial deposits	0.6—0.9	1926—1950	Maps	132
Covehithe, Suffolk	Glacial deposits	5.1	1925—1950	Maps	132
Southwold, Suffolk	Glacial deposits	3—3.3	1925—1950	Maps	132
Southwold, Suffolk	Glacial deposits	4.5—13.5	—	—	91
Dunwich, Suffolk	Glacial deposits	1.6	1589—1753	Maps	
		0.9	1753—1824		
		1.5	1824—1884		
		1.1	1884—1925		
		0.15	1925—1977	Maps, surveys	
Studd Hill, Kent	London clay	1.5	1872—1898	Maps	129
		2	1898—1931		
		3.4	1931—1939		
		1.2	1939—1961		
Beltinge, Kent	London Clay	0.7	1872—1907	Maps	129
		0.9	1907—1933		
		1.1	1933—1939		
		0.9	1939—1959		
Grenham Bay, Birchington, Kent	Chalk	0.08—0.4	1872—1932	Maps	173
South Foreland, Kent	Chalk	0.19	1878—1962	—	92
Seven Sisters, Sussex	Chalk	0.51	1873—1962	—	92
Birling Gap, Seven Sisters, Sussex	Chalk	0.91	1875—1916	—	92
		0.99	1950—1962	Surveys	
Middle Bottom, Dorset	Chalk	0.05	1882—1962	—	92
White Nothe to Hamburg Tout, Dorset	Chalk	0.22	1882—1962	—	92

Table 1 (continued)
WORLDWIDE COASTAL CLIFF EROSION RATES

Location	Lithology	Erosion rate (m/year)	Interval	Method	Ref.
Ballard Down, Dorset	Chalk	0.23	1882—1962	—	92
Fairy Dell, Dorset	Marls	0.4—0.5	1887—1969	Maps, air photos	17
Vale of Glamorgan, Wales	Lower Lias limestone, shale	0.008—0.099	1967—1969	Cliff-top stakes	159
Southeast County Down, N. Ireland	Glacial deposits	0.21—0.84	1834—1962	Maps, air photos	96
U.S.A.					
Martha's Vineyard, Mass.	Glacial deposits	1.7	1846—1886	Surveys	71
Boston Harbor, Mass.	Glacial deposits	{0.23	1860—1908	Surveys }	71
		0.31	1903—1905	Maps }	
Boston Harbor, Mass.	Glacial deposits	0.1}	—	—	40
Outer Cape Cod, Mass.	Glacial deposits	1	1848—1888	—	31
Outer Cape Cod, Mass.	Glacial deposits	0.96	1848—1888	Surveys	71
Outer Cape Cod, Mass.	Glacial deposits	0.9—1.5+	(Long continued observations)		124
Outer Cape Cod, Mass.	Glacial deposits	0.8	1879—1959	Surveys	181
Outer Cape Cod, Mass.	Glacial deposits	0.12—2.2	1938—1974	Air photos	48
Highland, Outer Cape Cod, Mass.	Glacial deposits	0.3	1953—1958	Surveys	180
North Shore, Long Island, N.Y.	Glacial depsoits	0—0.5	1933—1966	—	29
North Shore, Long Island, N.Y.	Glacial deposits	0.5	(80 yr)	Charts, air photos	15
East Fort Point, Long Island, N.Y.	Glacial deposits	0.9+	1833—1883	—	71
Lake County, Ohio	Glacial deposits	{0.3—0.9	1876—1937	Maps, air photos }	22
		{0.3 (max.)	1937—1973	Air photos }	
Berrien County, Mich.	Glacial deposits	3.8	1971—1974	Air photos	13
Eastern shore, Lake Michigan, Mich.	Glacial deposits	2.5	1970—1974	Surveys	14
Shannon Point, Wash.	Glacial deposits	0.2	1893—1969	Maps	125
North of Yaquina Bay, Ore.	Miocene sandstone	0.6	1880—1963	Photos	19

Rockaway Beach, Calif.	Franciscan volcanic and metavolcanic rocks	0.15	—	Maps	130
Montara, Calif.	Miocene conglomerate	0.26—0.29	1912—1965	Maps	130
Santa Cruz, Calif.	Mio-Pliocene mudstone	0—0.2	1943—1963	Air photos	56
Santa Cruz, Calif.	Miocene mudstone with shale	0.4	(80 yr)	Maps, surveys	130
Santa Barbara, Calif.	Miocene shale	0.2	1951—1965	Surveys	104
	Mio-Pliocene shale	0.08	1923—1967		
	Pliocene siltstone	0.3	1927—1947		
Isla Vista, Calif.	Mio-Pliocene shale	0.05—0.4	1966—1973	Maps, surveys	28
Leucadia to Pacific Beach, Calif.	Sandstone, claystone	0—0.5	1970—1975	Nails	87
La Jolla, Calif.	Cretaceous-Eocene sandstone	0.0003—0.0006	—	Dated inscriptions and graffiti	36, 40
La Jolla, Calif.	Cretaceous sandstone, Eocene shale	0.01—0.2	1940—1979	Photos	40
La Jolla, Calif.	Alluvium	0.5	(11 yr)	Photos	127
La Jolla, Calif.	Alluvium	0.28—0.50	1918—1930	Measurements	165
La Jolla, Calif.	Alluvium	0.09—0.3	1912—1975	Maps	40
Sunset Cliffs, San Diego, Calif.	Upper Cretaceous sandstone, siltstone	0.012	(75 yr)	Photos, observations	77
Sunset Cliffs, San Diego, Calif.	Pre-Miocene sandstone, shale	0.01—0.3	—	Photos, maps	86
U.S.S.R.					
Barrents Sea coast	Granitic rocks	0.001—0.002	—	—	183
Laptev Sea coast	Glacial clay	4—6	—	—	183
Laptev Sea coast	Glacial clay	4	—	—	183
Laptev Sea coast	Glacial clay	50 (max.)	—	—	183
Laptev Sea coast	Glacial clay	3.5—4	—	—	183
Bering Sea coast	Moraine deposits	1—2 (max.)	—	—	183
Okhotsk Sea coast	Volcanic ash	50 (max.)	—	—	183
Okhotsk Sea coast	Quaternary brown loam and clay	40 (max.)	—	—	183

Table 1 (continued)
WORLDWIDE COASTAL CLIFF EROSION RATES

Location	Lithology	Erosion rate (m/year)	Interval	Method	Ref.
Baltic Sea coast	Granitic rocks	0.001—0.002 (max.)	—	—	183
Baltic Sea coast	Sandstone	3 (max.)	—	—	183
Baltic Sea coast	Sandy deposits	20 (max.)	—	—	183
Baltic Sea coast	Sandy deposits	2	—	—	183
Baltic Sea coast	Sandy deposits	1—1.5	—	—	183
Azov Sea coast	Quaternary brown loan and clay	3	—	—	183
Black Sea coast	Flysh	0.02—0.03	(20 yr)	Surveys, photos	183
Black Sea coast	Flysh, shale	0.01—0.02	—	—	183
Black Sea coast	Coquinite	0.002—0.005	—	—	183
Black Sea coast	Crystallized limestone	0.003	—	—	183
Black Sea coast	Massive limestone	0.3—0.5 (max.)	—	—	183
Black Sea coast	Limestone with loess	2—3	—	—	183
Black Sea coast	Limestone with loess	0.61	—	—	183
Black Sea coast	Quaternary loess	0.5—1.0	—	—	183
Black Sea coast	Quaternary conglomerate	12	—	—	183
Black Sea coast	Quaternary brown loam and clay	1 (max.)	—	—	183
Black Sea coast	Quaternary clay	2—3	—	—	183
Black Sea coast	Diluvial deposits	0.11	—	—	183

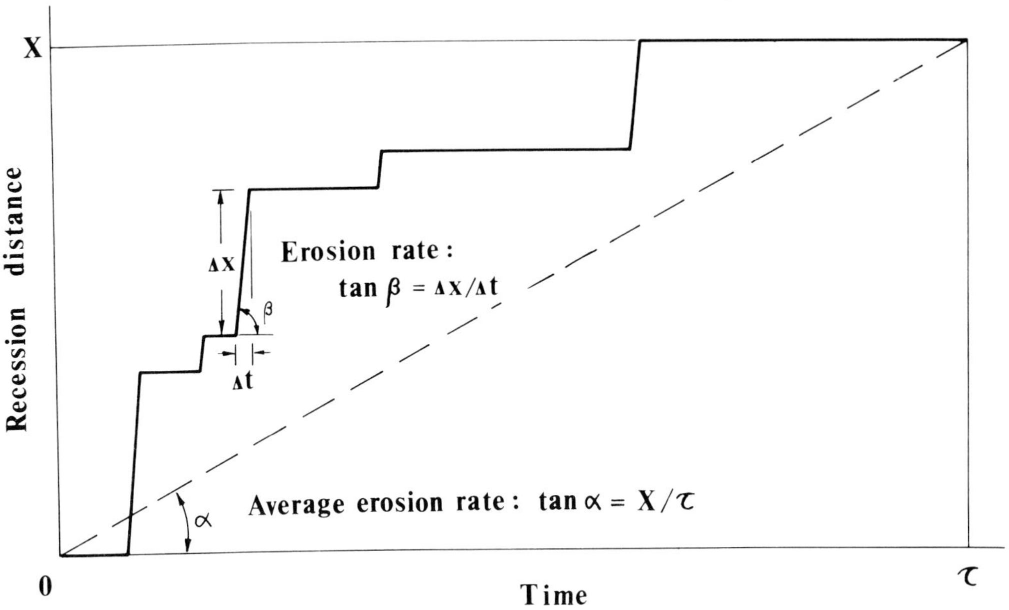

FIGURE 2. Schematic representation of temporal change of cliff recession distance. A very short-term erosion rate, $\Delta X/\Delta t$, is quite different from the long-term average value, X/τ.

waves, f_w, while the following three factors indirectly influence f_w: (1) water level or tide,[8,13,14,82,95,112,172] (2) shallow-water bottom and beach topography,[78,95,100,117,176,178] and (3) beach sediment.[22,58,95,116,120] Interrelations among the three factors and f_w are shown in Figure 3. The water level influences the elevation of maximal wave attack. A combination of water level and topographic factors controls the wave height near the cliff foot, while the wave period remains the same as in the deep-water region because the constancy of wave period can be assumed to hold (e.g., Eagleson and Dean[33]). The local topography and water level plus deep-water wave characteristics determine the type of waves immediately in front of the cliff. There are three distinct types of waves: standing, breaking, and broken; and for the last case the distance from the breaking point to the cliff base is of importance in the evaluation of f_w.[112]

Waves always exert hydraulic action on a cliff face. This action consists mainly of compression, tension, and shearing. When waves hit a cliff, the air in the cracks or fissures is suddenly compressed. At wave recession, the compressed air expands with explosive force. This is known as wedge action.[64,79,89,155] Immediately after a wave hits a cliff, the water mass climbs the cliff face exerting a shearing force. Cavitation[4] probably occurs when the up-rush is of high velocity.

Waves armed with beach sediment exert mechanical action which comprises abrasive and impact forces produced on the cliff face by solid particles. An extensive literature has stressed the importance of mechanical action in wave erosion of rocky coasts.[9,12,26,30,41,43,54,57,83,84,89,131,134,155,182]

Both hydraulic and mechanical actions characterize the wave assailing force, f_w. However, no direct field measurements of f_w have been performed, due to the complexity of the phenomena involved and the difficulty of measurement. Therefore, the most suitable physical and quantitative index for f_w has not yet been determined in actual field situations. This is one of the important problems which should be solved for furthering the development of dynamical studies of wave-induced rocky coast erosion.

Table 2
VERY SHORT-TERM CLIFF RECESSION DISTANCES

Location	Lithology	Recession distance (m)	Interval	Remarks	Ref.
Iceland					
Surtsey Island	Lava	100	2/1—2/5/1966	Storm	106
		140 (max.)	Winter of 1967/68	Storms	105
Japan					
Nishinoshima Island	Volcanic ejecta	120 (max.)	8/3—10/29/1974	Storms	100
Mexico					
Isla San Benedicto	Lava	18	4/16—9/12/1953	Storms	114
U.K.					
Covehithe, Suffolk	Glacial deposits	9	1/31—2/1/1953	Storm	171
Southeast County Down, . Ireland	Glacial deposits	0.03—0.28	1/11—1/13/1974		95
U.S.A.					
Outer Cape Cod, Mass.	Glacial deposits	1.2—15	April, 1958	Storm	180
North Shore, Long Island, N.Y.	Glacial deposits	12	9/14/1944	Storm	15

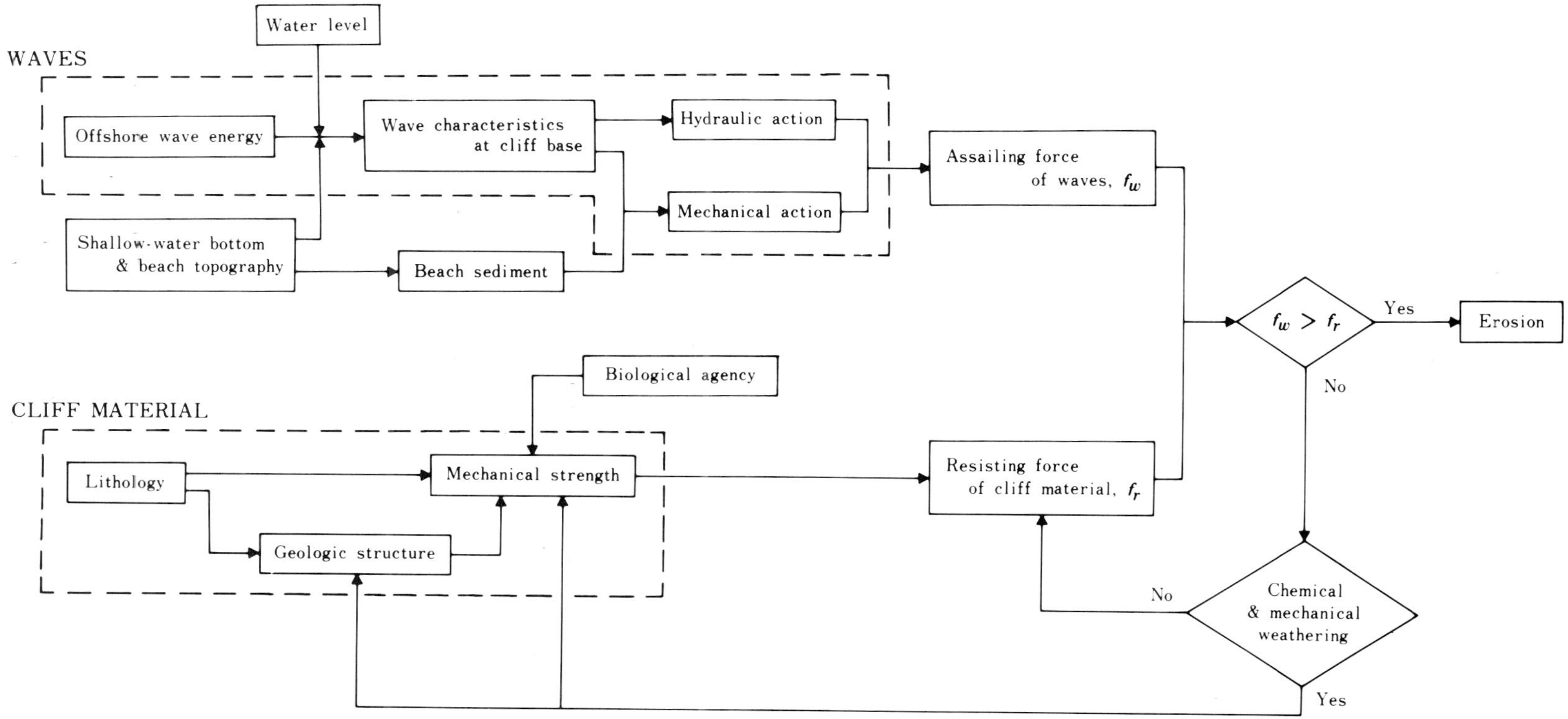

FIGURE 3. Factors affecting wave-induced cliff erosion. Ultimate factors are the assailing force of waves and the resisting force of cliff material. Cliff erosion is determined by the relative intensity of the two.

Resisting Force of Cliff Material

Lithology principally determines the resisting force of cliff material f_r (Figure 3). Mechanical strength of the cliff material, and geologic structure such as jointing, faulting, folding, and stratifications, are contributing factors. The former is directly related to f_r, while the latter acts to reduce the former. In some areas, the geologic structure is a dominant controlling factor for cliff recession.[20,107]

In order to examine the relationship between erosion and cliff resistance, the strength of cliff material has been tested: compressive strength,[3,65,107,135] tensile strength,[107,135] cohesive strength,[95] and penetration strength.[177,178] Of these indexes, the most appropriate one for expressing the resisting force, f_r, has not yet been fully determined. However, since these indexes are closely related and not independent of one another, it is possible to use one of them as representative of f_r. Compressive strength has been employed for the studies of eroding Tertiary cliffs in Japan,[67,135,143,144] simply because this is a widely used index in the fields of soil sciences and rock mechanics, with the testing criteria having been well established. A sound physical explanation for the application of this index to cliff erosion studies is required for greater progress in this field. Another problem to be solved is how to quantify the effect of geologic structure upon mechanical strength, and how to relate this to f_r. This problem becomes serious when cliffs are composed of densely jointed rocks. Suzuki,[147] who studied the erosion of channel bedrock by fluvial forces, has attempted to quantify the effect of jointing using a parameter, V_{pf}/V_{pc}, in which V_{pf} is the velocity of longitudinal waves propagating through the bedrock *in situ* and V_{pc} is the longitudinal wave velocity measured for cylindrical specimens without visible cracks. This parameter could also be applied to the coastal environment.

A reduction in mechanical strength of cliff material is brought about by both chemical and mechanical weathering (Figure 3). Chemical weathering encompasses hydration and solution, which are produced by the interaction between cliff material and sea water.[39] The effect of solution upon cliff retreat is marked for limestone areas, especially in the tropical environment.[63,74,75,163] Mechanical weathering comprises (1) the wetting and drying process in the intertidal zone,[38] (2) generation of repeated stresses in cliff material by periodic wave action, and (3) the frost effect in cold latitudes.[30,182]

Two contrasting effects by biological agencies upon cliff strength are found on calcarious rocky coasts. One is the well-known negative effect by rock-boring or rock-snapping organisms,[2,42,88,101,102,115,163,166] i.e., a reduction in mechanical strength due to many burrows on the cliff surface. The other is a positive effect by encrusting organisms which lithify tropical limestone cliffs,[46,47] i.e., an increase in mechanical strength due to lithification.

LABORATORY EXPERIMENTS AND A BASIC RELATION FOR CLIFF EROSION

A fundamental relation of cliff erosion by waves can be expressed as

$$X = \phi(f_w, f_r, t) \tag{1}$$

where X is the eroded distance and t is the time. As shown in Figure 3, erosion occurs when the wave assailing force is greater than the cliff resisting force, i.e., when $f_w > f_r$; otherwise, no erosion takes place. No field studies have been performed to determine the functional form of Equation 1. The absence of such studies is due to the fact that complicated and severe natural environments have discouraged actual measurements of erosional events occurring at cliff bases. To overcome this difficulty, laboratory approaches have been taken.

Horikawa and Sunamura[66] and Sanders[123] performed wave-tank experiments on cliff erosion in response to wave action. Sanders used a model cliff made of a mixture of plaster,

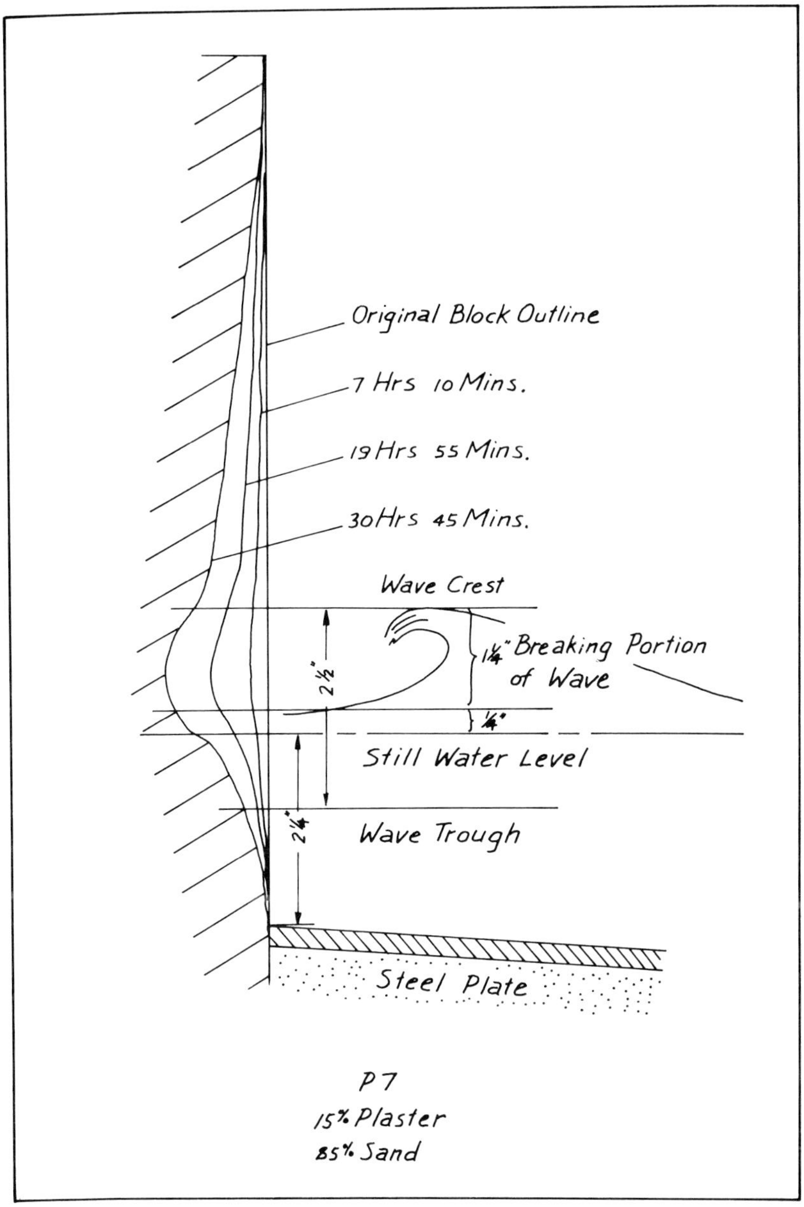

FIGURE 4. Profile change of plaster:sand cliff. Erosion is accelerated and it extends to the cliff base. (From Sanders, N. K., *Pap. Proc. R. Soc. Tasmania,* 102, 11, 1968. With permission of the Royal Society of Tasmania and the author.)

sand, and water, with the plaster used as a cementing material. Since plaster and such substances as bentonite,[23] diatomite, lime, and liquid glass are soluble in water, the strength of model cliffs cemented by one of these substances decreases with time due to solution after the cliffs are exposed to waves. This temporal reduction in mechanical strength has not yet been quantitatively investigated. These materials seem to be appropriate for the modeling of limestone cliffs. Figure 4 shows the profile change of a plaster:sand cliff caused by 0.65-sec breaking waves with a height of 6.4 cm.[123] Cliff erosion processes observed in such laboratory experiments would be helpful for an understanding of notch development occurring on actual calcarious rocky coasts.[63,154,163]

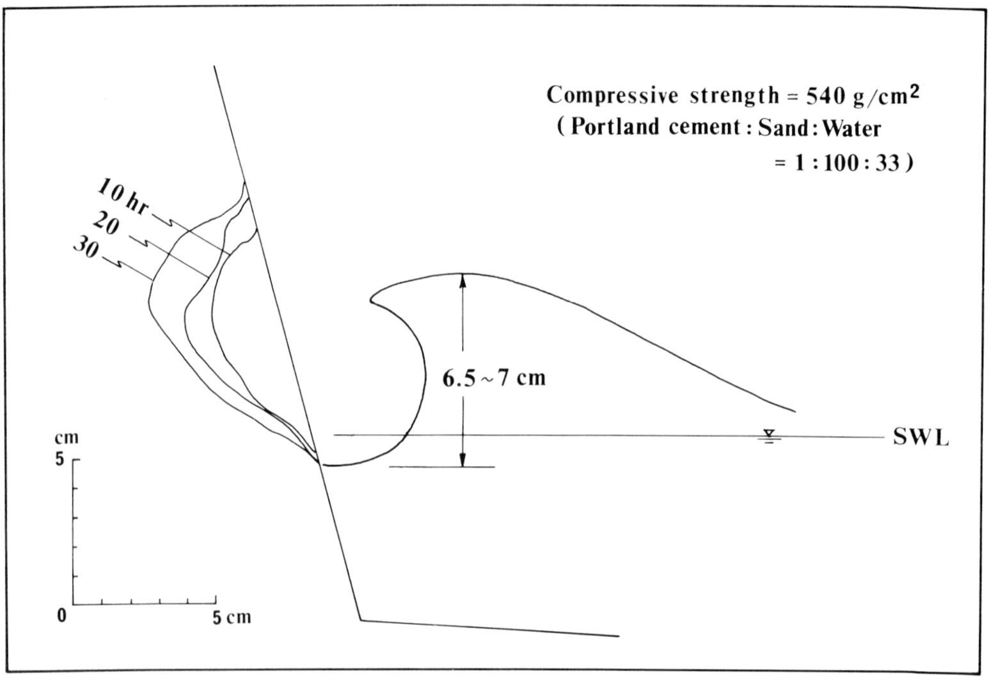

FIGURE 5. Erosion of cement:sand cliff caused by breaking waves. Limit of downward erosion is located slightly below SWL.

With the purpose of modeling noncalcarious rocks, Horikawa and Sunamura,[66] Sunamura and Horikawa,[145] and Sunamura[135] constructed cliffs with a mixture of portland cement, well-sorted fine sand (0.2 mm in diameter), and water. These model cliffs, made of homogeneous, unbedded, and unjointed material, have the desirable properties of being (1) insoluble in water, (2) low in compressive strength, but capable of standing in a wave tank, and (3) temporally constant in strength. By changing the cement:sand ratio, cliffs of arbitrary compressive strength can be made. A nonlinear stress-strain relationship for this kind of material was found,[135] which is analogous to the behavior of rocks in nature. Cliffs formed of portland cement require a long curing time before exposure to laboratory waves: at least a one-month water curing is desirable to obtain constant strength.

Figure 5 shows an example of the profile change of a cement:sand cliff due to breaking waves 6.5 to 7.0 cm in height and 1.2 sec in period. This cliff profile may be compared to the case of the plaster:sand cliff (Figure 4). In both experiments, breaking waves of approximately the same height were allowed to act on the cliffs. In the plaster:sand cliff experiment, erosion is seen to extend to the base of the model cliff, and the eroded area increased with time; while for the cement:sand cliff, no erosion occurred below a certain level, and no significant temporal increase in the eroded area is found. These distinctive differences reflect the chemical and mechanical properties of the cementing material used for making the model cliffs.

In order to determine the functional form of Equation 1, a wave-tank test was performed.[137] Broken waves were exerted on a cliff made of a cement:sand:water mixture in an 1:150:50 weight ratio, with a compressive strength of 340 g/cm^2. Figure 6 shows the resulting cliff erosion and sand beach formed in front of the cliff. As sand grains were increasingly deposited at the cliff base the input waves used them as an abrasive, so that the erosive force was intensified and the erosion was accelerated. In time, however, an excess of sand deposits brought about a dissipation of wave energy. Waves gradually lost their erosive force and

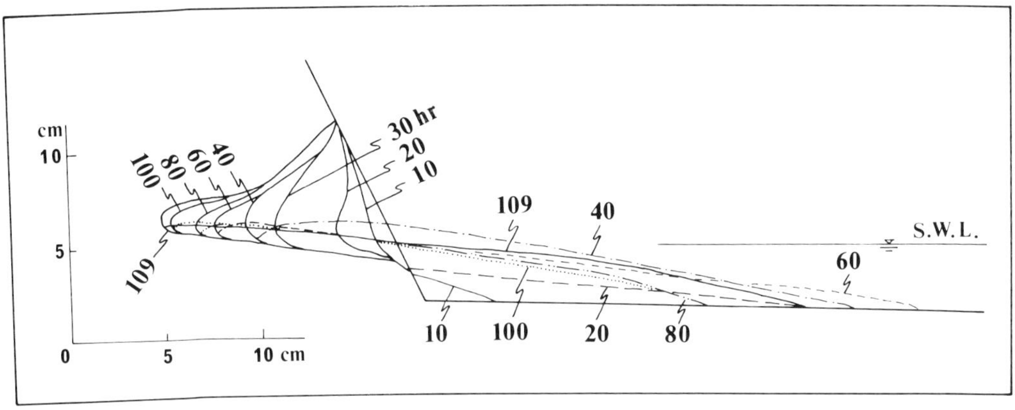

FIGURE 6. Erosion of cement:sand cliff caused by broken waves and development of beach in front of cliff. The notch develops obliquely upward. (From Sunamura, T., *J. Geol.*, 84, 427, 1976. Copyright© 1976, The University of Chicago. With permission.)

the erosion rate was reduced. Although the erosive force remained time-independent when no sand was deposited, the force became automatically time-dependent, receiving feedback from the erosional process due to the action of sand produced by cliff erosion itself.

Based on the result of the application of linear automatic control theory to this erosional process,[137] a basic equation for cliff erosion was found to obey the form[138]

$$\frac{dX}{dt} \propto \ln \frac{f_w}{f_r} \tag{2}$$

This relation holds only when $f_w > f_r$; otherwise $dX/dt = 0$. By physical intuition and dimensional considerations, it is assumed that,

$$f_w = A\rho gH \tag{3a}$$

$$f_r = B\ S_c \tag{3b}$$

where H is the wave height at the cliff base, S_c is the compressive strength of cliff material, ρ is the density of water, g is the gravitational acceleration, and A and B are nondimensional constants. Then Equation 2 becomes

$$\frac{dX}{dt} = k\left(C + \ln \frac{\rho gH}{S_c}\right) \tag{4}$$

where C is a nondimensional constant ($C = \ln [A/B]$) and k is a constant with units of $[LT^{-1}]$. Equation 4 relates the rate of wave-induced cliff erosion to the two major controlling factors: (1) wave height at the cliff foot and (2) compressive strength of the cliff-forming material.

SHORT-TERM EROSION PROBLEMS

Wave Height vs. Erosion

Equation 4 indicates that there exists a critical height of waves at a cliff base for initiating cliff erosion. This critical wave height, $H_{crit.}$, is obtained by setting $dX/dt = 0$ in Equation 4, to yield

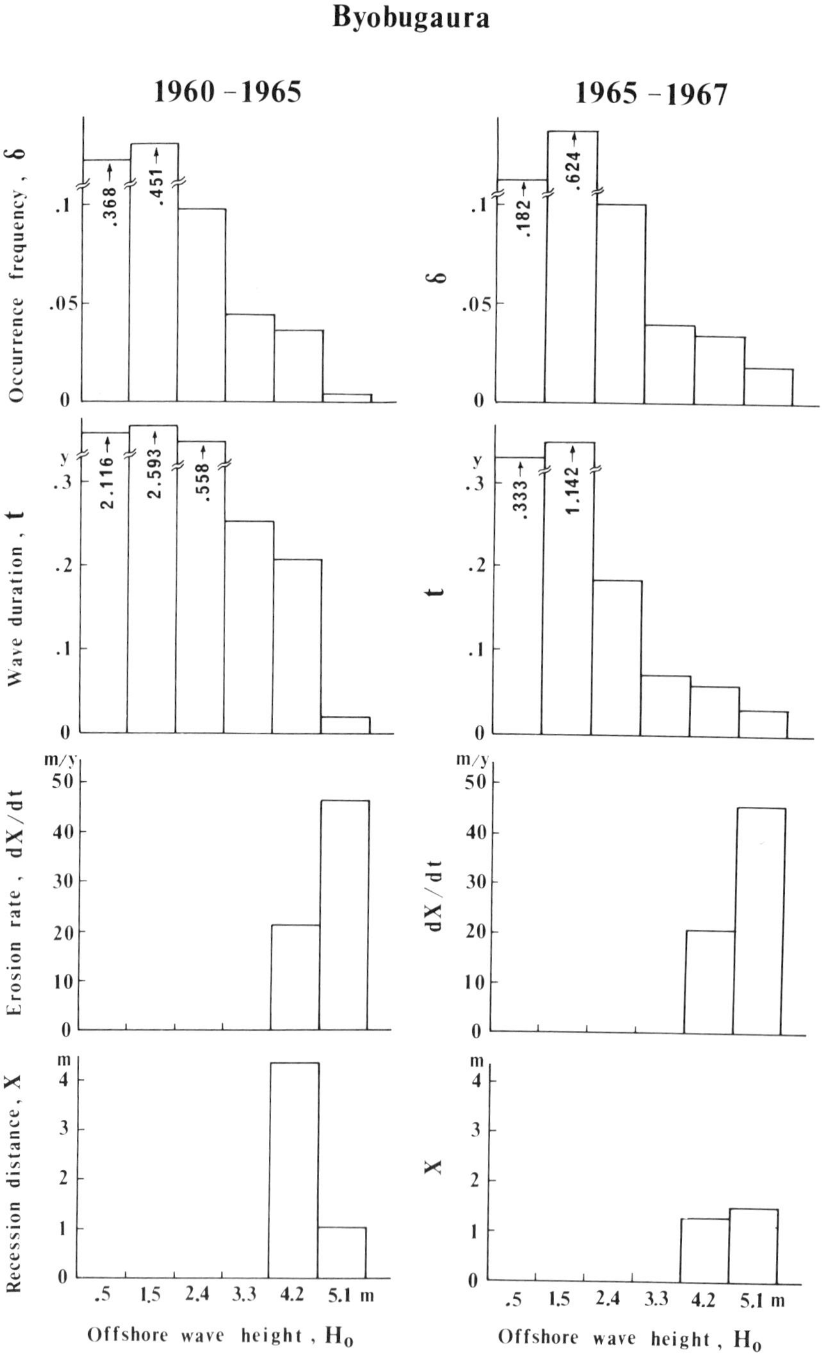

FIGURE 7. Wave occurrence frequency, wave duration, cliff erosion rate, and recession distance vs. offshore wave height, at Byobugaura, Japan. (Modified from Sunamura, T., *J. Geol.*, 90, 167, 1982. Copyright© 1982, The University of Chicago. With permission.)

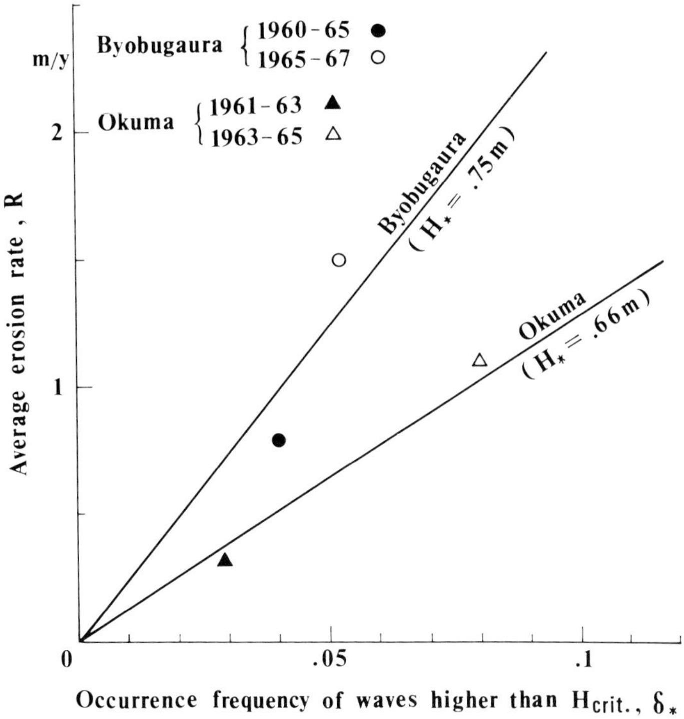

FIGURE 8. Relationship between average erosion rate and occurrence frequency of erosion-causing waves for Byobuguara and Okuma cliffs, Japan. (From Sunamura, T., *J. Geol.*, 90, 167, 1982. Copyright© 1982, The University of Chicago. With permission.)

$$H_{crit.} = \frac{S_c}{\rho g} e^{-C} \tag{5}$$

On a coast with little significant alongshore variations of quantities (1) the assailing force of waves and (2) the resisting force of cliffs, $H_{crit.}$ as obtained from Equation 5 gives one representative value for the overall coast.

The foregoing conditions are found to be satisfied on the Byobugaura cliff located on the Pacific coast of Japan. This cliff is 10 to 60 m high, and forms a nearly straight coastline of 9 km in length. Pliocene mudstone is exposed on the lower half of the cliff. Notable mass movements such as slumps, mudflows, and mudslides do not occur for this cliff. Using the data of (1) cliff strength, (2) wave climate, and (3) erosion rates of two time intervals (1960 to 1965 and 1965 to 1967), the constants C and k in Equation 4 were determined for this area, and $H_{crit.}$ was calculated from Equation 5.[143] Analysis of these results enabled plots of the relationships between offshore wave height, H_o, and the erosion rate, dX/dt, or the eroded distance X. These two relationships are respectively shown in the lower two graphs in Figure 7 for each interval considered. The plot of dX/dt vs. H_o shows that dX/dt abruptly increases as H_o increases beyond some critical value. The sum of X in the plot of X vs. H_o gives the total recession distance for each interval. The bottom graphs show that the larger but rarer waves are solely responsible for the cliff erosion, as compared with the smaller but more frequent waves.

Temporal Variations in Erosion Rates

It has been reported that considerable temporal variations in erosion rates exist for the short term of 10^0 to 10^1 years.[21,95,135] One would anticipate that the more frequently the

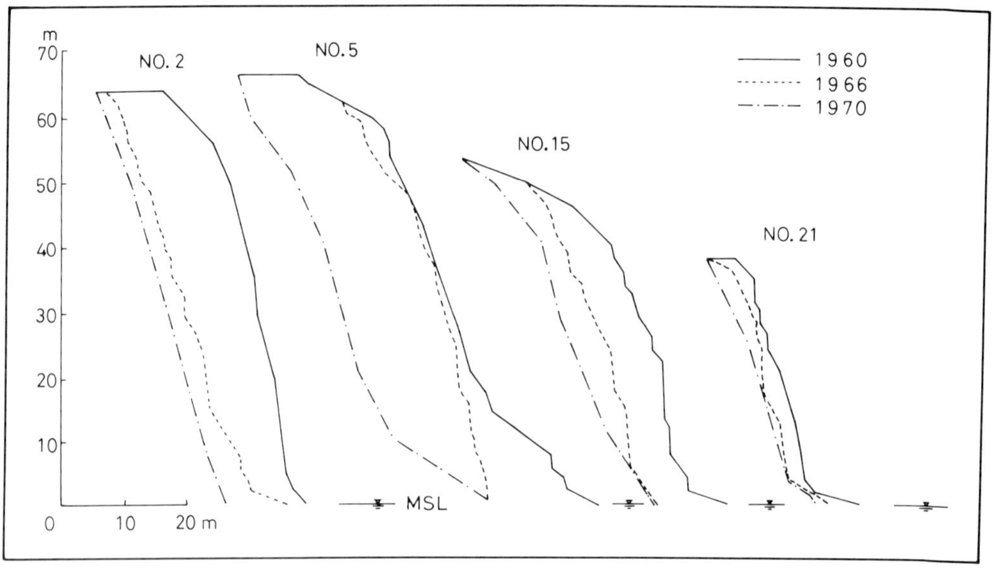

FIGURE 9. Some examples showing profile change of Taito-misaki cliff, Japan. The cliff profiles are based on air-photogrammetric maps. (From Sunamura, T., *J. Fac. Eng. Univ. Tokyo*, 32, 1, 1973. With permission.)

larger waves attack a coast during a certain period of time the faster the cliff would recede, thus producing a higher erosion rate. Since only the waves that exceed a threshold value (Equation 5) can cause erosion, the erosion rate should be related to the occurrence frequency of waves higher than $H_{crit.}$.

Figure 8 illustrates this relation using the data obtained from the Byobugaura and Okuma coasts in Japan.[143] Short-term annual erosion rates are, thus, strongly influenced by the occurrence frequency of the erosion-causing waves. If we take a longer time span of, say, more than 20 or 30 years for a coast, then the average erosion rate would mask the influence of the variation of wave occurrence frequency and show a characteristic value for the location, i.e., a time-average value of the relative intensity of wave action to the cliff resistance. The masking effect due to the length of the time-scale has not been quantitatively investigated because of the lack of long-term daily wave data.

Temporal Changes in Cliff Profile

Slope instability produced by cliff-base wave erosion gives rise to mass movement, which renders the cliff face profile stable, i.e., gently sloping, and simultaneously supplies debris masses to the foot of the cliff. Waves remove the debris to again undercut the cliff. As seen in this circular relationship, depicted in Figure 1, a cyclic change between steep and gentle cliff profiles occurs during cliff recession processes.

To illustrate this, an example is taken from the area of Taito-misaki on the Pacific coast of Japan, where a Pliocene mudstone bluff of 2.5-km alongshore length is found.[135] Based on the cliff profiles of 1960, 1966, and 1970, which were obtained from air-photogrammetric maps (Figure 9), the correlations between cliff recession rates at two different heights were examined (Figure 10). In this figure, the two quantities, $\dot{X}_B$ and $\dot{X}_T$, denote, respectively, the recession rates obtained at levels of 1/4 and 3/4 of the cliff height measured from MSL. In Figure 10a, most of the data points for 1960 to 1966 are located in the area $\dot{X}_T < \dot{X}_B$, but the reverse tendency is seen for 1966 to 1970 (Figure 10b). For the whole interval, 1960 to 1970, however, the data points lie with a smaller scatter around the line $\dot{X}_T = \dot{X}_B$ (Figure 10c). These results suggest the presence of a cyclic cliff profile change (Figure 11), with a possible period of ten years for this coast. Figure 11 also shows that a cliff profile with the

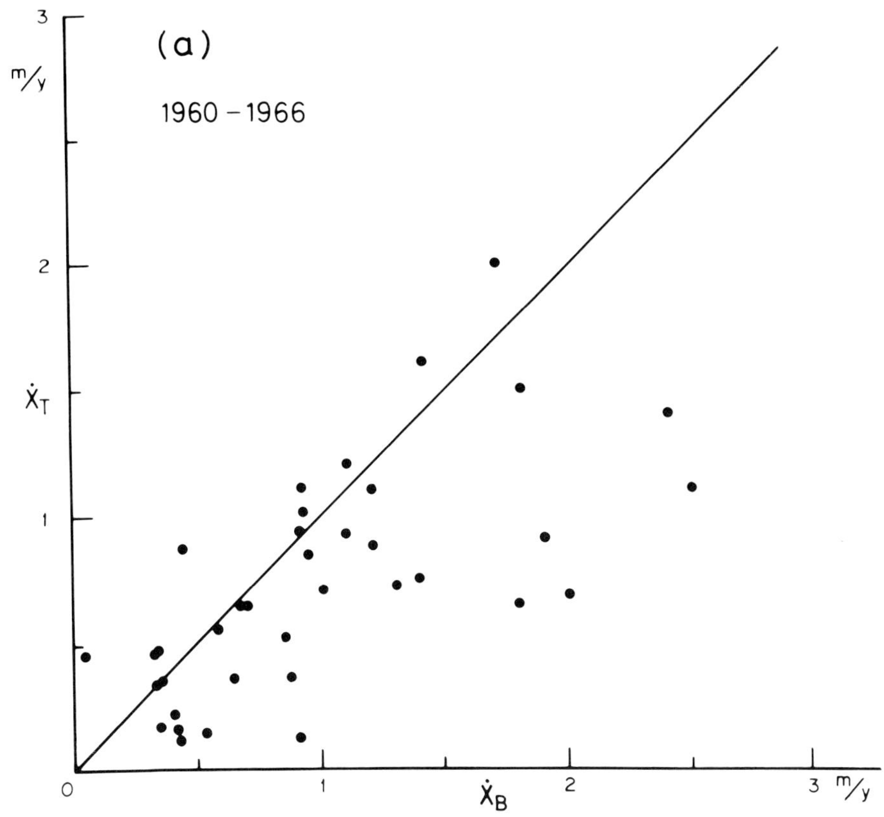

FIGURE 10. Relationship between cliff erosion rates at the lower and upper levels of a cliff profile for different intervals: (a) 1960 to 1966, (b) 1966 to 1970, and (c) 1960 to 1970. $\dot{X}_B$ and $\dot{X}_T$ denote the mean annual recession rates at a level of 1/4 and 3/4 of cliff height, respectively.

same slope angle appears repeatedly as the cliff recedes (e.g., profiles AA, A′A′, and A″A″ in Figure 11). Such a parallel retreat of the cliff profile has been found on Lake Erie northshore bluffs,[111,113] and in McGreal's[95] model of coastal slope recession.

Spatial Variations in Erosion Rates

If the respective alongshore variations of the wave assailing force, f_w, and the cliff resisting force, f_r, are uniform on a coast, then Equation 2 gives the same erosion rate at any location along the coast, i.e., there is no spatial variation in the erosion rate. Hence, the coefficient of variation, σ/R, should be zero, in which R is the space-average value of erosion rates and σ is the standard deviation of R. This implies the existence of a parallel recession of the cliffline.

Figure 12 is a plot of σ/R against τ, the length of interval, using the data from Byobugaura, Japan, where both f_w and f_r do not show notable alongshore variations.[135] This graph illustrates that σ/R decreases with increasing τ, suggesting that the degree of erosion-rate variation becomes smaller as a time scale is taken longer, as Cambers[21] has stated. On a coast with little alongshore variation of input wave height and cliff strength, parallel recession of the coastline will occur over a long time period.

Consider several cliffs with differing resistances that are exposed to a similar wave climate. Since f_w is space-constant and f_r is variable in this case, Equation 2 is rewritten using Equation 3b as

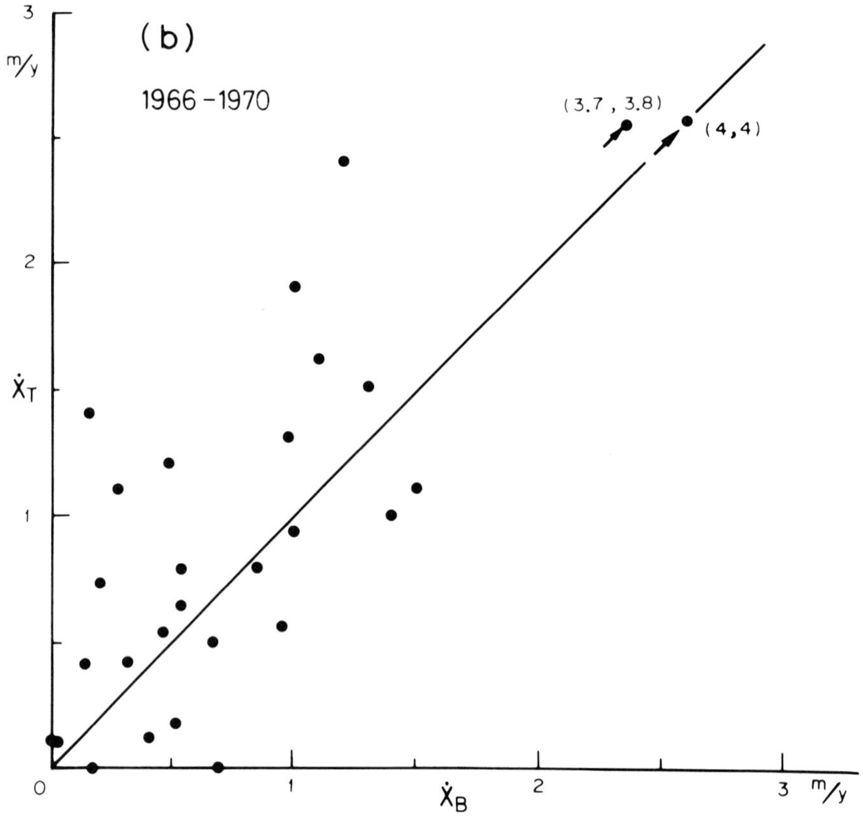
(b)
1966 – 1970
(3.7 , 3.8)
(4,4)
3
m/y
2
$\dot{X}_T$
1
0
1
$\dot{X}_B$
2
3 m/y

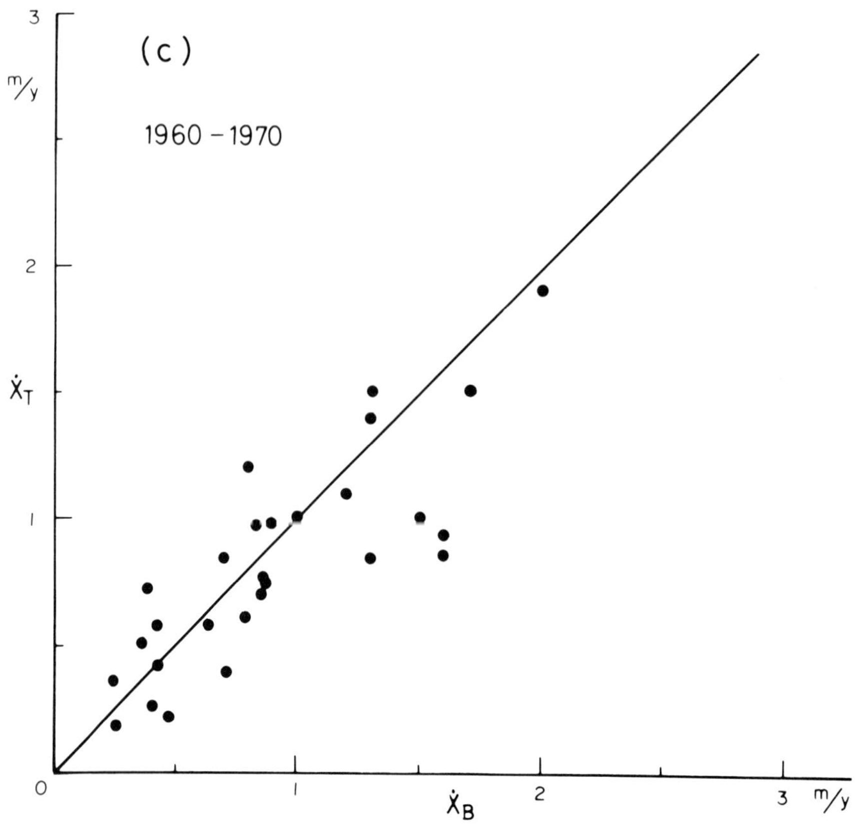
(c)
1960 – 1970
3
m/y
2
$\dot{X}_T$
1
0
1
$\dot{X}_B$
2
3 m/y

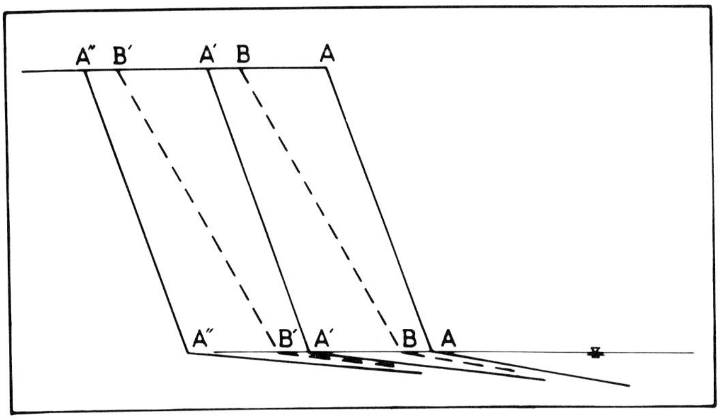

FIGURE 11. Schematic diagram showing cliff recession with an alternation of steep and gentle slopes. Parallel retreat of cliff face occurs. (From Sunamura, T., *J. Fac. Eng. Univ. Tokyo*, 32, 1, 1973. With permission.)

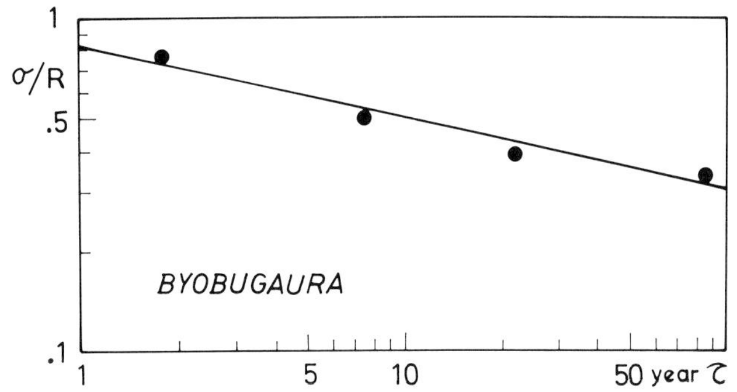

FIGURE 12. Variation coefficient of erosion rate, σ/R, vs. length of interval, τ, for Byobugaura cliff, Japan. R is the erosion rate and σ is the standard deviation. (From Sunamura, T., *J. Fac. Eng. Univ. Tokyo*, 32, 1, 1973. With permission.)

$$\frac{dX}{dt} \propto (C_1 - \ln S_c) \qquad (6)$$

where $C_1 = \ln (f_w/B) =$ const. This equation indicates that a linear relation should be found between the erosion rate and compressive strength in a semilogarithmic expression. This relation is clearly shown in Figure 13. The data plotted here were collected from the Pacific coasts of eastern Honshu, Japan, where the spatial variation of f_w appears to be small. As discussed before, the short-term erosion rate is strongly affected by temporal changes of wave climate. In order to eliminate this influence from the erosion rate, average values over a fairly long term (20 to 90 years) were employed. Figure 13 shows that the erosion rate decreases with an increase of compressive strength, i.e., the so-called "harder" cliffs exhibit lower erosion rates.

Cliff Height vs. Erosion Rate

If a cliffline with similar lithology but varying height is subjected to the same wave conditions alongshore, then the higher cliffs supply a larger amount of debris to the cliff foot for a given toe erosion by waves. Cliff undercutting is halted until the debris is removed by wave and current action. More time is needed for the removal of larger debris masses.

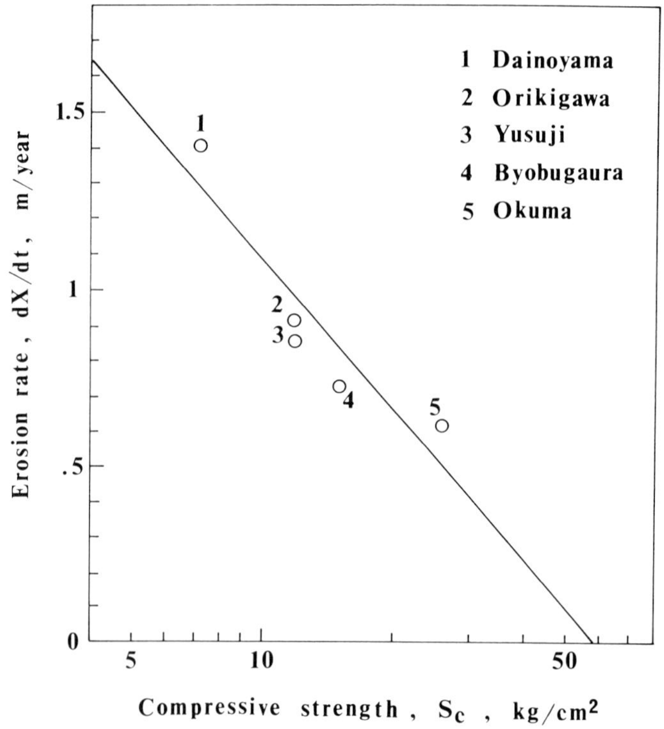

FIGURE 13. Relationship between erosion rate and rock strength of Tertiary cliffs on the Pacific coasts of eastern Honshu, Japan. (From Sunamura, T., *J. Geol.*, 85, 613, 1977. With permission.)

This leads to the concept that higher cliffs have smaller erosion rates than lower cliffs.[76,127,131] This relation between cliff height and erosion rate does not seem to hold, however, because a cliffline with varying height and similar lithology has a tendency toward parallel retreat with increasing time span, as already illustrated by the case of Byobugaura cliff which ranges from 10 to 60 m in height. A plot of the data taken for Holderness cliff (4 to 43 m high), Humberside, England, indicates no correlation between cliff height and erosion rate during the period of 100 years from 1852 to 1952.[32] Even for a short period of time, 7 years from 1960 to 1967, no significant correlation is present for Byobugaura.[135] In a very short interval, however, this relation may hold as suggested from Williams' record of the erosional events caused by the 1953 storm surge in England: on the coast south of Lowestoft, Suffolk, a glacial-sand cliff 12.2 m high was eroded 12.2 m overnight, while a cliff 1.8 m high was cut back 27.4 m.[79]

Effect of Beach Sediment Upon Erosion

Beach sediment residing at the foot of a cliff comprises (1) material produced through cliff erosion by waves, (2) material derived from disintegrated debris masses, and (3) material transported from adjacent coasts by longshore currents and/or from the offshore area by constructive wave action. Beach sediment at a cliff toe can either intensify the wave assailing force, f_w, by acting as an abrasive, or reduce f_w by working as a wave-energy dissipator. The control of f_w by the beach material depends upon (1) the wave energy level at the cliff base, (2) the wave direction, and (3) the size and amount of beach sediment.

The effect of the amount of beach sediment, or the beach elevation, upon cliff erosion rate has been reported from some locations. Carter[22] investigated cliff erosion along the Ohio shore of Lake Erie. He found that the bluff made of glacial deposits receded more

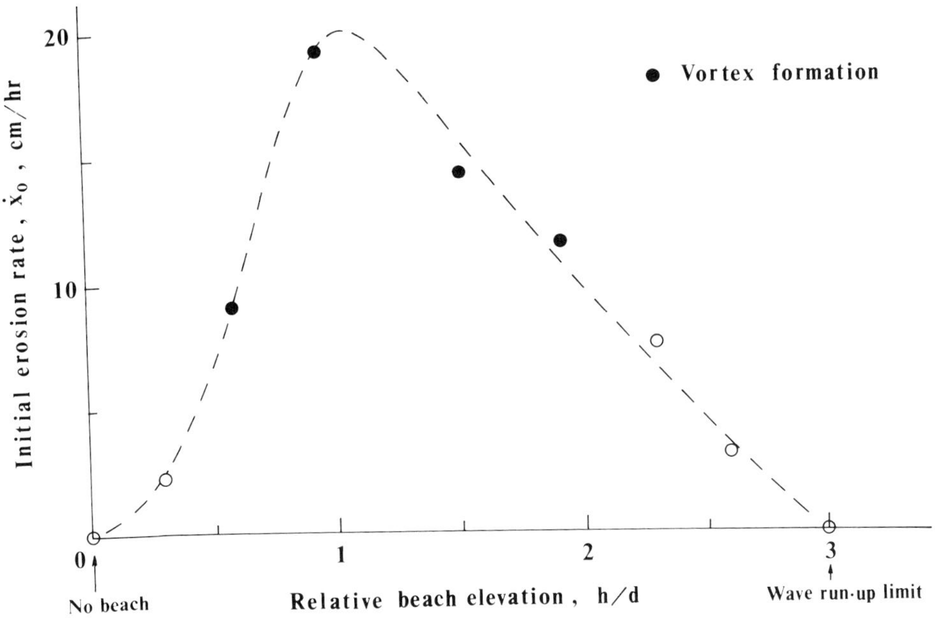

FIGURE 14. Experimental results showing the effect of the amount of beach sand upon cliff erosion rate. In this figure, $\dot{X}_o$ is the erosion rate immediately after wave action, h is the height of the cliff-beach junction, and d is the water level at the base of the cliff with no beach. The greatest erosion takes place when the beach top is located at about SWL. (From Sunamura, T., *Earth Surface Processes Landforms*, 7, 333, 1982. Copyright© 1982. Reprinted by permission of John Wiley & Sons, Ltd.)

rapidly on the downdrift side of artificial structures such as groins and jetties, where the beach is poorly developed at the cliff base, as compared with the updrift side having a well-developed beach. At Hallsands in South Devon, England, during the period of 5 years from 1897 to 1902, approximately 500,000 yd^3 of shingle were removed for construction material from the foreshore of a micaschist cliff on which the fishing village of Hallsands was located. As a result, the entire beach was lowered by as much as 3.6 m.[116] This shingle removal caused rapid cliff erosion, leading to the ruin of the fishing village.[58,116] Near Kilkeel, Northern Ireland, change in beach elevation is one of the main factors responsible for temporal variations in erosion rates of the glacial-deposit cliff.[95] Based on precise measurements using a microerosion meter, Robinson[120] reported that on the northeast Yorkshire coast of England, erosion rates of shale cliffs fronted by beaches were 15 to 18.5 times higher than those with no beaches.

Figure 14 is a plot of the erosion rate vs. relative beach elevation, as obtained from a wave-tank experiment.[142] The experiment was performed under constant wave conditions, by systematically changing the amount of beach sand at the foot of steep model cliffs of the same slope and strength. Duration of wave action was 2 hr for each run. Cliff erosion occurred when the beach material at the cliff-beach junction was moved by waves. Figure 14 shows that (1) the greatest erosion occurred when $h/d \approx 1$, viz., when the beach top abutting the cliff was located at about the SWL, and (2) formation of a vortex containing sand grains was closely related to severe cliff erosion, as previously noted for another laboratory experiment.[135,137]

Turbulence created by bores running up on the model beach mobilized the sand and exerted a mechanical shearing force on the cliff face using the sand as an abrasive. An analysis of the experimental results showed that the assailing force of the sediment-laden water masses is proportional to the square of the bore speed immediately in front of the cliff face, and the factor of proportionality is related to the quantity of beach sand entrapped in

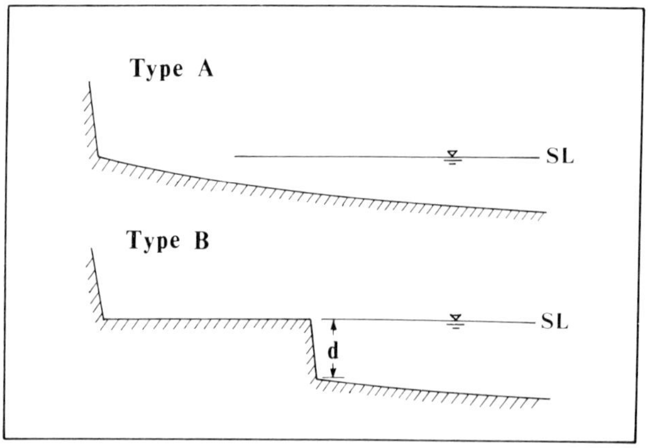

FIGURE 15. Diagrammatic representation of two types of shore platforms. Type A is characterized by a gently seaward-sloping bottom, and Type B is marked by a nearly horizontal surface with a seaward scarp.

the turbulent fluid.[142] For the evaluation of wave assailing force in the full-scale environment, the bore speed and the size and amount of beach sediment mobilized are of vital importance.

LONG-TERM EROSION PROBLEMS: SHORE PLATFORM FORMATION

Problems in Platform Studies

The recession of sea cliffs results in shore platforms at their feet; the platforms are gently seaward-sloping rock surfaces — or nearly horizontal ones depending upon location. Since Dana first described in 1849 the formative processes of shore platforms[6] numerous researchers have attempted to explain platform formation. In the last decade many studies on this point were performed in various parts of the world: Australia,[50,51,53,61,62] Canada,[161] Japan,[70,90,140,153] New Zealand,[81] U.K.,[119,121,158-160,175] and U.S.A.[16,40] However, an explicit explanation has not yet been provided to the following basic question: at what level, by what kind of waves, and how fast are platforms originated at a place with given topographical, geological, and tidal conditions?

Multiple eustatic changes plus local crustal movement make this kind of investigation very difficult. As Bird and Dent[10] have stated, it is probably misleading to attempt to explain the formation of platforms entirely in terms of agents visibly at work on them at present. In tectonically unstable regions, the occurrence of recent crustal movements should be scrutinized before exploring the formative mechanisms.[140]

Some existing studies have stressed the importance of the weathering effect upon platform development (e.g., Bartrum,[6] Wentworth,[167,168] Tricart[162]), while others have placed emphasis on the role of storm-wave action (e.g., Bartrum,[6,7] Edwards,[34,35] Jutson,[72] So[128]). Water level effect (tidal range) upon platform elevation or gradient has been recently investigated.[159,160,174]

Basically, the same conditions as shown in Figure 3 can be applied to this problem. Platform cutting takes place when the wave assailing force, f_w, is greater than the rock resisting force, f_r. Since f_w and f_r are crucial factors, platform studies require an understanding of both. It is clear that knowledge of nearshore wave dynamics and lithological factors, inclusive of the weathering effect, is essential for the solution of the problem.

Development of Platforms

Shore platforms found around the present sea level[9,62,157,174] are largely classified into two types (Figure 15): one is a platform that slopes seaward without significant topographic break extending to the nearshore bottom — Type A; and the other is a platform that has a nearly horizontal surface with marked topographic break, a scarp, at the seaward margin — Type B. The surface of Type B platforms shows a slightly seaward gradient in some places or a reverse inclination in other locations. The occurrence of the two distinctive types has not been dynamically explained.

Type-A platform development — Flemming,[45] in 1965 was the first researcher to mathematically treat Type-A platform evolution with a static sea level. His model, based on the assumptions of (1) a cliff made of uniform rock with no subaerial weathering, and (2) no beach formation, indicates that the rate of cliff recession decreases with time, but does not reach zero. Under the same assumptions as Flemming's, a physical model was constructed using Equation 4.[141] On this type of platform, all waves break before approaching the cliff face and a surf zone is formed, due to (1) very shallow water at the cliff foot, and (2) the seaward sloping bottom. Transformation of Equation 4 using a relation for the surf-zone wave height attenuation led to

$$\frac{dX}{dt} = Kke^{-\gamma kt} \tag{7}$$

where dX/dt is the rate of platform development, i.e., the long-term cliff erosion rate, $K = C - \gamma W_o + \ln(\rho gH_b/S_c)$. $\gamma = \epsilon\sqrt{h_b}/\sqrt{gh_a}L_b^2$, H_b is the breaker height, h_b is the breaking depth, L_b is the wavelength at the breaking point, ϵ is a wave height attenuation coefficient, h_a is the maximum water depth beyond which no bedrock abrasion can extend,[139] and W_o is the initial platform width. Equation 7 indicates that $dX/dt \rightarrow 0$ as $t \rightarrow \infty$. While waves at the cliff foot have sufficient force to cause cliff erosion, a platform continues to grow with a rapid rate at the early stage. In time, however, the growth rate decreases due to the interaction between waves and platform width, i.e., due to negative feedback. Finally, platform development halts when the wave assailing force becomes insufficient to further cut back the cliff. No effect of weathering was included in this modeling. The reduction in f_r (here represented by S_c) due to weathering should be quantified for the construction of more realistic models.

Figure 16 shows the elevation of platforms developing in front of rapidly eroding chalk cliffs near Minnis Bay, Kent, England.[173] The platform height in the bays is greater than that off headlands. A similar relationship is found in the adjacent area, near Grenham Bay.[175] According to Wood,[173] platforms in the bays are covered by beach sand, while platforms around the headlands consist of bare bedrock. As he suggested, waves exert a force intensified by the abrasive action of sand on a bay-head cliff at a higher level when they hit the cliff after running up on the beach. Such wave action develops caves, the deepest parts of which are located at a much higher elevation (Figure 16). This cave development is probably similar to the notch development observed in a wave-tank experiment (Figure 6).

No accurate and consecutive measurements of topographic change had been made in connection with platform development until Robinson[119] attempted to perform such investigations using the microerosion meter technique on the northeast Yorkshire shore platforms composed of Lower Jurrassic rocks. He found that the debris resting on the platform surface greatly influences the temporal change of the platform profile.

Recent studies conducted in the U.K.[119,173] treat the processes of platform growth in the contemporary littoral environment. Results of such work enable us to gain an insight into Type A platform evolution. This kind of study, preferably based on considerations of f_w and f_r, such as in Figure 3, should be continued.

Type-B platform development — Most of the field investigations reported in the literature have dealt with this type of platform, but no satisfying explanation has yet been given as

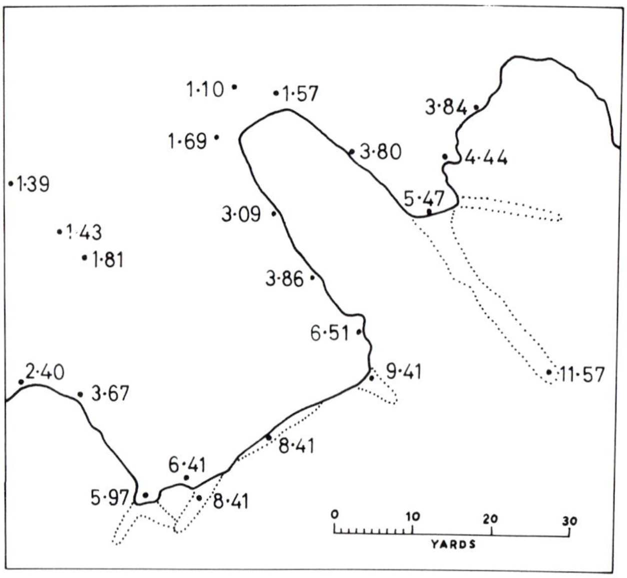

FIGURE 16. Platform elevation, measured in feet above Ordinance Datum (Newlyn), east of Minnis Bay, Kent, England. The dotted lines show caves below the cliffs, and heights of cave floors are given. (From Wood, A., *Z. Geomorphol.*, 12, 107, 1968. With permission.)

to the origin. In order to explore this basic problem, laboratory experiments have been performed under simplified conditions.[123,136] In these experiments, steep model cliffs were built in a wave tank. Sanders[123] used plaster as a cementing material for the model cliff. His study is useful when we consider the formative processes of platforms made of calcarious rocks in the real world.[52] Because plaster is soluble in water, the results should be analyzed considering the effect of solution upon the temporal reduction in mechanical strength of the cliff. To eliminate this chemical effect, portland cement was used in Sunamura's[136] experiment.

A seaward bluff characterizing Type B platforms (Figure 15) might be the initial profile of a steep drowned coast.[27] Due to the presence of the bluff, three different types of waves occur in front of this scarp: standing waves when $d > h_b$, breaking waves when $d = h_b$, and broken waves when $d < h_b$, where d is the water depth at the scarp, and h_b is the breaking depth of the incoming waves. A steep homogeneous model cliff made of a cement:sand mixture was exposed to the three types of waves.[136] Standing waves exerted little assailing force on the cliff and created no platforms. In the natural environment, platforms are not developed on plunging cliff coasts.[9,25] Since standing waves are likely to occur on such coasts due to reflection from the bluff, not only the rock factor, the resistance of cliff rocks,[9] but also the wave factor is unfavorable for producing platforms.

In the experiments, breaking and broken waves produce platforms. The platform created by breaking waves was located lower than that made by broken waves. The assailing force of breaking waves, which is marked by impulsive hydrodynamic pressure, is much greater than that of broken waves. Broken waves suffer a rapid reduction in energy with increase of distance from the breaking point in the surf zone due to dissipation by turbulence, bottom friction, etc. Consequently, breaking waves immediately in front of bluffs can be regarded as the most effective agent in platform initiation.

Even in a simplified laboratory environment, however, no generalized relationship describing the origin of platforms has been obtained. More model tests should be carried out selecting major factors considered to be responsible for platform origination. However, it is virtually impossible for laboratory experiments to include all the factors involved in this problem.

On the coast east of Warrnambool, Victoria, Aust., narrow Type B platforms are cut by present-day processes into huge blocks made of aeolianite, which have fallen from adjacent high cliffs.[50,52,53] The maximum rate of development for one of these platforms is estimated at 1.4 cm/year.[53] Since the initial boundary condition is clear, such contemporary platforms could serve best as a prototype for small-scale model studies.

Platform Surface Lowering

Shore platforms grow with cliff retreat, lowering the bedrock surface at the cliff-platform junction, especially for Type A platforms. Zenkovich[182] related the lowering rate of platform surface, dZ/dt, to the rate of cliff recession dX/dt, by

$$\frac{dZ}{dt} = \frac{dX}{dt} \tan \beta \qquad (8)$$

where tanβ is the platform gradient. Calculations using this relation showed that platforms of Neogene calcarious rocks are being lowered on the order of 0.1 mm/year on the coast of the Black Sea.[182] Another application of Equation 8 was made to Byobugaura on the Pacific coast of Japan. The result showed that Pliocene mudstone platforms are being lowered at an average rate of 2 cm/year.[67] This lowering rate is 3% of the cliff recession rate. Based on a similar relation to Equation 8, Trenhaile[160] estimated lowering rates on the coasts of England and Wales: for chalk platforms the lowering rates were about 2% of the growth rate, while for Lias platforms they were approximately 5%. A value of 3% was obtained for Tertiary mudstone and limestone platforms at the Kaikoura Peninsula. N.Z.[81] An attempt to check the validity of Equation 8 was also made by Kirk[81] using direct measurements.

Apart from such primary surface-lowering inherent in platform development, there is always a secondary lowering of the surface of platforms originally cut by wave action. Secondary processes fall into three categories: mechanical, chemical, and biological. It is difficult to treat them in a perfectly isolated manner, because they partially interact with one another. Mechanical processes include (1) abrasive action of wave-moved sediment on platforms,[81,121,173] (2) potholing,[150,169] and (3) disintegration of rocks forming the platform surface due to the alternation of wetting and drying.[148,149,151,152] Based on rock mechanics, a new approach to the wetting and drying weathering processes has been made by Suzuki et al.[148,149] and Takahashi.[151,152] Chemical processes are characterized by solution of platform-forming rocks.[37,40,99] Biological factors are the action of rock-boring and rock-browsing organisms.[59]

Direct measurements of the rate of secondary lowering have been made by Robinson[121] using the microerosion meter (MEM) technique on the northeast Yorkshire coast. The average lowering rate of shale platforms was 0.1 cm/year on the horizontal portion of the platforms where the dominant erosional process is the removal by waves of shale fragments produced by the alternate wetting and drying. On the sloping portion of the platforms near cliff bases where the abrasive action by sand and pebbles is pronounced, the lowering rates were different according to differing thickness of the beach sediment. Kirk's[81] study applying the MEM technique to the Kaikoura Peninsula coast, N.Z., indicated that the average lowering rate is 0.15 cm/year on Tertiary mudstone and limestone platforms, where the lowering is caused mainly by abrasion. Comparing old with new photographs of the surfaces of Cretaceous sandstone platforms at La Jolla, Calif., Emery and Kuhn[40] estimated the rate of lowering due to abrasion combined with solution. The results ranged from 0.003 to 0.06 cm/year.

These quantitative studies surely provide us with a key for the estimation of the altitude of platforms originally created. Such investigations will hopefully be advanced through a deeper understanding of the wave assailing force causing the surface lowering.

CONCLUDING REMARKS

Considerable progress in our understanding of processes of sea cliff and platform erosion has been made during the last decade. This has been in step with the development of measuring techniques, accumulation and analysis of field data, and studies through laboratory experiments. However, the evolutionary process of rocky coasts has not yet been thoroughly explained. To elucidate this we must make further effort to (1) evaluate the wave assailing force, (2) quantify the rock resisting force, and (3) measure resulting topographic changes. Finally, the results of these three basic investigations must be incorporated on a physical basis.

ACKNOWLEDGMENTS

Most of the author's original research was performed when he was at the Coastal Engineering Laboratory, University of Tokyo. Dr. K. Horikawa, the director of the CEL, afforded the opportunity to study rocky coasts from the dynamical point of view. Many people have provided much input through discussions, especially Drs. Y. Nishimatsu, T. O. Sasaki, T. Suzuki, K. Takahashi, and E. Waddell. Dr. N. Kraus reviewed a preliminary version of the manuscript. Financial support has been provided through the Science Research Fund of the Ministry of Education (975242, 075112, 175272, and 365180).

REFERENCES

1. **Agar, R.,** Coastal cliffs: Report of a symposium — Discussion, *Geogr. J.*, 128, 303, 1962.
2. **Ahr, W. M. and Stanton, R. J., Jr.,** The sedimentologic and paleoecologic significance of *Lithotrya*, a rock-boring barnacle, *J. Sediment. Petrol.*, 43, 20, 1973.
3. **Aramaki, M.,** Erosion of coastal cliff at Iwaki coast in Fukushima prefecture, Japan, (Engl. abstr.), *Bull Assoc. Nat. Sci. Senshu Univ.*, 11, 5, 1978.
4. **Barnes, H. L.,** Cavitation as a geological agent, *Am. J. Sci.*, 254, 493, 1956.
5. **Barnes, R. S. K., Ed.,** *The Coastline*, John Wiley & Sons, New York, 1977.
6. **Bartrum, J. A.,** "Abnormal" shore platforms, *J. Geol.*, 34, 793, 1926.
7. **Bartrum, J. A.,** Shore platforms: discussion, *J. Geomorphol.*, 1, 266, 1938.
8. **Benson, D. J.,** Shore erosion along a low-lying shoreline, western Lake Erie, Abstr. 17th Conf. Great Lakes Res., McMaster University, Hamilton, Ontario, 1974, 219.
9. **Bird, E. C. F.,** *Coasts*, MIT Press, Cambridge, 1969.
10. **Bird, E. C. F. and Dent, O. F.,** Shore platforms on the south coast of New South Wales, *Aust. Geogr.*, 10, 71, 1966.
11. **Bird, J. B., Richards, A., and Wong, P. P.,** Coastal subsystems of Western Barbados, West Indies, *Geogr. Annaler*, 61A, 221, 1979.
12. **Birkeland, P. W. and Larson, E. E.,** *Putnam's Geology*, Oxford University Press, New York, 1978.
13. **Birkemeier, W. A.,** The effect of structures and lake level on bluff and shore erosion in Berrien County, Michigan, 1970-74, *U.S. Army Coastal Eng. Res. Cent. Misc. Rep.*, No. 80-2, 1980.
14. **Birkemeier, W. A.,** Coastal changes, Eastern Lake Michigan, 1970-74, *U.S. Army Coastal Eng. Res. Cent. Misc. Rep.*, No. 81-2, 1981.
15. **Bokuniewicz, H. and Tanski, J.,** Managing localized erosion of coastal bluffs, *Coastal Zone '80*, Vol. 3, American Society of Civil Engineers, New York, 1980, 1883.
16. **Bradley, W. C., and Griggs, G. B.,** Form genesis, and deformation of central California wave-cut platforms, *Geol. Soc. Am. Bull.*, 87, 433, 1976.
17. **Brunsden, D. and Jones, D. K. C.,** Relative time scales and formative events in coastal landslide systems, *Z. Geomorphol.*, 34(Suppl. Bd.), 1, 1980.
18. **Bryan, R. B. and Price, A. G.,** Recession of the Scarborough bluffs, Ontario, Canada, *Z. Geomorphol.*, 34(Suppl. Bd.), 48, 1980.
19. **Byrne, J. V.,** Coastal erosion, Northern Oregon, in *Essays in Marine Geology in Honor of K. O. Emery*, Clements, T., Ed., University of Southern California Press, Los Angeles, 1963, 11.

20. **Byrne, J. V.,** An erosional classification for the northern Oregon coast, *Ann. Assoc. Am. Geogr.*, 54, 329, 1964.
21. **Cambers, G.,** Temporal scales in coastal erosion systems, *Trans. Inst. Br. Geogr. New Ser.*, 1, 246, 1976.
22. **Carter, C. H.,** Lake Erie Shore Erosion, Lake County, Ohio: Setting, Processes, and Recession Rates from 1876 to 1973, Div. Geol. Survey, Dept. Nat. Res. Invest. Rep. no. 99, State of Ohio, Columbus, 1976.
23. **Chien, N.,** Sediment motion at the vicinity of littoral barrier, U.S. Army Beach Erosion Board, Bull. 10, 21, 1956.
24. **Clark, M. J., Ricketts, P. J., and Small, R. J.,** Barton does not rule the waves, *Geogr. Mag.*, 48, 580, 1976.
25. **Cotton, C. A.,** Plunging cliffs, Lyttelton Harbour, *N.Z. Geogr.*, 2, 130, 1949.
26. **Cotton, C. A.,** *Geomorphology*, Whitcombe and Tombs, London, 1960.
27. **Cotton, C. A.,** Levels of planation of marine benches, *Z. Geomorphol.*, 7, 97, 1963.
28. **Cottonaro, W. F.,** Sea cliff erosion, Isla Vista, California, *Calif. Geol.*, June, 140, 1975.
29. **Davies, D. S., Axelron, E. W., and O'Conner, J. S.,** *Erosion of the North Shore of Long Island*, Mar. Sci. Res. Cent., Tech. Rep. 18, State University New York, Stony Brook, 1972.
30. **Davies, J. L.,** *Geographical Variation in Coastal Development*, Oliver and Boyd, Edinburgh, 1972.
31. **Davis, W. M.,** The outline of Cape Cod, in *Geographical Essays*, Johnson, D. W., Ed., Dover Publ., New York, 1954, 690.
32. **Doornkamp, J. C. and King, C. A. M.,** *Numerical Analysis in Geomorphology, An Introduction*, Edward Arnold, London, 1971.
33. **Eagleson, P. S. and Dean, R. G.,** Small amplitude wave theory, in *Estuary and Coastline Hydrodynamics*, Ippen, A. T., Ed., McGraw-Hill, New York, 1966, 1.
34. **Edwards, A. B.,** Storm-wave platforms, *J. Geomorphol.*, 4, 223, 1941.
35. **Edwards, A. B.,** Wave action in shore-platform formation, *Geol. Mag.*, 88, 41, 1951.
36. **Emery, K. O.,** Rate of surface retreat of sea cliffs based on dated inscriptions, *Science*, 93, 617, 1941.
37. **Emery, K. O.,** Marine solution basins, *J. Geol.*, 54, 209, 1946.
38. **Emery, K. O.,** *The Sea off Southern California*, John Wiley & Sons, New York, 1960.
39. **Emery, K. O. and Foster, H. L.,** Shoreline nips in tuff at Matsushima, Japan, *Am. J. Sci.*, 254, 380, 1956.
40. **Emery, K. O. and Kuhn, G. G.,** Erosion of rock shores at La Jolla, California, *Mar. Geol.*, 37, 197, 1980.
41. **Emmons, W. H., Thiel, G. A., Stauffer, C. R., and Allison, I. S.,** *Geology*, McGraw-Hill, New York, 1955.
42. **Evans, J. W.,** The role of *Penitella Penita* (Conarad 1837) (Family Pholadidae) as eroders along the Pacific Coast of North America, *Ecology*, 49, 156, 1968.
43. **Fairbridge, R. W.,** "Platforms — wave-cut", in *Encyclopedia of Geomorphology*, Fairbridge, R. W., Ed., Reinhold, New York, 1968, 859.
44. **Flemming, C. A.,** The Geology of Wanganui Subdivision, N.Z. Geol. Surv., Bull. No. 52, Wellington, 1953.
45. **Flemming, N. C.,** Form and relation to present sea level of Pleistocene marine erosion features, *J. Geol.*, 73, 799, 1965.
46. **Focke, J. W.,** Limestone cliff morphology and organism distribution on Curacao (Netherlands Antilles), *Leidse Geol. Meded.*, 51, 131, 1978.
47. **Focke, J. W.,** Limestone cliff morphology on Curacao (Netherlands Antilles), with special attention to the origin of notches and vermetid/coralline algal surf benches ("cornices", "trottoirs"), *Z. Geomorphol.*, 22, 329, 1978.
48. **Gatto, L. W.,** Shoreline changes along the outer shore of Cape Cod, U.S. Army Corps from Long Point to Monomoy Point, U.S. Army Corps Eng. Cold Reg. Res. Eng. Lab., CRREL Rep. 78-17, 1978, 49.
49. **Gelinas, P. J. and Quigley, R. M.,** The influence of geology on erosion rates along the north shore of Lake Erie, *Proc. 16th Conf. Great Lakes Res.*, McMaster University, Hamilton, Ontario, 1973, 421.
50. **Gill, E. D.,** The relationship of present shore platforms to past sea levels, *Boreas*, 1, 1, 1972.
51. **Gill, E. D.,** Ramparts on shore platforms, *Pac. Geol.*, 4, 121, 1972.
52. **Gill, E. D.,** Sanders' wave tank experiments and shore platforms, *Pap. Proc. R. Soc. Tasmania*, 106, 17, 1972.
53. **Gill, E. D.,** Rate and mode of retrogradation on rocky coasts in Victoria, Australia, and their relationship to sea level changes, *Boreas*, 2, 143, 1973.
54. **Gilluly, J., Waters, A. C., and Woodford, A. O.,** *Principles of Geology*, 2nd ed., W. H. Freeman, San Francisco, 1959.
55. **Goldsmith, F. B.,** Rocky cliffs, in *The Coastline*, Barnes, R. S. K., Ed., John Wiley & Sons, Chichester, 1977, 237.
56. **Griggs, G. B. and Johnson, R. E.,** Coastal erosion, Santa Cruz County, *Calif. Geol.*, April, 67, 1979.

57. **Guilcher, A.,** *Coastal and Submarine Morphology,* Methuen, London, 1964.
58. **Hails, J. R.,** Sediment distribution and Quaternary history, *J. Geol. Soc. London,* 131, 19, 1975.
59. **Healy, T. R.,** Bioerosion on shore platforms in the Waitemata Formation, Auckland, *Earth Sci. J.,* 2, 26, 1968.
60. **High, C. and Hanna, F. K.,** A Method for the Direct Measurement of Erosion on Rock Surfaces, Br. Geomorphol. Res. Group, Tech. Bull. no. 5, Geo. Abstracts, University of East Anglia, Norwich, England, 1970.
61. **Hills, E. S.,** A study of cliffy coastal profiles based on examples in Victoria, Australia, *Z. Geomorphol.,* 15, 137, 1971.
62. **Hills, E. S.,** Shore platforms and wave ramps, *Geol. Mag.,* 109, 81, 1972.
63. **Hodgkin, E. P.,** Rate of erosion of intertidal limestone, *Z. Geomorphol.,* 8, 385, 1964.
64. **Holmes, A.,** *Principles of Physical Geology,* 3rd ed., Thomas Nelson and Sons, Middlesex, 1978.
65. **Horikawa, K. and Sunamura, T.,** A study on erosion of coastal cliffs by using aerial photographs, *Coastal Eng. Jpn.,* 10, 67, 1967.
66. **Horikawa, K. and Sunamura, T.,** An experimental study on erosion of coastal cliffs due to wave action, *Coastal Eng. Jpn.,* 11, 131, 1968.
67. **Horikawa, K. and Sunamura, T.,** A study on erosion of coastal cliffs and of submarine bedrocks, *Coastal Eng. Jpn.,* 13, 127, 1970.
68. **Hutchinson, J. N.,** A reconsideration of the coastal landsides at Folkestone Warren, Kent, *Géotechnique,* 19, 6, 1969.
69. **Hutchinson, J. N.,** A coastal mudflow on the London Clay cliffs at Beltinge, north Kent, *Géotechnique,* 20, 412, 1970.
70. **Imanaga, I.,** On the coastal erosive features of Enoshima island, (Engl. abstr.), *Bull. Kanagawa Prefectural Mus.,* 8, 79, 1975.
71. **Johnson, D. W.,** *New England — Acadian Shoreline,* John Wiley & Sons, New York, 1925.
72. **Jutson, J. T.,** Shore platforms near Sydney, New South Wales, *J. Geomorphol.,* 2, 237, 1939.
73. **Kanazaki, H.,** Shoreline recession on the western Kanazawa coast, Japan, in *Dr. Tsujimura Commemorative Volume of Geographical Treatises,* (Engl. abstr.) Tada, F., Ed., 1961, 145.
74. **Kaye, C. K.,** The effect of solvent motion on limestone solution, *J. Geol.,* 65, 35, 1957.
75. **Kaye, C. K.,** Shoreline features and Quarternary shoreline changes, Puerto Rico, U.S. Geol. Surv. Prof. Pap. 317B, Department of Commerce, Washington, D.C., 1959, 49.
76. **Kawasaki, I.,** Geomorphological study of the Byobugaura sea-cliff in the vicinity of Iioka-machi, Chiba prefecture, (Engl. abstr.) *Geogr. Rev. Jpn.,* 27, 213, 1954.
77. **Kennedy, M. P.,** Sea-cliff erosion at Sunset Cliffs, San Diego, *Calif. Geol.,* February, 27, 1973.
78. **King, C. A. M.,** Coastal cliffs: Report of a symposium — Discussion, *Geogr. J.,* 128, 303, 1962.
79. **King, C. A. M.,** *Beaches and Coasts,* 2nd ed., Edward Arnold, London, 1972.
80. **Kirk, R. M.,** Coastal changes at Kaikoura, 1942-74, determined from air photographs, *N.Z. J. Geol. Geophys.,* 18, 788, 1975.
81. **Kirk, R. M.,** Rates and forms of erosion on intertidal platforms at Kaikoura Peninsula, South Island, New Zealand, *N.Z. J. Geol. Geophys.,* 20, 571, 1977.
82. **Kolberg, T. O.,** Evaluation of water levels effect on erosion rate, *Summaries,* 14th Int. Conf. Coastal Eng., American Society of Civil Engineers, New York, 1974, 413.
83. **Komar, P. D.,** *Beach Processes and Sedimentation,* Prentice-Hall, Englewood Cliffs, N. J., 1976.
84. **Kuenen, Ph. H.,** *Marine Geology,* John Wiley & Sons, New York, 1960.
85. **Kuhn, G. G. and Shepard, F. P.,** Coastal erosion in San Diego, California, *Coastal Zone '80,* Vol. 3, American Society of Civil Engineers, New York, 1980, 1899.
86. **Lee, L. J.,** Sea cliff erosion in Southern California, *Coastal Zone '80,* Vol. 3, American Society of Civil Engineers, New York, 1980, 1919.
87. **Lee, L. J., Pinckney, C., and Bemis, C.,** Sea cliff erosion, *Natl. Water Res. Ocean Eng. Convention,* San Diego, Calif., American Society of Civil Engineers, New York, 1976, preprint 2708.
88. **Lewis, J. R.,** *The Ecology of Rocky Shores,* English University Press, London, 1964.
89. **Longwell, C. R., Flint, R. F., and Sanders, J. E.,** *Physical Geology,* John Wiley & Sons, New York, 1969.
90. **Machida, T.,** Geomorphology and bottom materials of Nabeta Bay, Izu peninsula, central Japan, (Engl. abstr.) *Tokyo Geogr. Pap.,* 16, 59, 1972.
91. **Matthews, E. R.,** *Coast Erosion and Protection,* Charles Griffin, London, 1934.
92. **May, V. J.,** The retreat of chalk cliffs, *Geogr. J.,* 137, 203, 1971.
93. **May, V. J.,** Earth cliffs, in *The Coastline,* Barnes, R. S. K., Ed., John Wiley & Sons, Chichester, 1977, 215.
94. **McGreal, W. S.,** Factors promoting coastal slope instability in southeast County Down, N. Ireland, *Z. Geomorphol.,* 23, 76, 1979.

95. **McGreal, W. S.,** Marine erosion of glacial sediments from a low-energy cliffline environment near Kilkeel, Northern Ireland, *Mar. Geol.*, 32, 89, 1979.
96. **McGreal, W. S.,** Cliffline recession near Kilkeel, N. Ireland; an example of a dynamic coastal system, *Geogr. Annaler*, 61A, 211, 1979.
97. **McGreal, W. S. and Craig, D.,** Mass-movement activity: An illustration of differing responses to groundwater conditions from two sites in northern Ireland, *Ir. Geogr.*, 10, 28, 1977.
98. **McLean, R. F. and Davidson, C. F.,** The role of mass-movement in shore platform development along the Gisborne coastline, *N.Z. Earth Sci. J.*, 2, 15, 1968.
99. **Mii, H.,** Coastal geology of Tanabe Bay, *Sci. Rep. Tohoku Univ. Ser. 2*, 34, 1, 1962.
100. **Mogi, A., Tsuchide, M., and Fukushima, M.,** Coastal erosion of new volcanic island Nishinoshima, (Engl. abstr.), *Geogr. Rev. Jpn.*, 53, 449, 1980.
101. **Neumann, A. C.,** Observations on coastal erosion in Bermuda and measurements of the boring sponge, *Cliona lampa*, *Limnol. Oceanogr.*, 11, 92, 1966.
102. **Neumann, A. C.,** Biological erosion of limestone coasts, in *Encyclopedia of Geomorphology*, Fairbridge, R. W., Ed., Reinhold, New York, 1968, 75.
103. **Noda, H.,** *Beach Processes*, Summer Semin. Hydraul. Eng., Japan Society of Civil Engineers, Tokyo, 1971, 1. (In Japanese.)
104. **Norris, R. M.,** Sea Cliff Retreat Near Santa Barbara, California, Mineral Information Service, Vol. 21 (No. 6), California Division of Mines and Geology, Sacramento, 87.
105. **Norrman, J. O.,** Trends in postvolcanic development of Surtsey Island, Surtsey Res. Prog. Rep., No. 5, Department of Geography, Uppsala University, Uppsala, Sweden, 1.
106. **Norrman, J. O.,** Coastal erosion and slope development in Surtsey Island, *Z. Geomorphol.*, 34(Suppl. Bd.), 20, 1980.
107. **Ohshima, H.,** Erosional pattern of sea cliffs, with special reference to geological aspects, *Proc. Symp. Coastal Cliff Recession*, Japan Society of Civil Engineers, Tokyo, 1974, 9. (In Japanese.)
108. **Prêcheur, C.,** Le littoral de la Manche, de Sainte-Adresse à Ault, Étude morphologique, Special issue, Norois, Centre National de la Recherche Scientifique et du Conseil Général de la Seine-Maritime, Poitiers, 1960.
109. **Prior, D. B. and Renwick, W. H.,** Landslide morphology and processes on some coastal slopes in Denmark and France, *Z. Geomorphol.*, 34(Suppl. Bd.), 63, 1980.
110. **Prior, D. B., Stephens, N., and Archer, D. R.,** Composite mudflows on the Antrim coast of north-east Ireland, *Geogr. Annaler*, 50A, 65, 1968.
111. **Quigley, R. M. and Di Nardo, L. R.,** Cyclic instability modes of eroding clay bluffs, Lake Erie northshore bluffs at Port Bruce, Ontario, Canada, *Z. Geomorphol.*, 34(Suppl. Bd.), 39, 1980.
112. **Quigley, R. M. and Zeman, A. J.,** Strategy for hydraulic, geologic and geotechnical assessment of Great Lakes shoreline bluffs, in The Coastline of Canada, McCann, S. B., Ed., Pap. 80-10, Geological Survey Canada, Ottawa, 1980, 397.
113. **Quigley, R. M., Gelinas, P. J., Bou, W. T., and Packer, R. W.,** Cyclic erosion-instability relationships: Lake Erie northshore bluffs, *Can. Geotech. J.*, 14, 310, 1977.
114. **Richards, A. F.,** Rates of marine erosion of tephra and lava at Isla San Benedicto, Mexico, *Proc. 21st Int. Geol. Congress, Part 10, Submarine Geology*, Det Berlingske Bogtrykkeri, Copenhagen, 1960, 59.
115. **Ricketts, E. F. and Calvin, J.,** *Between Pacific Tides*, Stanford University Press, Stanford, Calif., 1968.
116. **Robinson, A. H. W.,** The hydrography of Start Bay and its relationship to beach changes at Hallsands, *Geogr. J.*, 127, 63, 1961.
117. **Robinson, A. H. W.,** Erosion and accretion along part of the Suffolk coast of East Anglia, England, *Mar. Geol.*, 37, 133, 1980.
118. **Robinson, L. A.,** The micro-erosion meter technique in a littoral environment, *Mar. Geol.*, 22, M51, 1976.
119. **Robinson, L. A.,** The morphology and development of the northeast Yorkshire shore platform, *Mar. Geol.*, 23, 237, 1977.
120. **Robinson, L. A.,** Marine erosive processes at the cliff foot, *Mar. Geol.*, 23, 257, 1977.
121. **Robinson, L. A.,** Erosive processes on the shore platform of northeast Yorkshire, England, *Mar. Geol.*, 23, 339, 1977.
122. **Rudberg, S.,** The cliff coast of Gotland and the rate of cliff retreat, *Geogr. Annaler*, 49A, 283, 1967.
123. **Sanders, N. K.,** Wave tank experiments on the erosion of rocky coasts, *Pap. Proc. R. Soc. Tasmania*, 102, 11, 1968.
124. **Schalk, M.,** A texture study of the outer beach of Cape Cod, Massachusetts, *J. Sediment. Petrol.*, 8, 41, 1938.
125. **Schwartz, M. L.,** Shannon Point cliff recession, *Shore Beach*, April, 45, 1971.
126. **Seibold, E.,** Geological investigation of nearshore sand-transport, examples of methods and problems from the Baltic and North Seas, in *Progress in Oceanography*, Sears, M., Ed., Pergamon Press, Oxford, 1963, 1.

127. **Shepard, F. P. and Grant, U. S.,** Wave erosion along the Southern California coast, *Bull. Geol. Soc. Am.*, 58, 919, 1947.
128. **So, C. L.,** Coastal platforms of the Isle of Thanet, Kent, *Trans. Inst. Br. Geogr.*, 37, 147, 1965.
129. **So, C. L.,** Some coastal changes between Whitstable and Reculver, Kent, *Proc. Geol. Assoc.*, 77, 475, 1967.
130. **Sorensen, R. M.,** Recession of marine terraces — with special reference to the coastal area north of Santa Cruz, California, *Proc. 11th Int. Conf. Coastal Eng.*, American Society of Civil Engineers, New York, 1968, 653.
131. **Sparks, B. W.,** *Geomorphology*, Longmans, London, 1960.
132. **Steers, J. A.,** Notes on erosion along the coast of Suffolk, *Geol. Mag.*, 88, 435, 1951.
133. **Steers, J. A.,** Coastal cliffs: Report of a symposium, *Geogr. J.*, 128, 303, 1962.
134. **Strahler, A. N.,** *The Earth Sciences*, Harper & Row, New York, 1963.
135. **Sunamura, T.,** Coastal cliff erosion due to waves — field investigations and laboratory experiments, *J. Fac. Eng. Univ. Tokyo*, 32, 1, 1973.
136. **Sunamura, T.,** A laboratory study of wave-cut platform formation, *J. Geol.*, 83, 389, 1975.
137. **Sunamura, T.,** Feedback relationship in wave erosion of laboratory rocky coast, *J. Geol.*, 84, 427, 1976.
138. **Sunamura, T.,** A relationship between wave-induced cliff erosion and erosive force of waves, *J. Geol.*, 85, 613, 1977.
139. **Sunamura, T.,** A model of the development of continental shelves having erosional origin, *Geol. Soc. Am. Bull.*, 89, 504, 1978.
140. **Sunamura, T.,** Mechanisms of shore platform formation on the southeastern coast of the Izu peninsula, Japan, *J. Geol.*, 86, 211, 1978.
141. **Sunamura, T.,** A mathematical model of submarine platform development, *Math. Geol.*, 10, 53, 1978.
142. **Sunamura, T.,** A wave tank experiment on the erosional mechanism at a cliff base, *Earth Surface Processes Landforms*, 7, 333, 1982.
143. **Sunamura, T.,** A predictive model for wave-induced cliff erosion, with application to Pacific coasts of Japan, *J. Geol.*, 90, 167, 1982.
144. **Sunamura, T. and Horikawa, K.,** A study on erosion of coastal cliffs by using aerial photographs (Rep. No. 2), *Coastal Eng. Jpn.*, 12, 99, 1969.
145. **Sunamura, T. and Horikawa, K.,** A quantitative study on the effect of beach deposits upon cliff erosion, *Coastal Eng. Jpn.*, 14, 97, 1971.
146. **Sunamura, T. and Horikawa, K.,** A study using aerial photographs of the effect of protective structures on coastal cliff erosion, *Coastal Eng. Jpn.*, 15, 105, 1972.
147. **Suzuki, T.,** Rate of lateral planation by Iwaki River, Japan, *Trans. Jpn. Geomorphol. Union*, 3, 1, 1982.
148. **Suzuki, T., Takahashi, K., and Sunamura, T.,** Rock control in coastal erosion at Arasaki, Miura peninsula, Japan, *International Geography*, Vol. 1, University of Toronto Press, Toronto, Canada, 1972, 66.
149. **Suzuki, T., Takahashi, K., Sunamura, T., and Terada, M.,** Rock mechanics on the formation of washboard-like relief on wave-cut benches at Arasaki, Miura peninsula, Japan, (Engl. abstr.) *Geogr. Rev. Jpn.*, 43, 211, 1970.
150. **Swinnerton, A. C.,** Observations on some details of wave erosion: wave furrows and shore potholes, *J. Geol.*, 15, 171, 1927.
151. **Takahashi, K.,** Differential erosion originating washboard-like relief on wave-cut bench at Aoshima Island, Kyushu, Japan, (Engl. abstr.), *Geogr. Rev. Jpn.*, 48, 43, 1975.
152. **Takahashi, K.,** Differential erosion on wave-cut bench, (Engl. abstr.) *Bull. Fac. Sci. Eng. Chuo Univ.*, 19, 253, 1976.
153. **Takahashi, T.,** Geomorphological study of shore platforms — analytical and genetical, *Sci. Rep. Tohoku Univ. Geogr. Ser.*, 24, 115, 1974.
154. **Takenaga, K.,** The classification of notch profiles and the origin of notches, (Engl. abstr.), *J. Geogr.*, 77, 329, 1968.
155. **Thornbury, W. D.,** *Principles of Geomorphology*, John Wiley & Sons, New York, 1960.
156. **Toyoshima, O.,** Sea cliff erosion and countermeasures, *Proc. Symp. Coastal Cliff Recession*, Japan Society of Civil Engineers, Tokyo, 1974, 1. (In Japanese.)
157. **Toyoshima, Y.,** A study on marine erosive features along San-In coast, (Engl. abstr.), *Sci. Rep. Fac. Educ. Tottori Univ.*, 18, 64, 1967.
158. **Trenhaile, A. S.,** Lithological control of high-water rock ledges in the Vale of Glamorgan, Wales, *Geogr. Annaler*, 53A, 59, 1971.
159. **Trenhaile, A. S.,** The shore platforms of the Vale of Glamorgan, Wales, *Trans. Inst. Br. Geogr.*, 56, 127, 1972.
160. **Trenhaile, A. S.,** The morphology and classification of shore platforms in England and Wales, *Geogr. Annaler*, 56A, 103, 1974.

161. **Trenhaile, A. S.,** The shore platforms of Gaspé, Québec, *Ann. Assoc. Am. Geogr.*, 68, 95, 1978.
162. **Tricart, J.,** Problèmes géomorphologiques du littoral oriental du Brésil, *Cah. Océanogr.*, 11, 276, 1959.
163. **Trudgill, S.,** The marine erosion of limestones on Aldabra Atoll, Indian Ocean, *Z. Geomorphol.*, 26(Suppl. Bd.), 164, 1976.
164. **Valentin, H.,** Der Landverlust in Holderness, Ostengland, von 1852 bis 1952, *Die Erde*, 6, 296, 1954.
165. **Vaughan, T. W.,** Rate of sea cliff recession on the property of the Scripps Institution of Oceanography at La Jolla, California, *Science*, 75, 250, 1932.
166. **Vita-Finzi, C. and Cornelius, P. F. S.,** Cliff sapping by molluscs in Oman, *J. Sediment. Petrol.*, 43, 31, 1973.
167. **Wentworth, C. K.,** Marine bench-forming processes: water-level weathering, *J. Geomorphol.*, 1, 6, 1938.
168. **Wentworth, C. K.,** Marine bench-forming processes. II. Solution benching, *J. Geomorphol.*, 2, 3, 1939.
169. **Wentworth, C. K.,** Potholes, pits, and pans: subaerial and marine, *J. Geol.*, 52, 117, 1944.
170. **Williams, A. T. and Davies, P.,** Man as a geological agent: the sea cliffs of Llantwit Major, Wales, U. K., *Z. Geomorphol.*, 34(Suppl. Bd.), 129, 1980.
171. **Williams, W. W.,** An East Coast survey: some recent changes in the coast of East Anglia, *Geogr. J.*, 122, 317, 1956.
172. **Williams, W. W.,** *Coastal Changes*, Routledge and Kegan Paul, London, 1960.
173. **Wood, A.,** Beach platforms in the chalk of Kent, England, *Z. Geomorphol.*, 12, 107, 1968.
174. **Wright, L. W.,** Some characteristics of the shore platforms of the English Channel Coast and the northern part of the North Island, New Zealand, *Z. Geomorphol.*, 11, 36, 1967.
175. **Wright, L. W.,** Variation in the level of the cliff/shore platform junction along the south coast of Great Britain, *Mar. Geol.*, 9, 347, 1970.
176. **Yajima, A.,** A Study on Recession of a Coastal Cliff at Habuseura, Niijima, Japan, Graduation thesis, Tokyo Kyoiku University, Japan, 1965. (In Japanese.)
177. **Yamanouchi, H.,** On the retreat of the sea cliffs along the Pacific coast at Haranomachi, Japan, (Engl. abstr.), *Geogr. Rev. Jpn.*, 37, 138, 1964.
178. **Yamanouchi, H.,** A geomorphological study about the coastal cliff retreat along the southwest coast of the Atsumi peninsula, central Japan, *Sci. Rep. Fac. Educ. Gunma Univ.*, 26(7), 95, 1977.
179. **Yoshikawa, T.,** Shore erosion and its bearing on topography of Kurobe alluvial fan, Toyama Prefecture, (Engl. abstr.), *Geogr. Inst. Bull., Univ. Tokyo*, 2, 92, 1952.
180. **Zeigler, J. M., Hayes, C. R., and Tuttle, S. D.,** Beach changes during storms on Outer Cape Cod, Massachusetts, *J. Geol.*, 67, 318, 1959.
181. **Zeigler, J. M., Tuttle, S. D., Giese, G. S., and Tasha, H. J.,** Residence time of sand composing the beaches and bars of Outer Cape Cod, *Proc. 9th Int. Conf. Coastal Eng.*, American Society of Civil Engineers, New York, 1964, 403.
182. **Zenkovich, V. P.,** *Processes of Coastal Development*, Oliver and Boyd, Edinburgh, 1967.
183. **Zenkovich, V. P., Ionin, A. S., and Kaplin, P. A.,** Abrasion as the resources of debris in the sea shore zone, (Engl. abstr.), *Inst. Oceanogr. Acad. Sci. U.S.S.R.*, 76, 103, 1965.
184. **Zeuner, F. E.,** *Dating the Past*, Methuen, London, 1952.

Chapter 13

BEACH PROCESSES AND SEA CLIFF EROSION IN SAN DIEGO COUNTY, CALIFORNIA

G. G. Kuhn and F. P. Shepard

In examining the history of beaches and sea cliff erosion it is important to include available information concerning the history of storms in the area, in that the present weather conditions may not be particularly representative. This has been demonstrated to be the case for the southern California region.[15] Records for the past two centuries clearly show that during the three or more most recent decades both rainfall and high storm winds have been far less critical than in most preceding periods, although there are indications that we are drawing to the end of this quiescent period. In this chapter we will examine the evidence for changing weather conditions in southern California and consider its effects on the beaches and sea cliff erosion.

Southern California also provides examples of man-induced effects on the beaches and sea cliff erosion. These include the impact of sea wall and culvert construction, the effects of increased ground water due to irrigation and septic tanks, and changes induced by bluff-top grading.

THE VARIABLE HISTORY OF SOUTHERN CALIFORNIA WEATHER

Several lines of evidence (Figure 1) demonstrate the variability and periodicity of weather conditions in southern California. Figure 1A shows the rainfall during 1880 to 1980 at Julian, located in the mountains east of San Diego. A sine wave variation is evident, showing periods of high and low rainfalls. The rainfall highs are separated by about 40 years and a high appears to be developing at present. Figure 1B is a combined core-varve index for the Santa Barbara basin expressed in varve thickness units which presumably represent the annual contribution of sediment during the rainy seasons.[23] There is a fairly consistent periodicity of about 25 years between highs, except during the years 1824 to 1859 when the sediments were highly bioturbated. Figure 1C shows the standard tree ring indexes for big cone spruce trees along the steep slope of the Santa Ana Mountains at an elevation of 1114 m for the years 1800 to 1974.[3,4] This shows a periodicity somewhat similar to that of the rainfall at Julian and clearly demonstrates the recent drought period from 1946 to the end of the record. Figure 1D shows the winter water temperature 5 m below the surface at La Jolla.[4] There seem to have been long-period trends of warm water with abrupt changes to cold periods. While the water temperature data show reversals, a periodicity is less well defined.

From 1883 to the present we find considerable correlation of high and low rainfalls with the other histograms. Thus periods of high rainfall from 1884 to 1891 correspond with thicker varve sediments, thicker tree rings, and warm winter sea surface temperatures. Lower rainfall from 1892 to 1905 corresponds with thinning of the varve sediments, thin tree rings, and cold water. From 1934 to 1945 we had relatively high rainfall and a warming trend of the sea surface toward the middle of this period. Low rainfall occurred during the years 1946 through 1977, along with thin varves, thin tree rings, and colder water temperatures. This long period of low rainfall is rather well reflected in the other histograms.

More intense weather conditions in the past are also suggested by various qualitative indicators. For example, the loss of many ships in the region is noted by the records of Lloyds of London. The logs of sea captains contain many entries concerning violent storms,

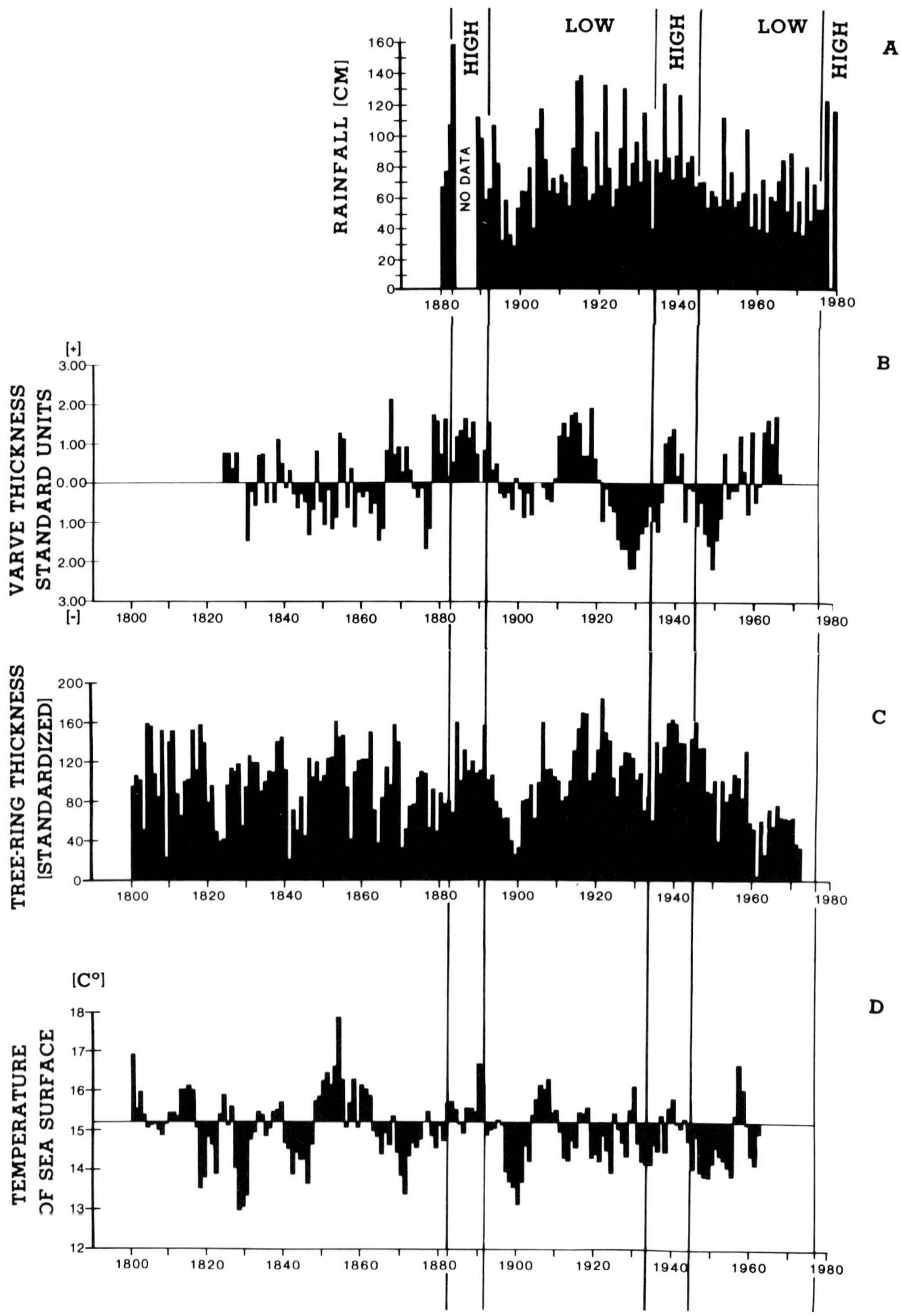

FIGURE 1. See caption opposite.

describing them as being worse than those encountered around Cape Horn. The waves were described as 50 to 60 ft high, much higher than have been seen as a regular occurrence after about 1856. The destructive storms apparently came from the southeast, which is also the case for the more recent storms arriving from Baja California, causing considerable damage to south-facing portions of the coast. The last severe storm of this type was in 1939.

The history of rainfall along the southern California coast, like the wind storms, is apparently quite different from that of recent decades. There were, undoubtedly, periods of intense rain having no equivalents in our present century. The most amazing downpours occurred in 1862 and again at intervals between 1884 and 1891. The floods of 1862 were not confined to the southern California region, but affected the entire western U.S. The Great Valley of California almost became one continuous lake with a length of about 480 km and maximum width of about 95 km. Sacramento was so flooded that the state government had to move to San Francisco. In San Diego and Oceanside (Figure 2) flooding also took place and the large valleys underwent extensive changes. Mission Valley became a rushing torrent, filled from side to side. The entire landscape of southern California must have been changed by this great flood. A long-time resident of San Diego noted that the hillsides at San Diego (New Town) were rounded prior to the flood, and that the gullies and canyons that now exist were formed during the 1862 flood. It is apparent that such flooding, at present, would cause billions of dollars in damage since no flood controls have been developed that could handle such volumes of flood waters.

Our recent intensive studies of weather records for more than the past century clearly show that each phreatic volcanic eruption has been followed by a period, lasting approximately three years, during which the weather conditions have become decidedly abnormal. These periods are charaterized by greatly increased storminess with record numbers of tornados, tremendous rain storms producing floods of exceptional proportions, hurricanes, and heavy snowfalls. The evidence suggests that these periods are related to the aerosols injected into the upper atmosphere by the volcanoes. The recent eruption of El Chichon provides a current example of the climatic effects of volcanism.

COASTAL CHANGES DUE TO LONG-TERM WEATHER VARIATIONS

Such major flooding in the past must also have contributed great quantitites of sediments to the coastal region, producing a filling of estuaries and widening the beaches. On the other hand, the greater storms and higher wave conditions would have intensified the nearshore processes of sediment movement and sea cliff retreat. There is evidence for these impacts, and of special interest, a contrast with conditions observed on the coast under the present more quiescent climate.

Prior to 1900 the sandy beach served as the most heavily traveled "highway" along the California coast. During the 1880s the coastal sand beaches varied in width from 30 to 80 m (Figure 3). The U.S. Coast and Geodetic Survey (1889) noted that " . . . from Mussel

FIGURE 1. (A) Histogram of annual rainfall data 1880 to 1980 at Julian, east of San Diego, Calif. High rainfall periods are separated by about 40-year intervals. (From Emery, K. O. and Kuhn, G. G., *Geol. Soc. Am. Bull.*, 93(7), 651, 1982. With permission.) (B) Histogram of the combined core-varve index for Santa Barbara Basin expressed in varve thickness units for the period 1824 to 1966. Fairly consistent periodicity of 24 to 26 years between highs is found, except during the years 1824 to 1859 when the sediments were highly bioturbated. (From Soutar, A. and Crill, P. A., *Geol. Soc. Am. Bull.*, 88, 1166, 1977. With permission.) (C) Histogram of standard tree ring indices for Big Cone Spruce trees located along the steep slopes of the Santa Ana Mountains north of San Diego at an elevation of 1114 m, for the period 1800 to 1974. Shows relationship to rainfall at Julian and clearly indicates recent drought period from 1946 to end of record. (From Douglas, A. V., Ph.D. thesis, University of Arizona, Tucson, 1976, 188.) (D) Histogram of winter water temperatures 5 m below the surface at La Jolla, Calif. Histogram was partly reconstructed from tree ring data using method explained by Douglas.[4]

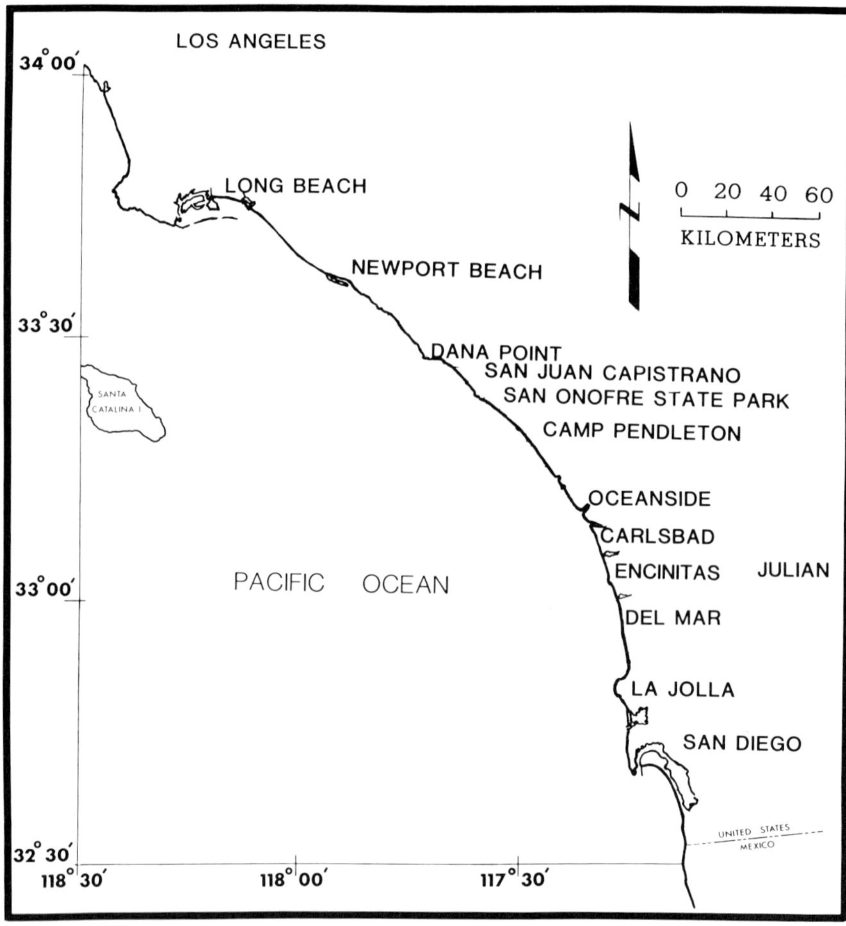

FIGURE 2. Map of locations cited in Southern California.

Rocks [Bathtub Rock at Torrey Pines] there is an unbroken sand beach for forty or fifty miles or as far north as the valley of San Juan Capistrano and in former times this stretch of beach was used whenever specially fast time [was] to be made on the route Los Angeles, via Capistrano, to San Diego.'' (With horse-drawn vehicles.) Wide, sandy beaches existed along this shore until the early 1940s.

Today the formerly wide sandy beaches south of Oceanside have, for the most part, been replaced by beach cobbles resting on exposed bedrock outcrops. This is certainly the result of numerous factors, both natural and man-induced. Of probable major importance is the fact that no flood of any significance has occurred during the past 40 years, contrasting with the large floods and sediment supply prevailing during the 19th century. What little sediment that would be derived at present from the rivers is further decreased by the agricultural dams that have been constructed on the principal rivers in San Diego County.[12]

During the latter part of the 19th century extensive beach cobble deposits were reported along the shoreline in San Diego County, derived from the rivers during flood stage, from sea cliff erosion, and perhaps transported along the shore by the longshore currents[6,7] or infrequently rafted onshore by kelp.[9] In 1874 Judge Benjamin Hayes (*San Diego Union*, September 19, 1874) described the large breakwater-like feature that he observed across the mouth of Penasquitos Lagoon (Sorrento Valley) south of Del Mar as follows: ''Perhaps the greatest curiosity is the high wall composed of small rocks, which at some day, long ago,

FIGURE 3. View in 1885 looking north along eroding bluffs and extremely wide, sandy beach at Oceanside, Calif. Oceanside Pier is presently located where building is seen in photo. Photo courtesy of Scripps Institution of Oceanography.

the dashing waves have built up, extending the whole length of the beach, across the mouth of the valley, making sort of a natural breakwater, through which now the sea can pass only at the northern side, in the Spring tides and storms to mingle with the fresh water of the creek.'' In 1885 at this same location a ''beach shingle railroad siding'' of the California Southern Railroad was constructed in order to mine the beach cobbles for both street paving materials for the city of San Diego and for providing rocks used as abrasives.[24]

Emery[6] described another locality at Carlsbad as the '' . . . longest gravel beach in Southern California.'' He noted that this was part of a depositional coast in which the majority of the cobble was derived by local cliff erosion. This site was also continually mined and the rocks were used for grinding mill pebbles and for filtering systems.[2,25]

These large cobble beaches may have been the product of more intense sea cliff erosion in the past. In the opposite direction, the presence of beach cobbles could have contributed to greatly accelerated erosion rates during the larger storms of the period. Playfair[19] and subsequent investigators have noted that waves armed with cobbles become a form of ''powerful artillery'' through which the ocean assails the land. Johnson[13] noted that ''It is perfectly evident that waves which are armed with cobblestones and enormous boulders must accomplish great erosive work when they beat against a cliff or artificial wall.''

This is further confirmed by our own observations in southern California. In 1963 approximately 2.9 million m^3 of sedimentary material was dredged from the Oceanside Harbor area and deposited on the beaches south of San Luis Rey River.[20] Much of this sediment consisted not only of mud and sand, but also contained a significant number of cobbles. Kuhn and Shepard[14] observed that, following the storms of December 1977, beach sand was rapidly removed by storm surf, the cobbles were withdrawn by the waves and in effect were thrown like artillery against houses along the strand at Oceanside. The following is a description of cobble abrasion and erosion during the storms of 1977 to 1978 and 1979 to 1980. During the last two weeks in December 1977 the offshore sand bars which buffered wave attack of the beaches at Oceanside were breached or removed by stormswell from the

FIGURE 4. (Before.) Top: September 1979 view looking north along The Strand at Oceanside, Calif. Building and road were constructed on former beach. Riprap fronts and protects concrete seawall and road. Photo: Sea Grant, G. Kuhn, September 22, 1979. (After.) Bottom: February 1980 view looking along same site as above. Road and seawall were undermined and collapsed from February 12 to 15, 1980, during a period with waves of less than 2-m height. Photo: Sea Grant, G. Kuhn, February 17, 1980.

FIGURE 5. (Before.) Top: A 1978 photo of beach cottages located south of Wisconsin Street in Oceanside. Photo: Sea Grant, G. Kuhn, January 4, 1978. (After.) Bottom: Same view in 1980 showing wave and beach cobble abrasion which damaged houses along the shore during storms in February 1980. Property damage occurred with only 1- to 2-m swell. Photo: Sea Grant, G. Kuhn, February 17, 1980.

FIGURE 6. An 1873 view of Cathedral Rock arch, off Point La Jolla, Calif. Photo: F. P. Shepard collection.

west-southwest. Subsequently, waves broke directly on the beach, removed the sand, and exposed the underlying cobbles. Distinct symmetrical cusps developed in the cobble beach. The cobbles were seen to have been moved as much as 6 m through the air by the waves, causing damage to beach-front cottages.

Again, in the winter of 1980 the cobbles were a destructive element, this time contributing to the undermining and collapse of the beach road (Figure 4) and damage to apartments and homes along the shore. At other locations roofs collapsed and windows were broken out as the cobbles became projectiles (Figure 5), despite the fact that the waves were consistently less than 2 m high.

In the past there must have been greatly accelerated coastal erosion, both due to the existence of extensive cobble beaches and because of larger sea storms. Even with relatively small present-day storms the cobbles are picked up by the waves and thrown through the windows and roofs of the cottages. It is alarming to contemplate what large storm waves could do with the cobble mass in attacking the shoreline, should such large storms return as a frequent occurrence. The large storm did return in early 1983.

Features of coastal erosion that appear to show some relation to the weather history of the southern California region are the arches and stacks that were so well developed along the coast of the La Jolla area, but have now largely disappeared. Old photographs show these arches and stacks to have been very common on the south side of Point La Jolla in the area now known as the Casa Manana Retirement Home near the Children's Pool. The beautiful arch called the Cathedral Rocks was a scenic marvel of the past century, but collapsed in 1906 (Figure 6), and the buttresses have since been entirely removed by the waves. During the same period other arches also collapsed and no remnant of them is now preserved. Arches to the north, around Alligator Head and at the cove, were more permanent, partly because they were reinforced with concrete. They did finally collapse, although one still stands at Goldfish Point and can be entered from either end by swimming.

The formation of the arches and stacks would appear to correspond to past periods of more intense sea cliff erosion, their destruction to the present period of more gradual cliff retreat. This is also indicated by the sea cliffs themselves, such as those at the southern end of La Jolla Shores, which give the appearance of having been actively eroded in the not too distant past, but have shown almost no changes in recent decades.[21]

MAN-INDUCED EROSION

As an indication of what can be found by a careful investigation of an area where casual observations suggested that there was little or no coastal erosion, we cite the results of work north of Del Mar, where many sections of the cliffs collapsed between 1973 and 1977. Failure involved both landslides and block falls. For the most part, collapse of the cliffs along this section occurred as instantaneous failures, where masses separated from the cliff face along fractures, joint sets, and contact planes. During the dry years from May 1973 to December 1977, at least 25 block falls occurred along the cliffs at Encinitas. Measuring from the cliff face landward, these blocks varied in dimensions from 0.3 to 3.7 m across, 1.5 to 33.5 m in length, and a maximum of 9.6 m in height. Many of these sea cliff failures can at least in part be attributed to man-induced effects.

Erosion Due to Increased Ground Water

These failures are partially related to the undercutting of the sea cliff, but a more significant factor is a recent steady rise in ground water levels along the sea cliffs. This rise is occurring commensurate with urbanization along, and inland of the sea cliff. It is a result of excessive watering of lawns and the attempt to grow nonnative vegetation, agricultural irrigation, septic tanks, leach lines, and cesspools. Sorben and Sherrod[22] indicate that the overirrigation of landscaping in residential subdivisions alone has been equivalent to 125 to 150 cm of precipitation per year.

This extensive watering in the coastal area has had at least three important effects: it has caused a slow but steady rise in the water table — even during the dry period preceding the heavy rains of 1978 — and this has added weight to the cliff material, which contributes to development of landslides, it has lubricated surfaces along which slides and block falls are initiated, and in some cases it has greatly increased solution in the underlying rocks.

Specific sites that experienced erosion during the drought years prior to 1978 have continued to collapse. Along the cliffs at the Self-Realization Fellowship property in Encinitas a large section of the cliff face separated along parallel joint sets and collapsed on April 26, 1978 (Figure 7). This recent failure measured 3.5 to 5 m landward, 34 m in width, and a maximum of 12 m in height. Following the collapse, springs were seen to flow from the cliff face. This is the same site where the cliff collapsed during the 1940s and the temple was lost (Figure 8).

Artificial Coastal Protection

Construction intended to protect the coast, such as groins, concrete and wooden sea walls, riprap and rock debris, as well as concrete lifeguard towers and buildings, are located along the cliffed shoreline of San Diego County. The erosive effects of improperly located artificial

FIGURE 7. (Before.) Top: 1978 photo looking southeast along base of cliff at Self Realization Fellowship at Encinitas, California. Photo: Sea Grant, G. Kuhn, February 3, 1978. (After.) Bottom: Same site following collapse, April 26, 1978.

protection and other structures have not been fully understood, especially during periods of large storms and high tides. It has been observed that where such protective measures project or extend seaward beyond adjacent unprotected lots, there is *immediate* erosion and notching of the latter. As beach sand levels fall, storm waves tend to converge on projecting structures and the waves refract toward unprotected lots. Recent examples were seen along the La

FIGURE 8. Collapse of Self Realization Fellowship temple along bluff at Encinitas following stormy period in winter of 1940 to 1941. Photo: Self Realization Fellowship, 1941.

Jolla Shores, Carlsbad, and Oceanside areas. During the storms of January through April 1978, beach sand levels dipped to a very low point, storm waves and high tides undercut the slope toe, and unprotected lots which were located adjacent to artificial projections were eroded landward in a short period of time (Figure 9). Since 1978 the sandy beach has continued to decrease in width and at many locations has been replaced by extensive cobble beaches as discussed above.

Because of the heavy rains in early 1978, cliffs that had been relatively stable were again subject to considerable retreat. A notable example is the low Quaternary alluvial cliffs south of Scripps Institution of Oceanography, which seemed to have stabilized during the dry period from 1947 to 1977. The series of houses built along the edge of these low cliffs, mostly without sea walls or any adequate protection, were subjected to erosion in January and again in February 1978. As the rains continued and the beaches were washed out by a combination of high waves and perigean spring high tides, as well as rain gullying, erosion again occurred along these cliffs. The residents were threatened with loss of their property and went to desperate measures to stop the erosion. Old car bodies were placed at the slope toe and later riprap was placed at these sites and cemented in place; temporary sea walls were installed. This was only partially successful as the houses had to be reinforced in various places to prevent their collapse. Adjacent unprotected lots were more severely eroded during the storm period. In 1979 a concrete seawall was constructed along most of this area.

Rapid Canyon Head Erosion South of San Onofre and Large Beach Accretion

Accelerated headward erosion is taking place along existing paleo-drainage avenues where culverts have been constructed under the railroad track and highways at San Onofre State Park and Camp Pendleton.[16] Between 1885 and 1980, normal erosional drainage avenues have been drastically altered by man. In 1885 the Atcheson-Topeka & Santa Fe Railroad Company constructed trestles across existing gullies and canyons. These structures were designed to span channels and did not appreciably alter the existing drainage patterns. During construction of California State Highway 101 in 1912 to 1918, culverts were installed under the highway in order to control surface runoff under the road and down the canyons during periods of heavy rainfall. In the wet years following installation of these drains, erosion of gullies accelerated where these structures concentrated the flow.

FIGURE 9. (Top) March 1978 view of eroded beach at Del Mar, Calif. Wooden seawall collapsed and adjacent unprotected lots were eroded. Photo: Sea Grant, G. Kuhn, March 5, 1978. (Bottom) February 1980 view of beach erosion at Del Mar, Calif. Concrete dumped on beach in front of house accelerated erosion of adjacent unprotected lots. Photo: Sea Grant, G. Kuhn.

FIGURE 10(A). (Before.) A 1968 vertical aerial photograph of canyon along bluffs at San Onofre State Park in northern San Diego County. Prior to this time only a notch existed perpendicular to shoreline at this site. Photo: U.S. Navy, 1968.

During construction of Interstate 5 in the late 1960s, large culverts and storm drains were installed directly below the San Onofre watershed and east of old Highway 101. These drains concentrated the surface runoff from numerous adjacent drainage avenues, directing it parallel to the highway and then west into culverts under Highway 101. No severe wet years occurred after completion of Interstate 5 until 1978. During January through April 1978, however, rainfall was greater than normal and headward erosion along these altered drainage avenues accelerated greatly. At Horno Canyon on Camp Pendleton the southbound lanes of Highway 101 collapsed, leaving a 23-m deep vertical scarp.

At one location no canyon existed perpendicular to the shoreline prior to the 1969 storms (Figure 10A). However, between 1968 and 1976 a new canyon began to be cut, reaching headward about 19 m. During January and February of 1978 approximately 49 m of headward erosion occurred. On February 14, 1980, another 2 m of erosion occurred, and on the 20th of February, during one day, the canyon actually eroded approximately 72 m landward, 15 to 30 m deep, and 18 to 27 m in width (Figure 10B). This amounted to approximately 38,000 m^3 of sedimentary material which moved immediately out onto the beach.

Dead Dog Canyon, south of Horno Canyon on Camp Pendleton, experienced approximately 170 m of headward erosion between 1932 and 1977. This same canyon eroded

FIGURE 10(B). (After.) March 1980 vertical aerial photograph showing development of 140-m canyon perpendicular to the shoreline. It was actually cut 133 m on February 20, 1980 alone, indicated between small arrows. Photo: U.S. Navy, March 1, 1980.

another 30 m during a 2-week period of January and February 1978[1] and approximately 30 m in February 1980. These are but two examples of the man-induced rapid headward erosion during wet years that can occur in canyons located perpendicular to the shoreline. Elsewhere during the same period of rapid erosion, degradation of the bluffs occurred where fences, stairways, lifeguard towers, and storm drain pipes were located along former drainage avenues. Surface water tends to adhere and flow along the underside of such structures, thus eroding the bluff face (Figure 11). At Carlsbad State Park numerous beach stair accesses have been undermined and storm drains collapsed between 1978 and 1980.

Bluff-Top Grading

The grading of the bluff tops, in many cases to the edge of the slope, initiates slope failures in the sea bluff and creates drainage avenues along which further surface runoff erodes the bluff-top edge and the bluff face. It has been observed that where the existing terrace deposits and soil profile are left intact and surface drainage is diverted, the bluff edge appears to erode very slowly. In 1971 to 1972 along south Solana Beach, the bluff top was lowered by excavation by 4.5 to 5.5 m and surface water was allowed to run over and down the bluff face. The upper slope continued to collapse during and following heavy rainfall. The base of the cliff along this site eroded 2 to 3 m between 1972 and 1978. A condominium homeowners association connected with the property threatened at this location has spent in excess of $1.5 million since 1978 in an attempt to forestall future erosion. A concrete seawall was constructed along the slope toe and concrete "cribbing" was installed along specific sites in the unconsolidated bluff face. In early May 1979, the week that the concrete "crib wall" was expected to be finished, the entire structure rotated and collapsed, leaving the debris and sediment resting on the seawall below.

Another example of cliff collapse following extensive grading of the bluff top occurred along the cliffs at Scripps Institution of Oceanography. The Quaternary cliffs in front of the Marine Biology building retreated 16 m between 1912 and 1975.[11] During extensive grading of the bluff top in 1975 the cliffs collapsed 2 to 3 m landward.

The importance of grading in relation to construction and the removal of protective vegetation and case-hardening surface materials has also had an important effect and is clearly indicated where areas have been used for construction along bluff margins.

EVALUATIONS OF CLIFF RETREATS BY OTHER METHODS

The popular concept of an average coastal-retreat rate for the evaluation of developmental projects is open to serious questioning, at least in the San Diego region and probably elsewhere. Sea-cliff and bluff retreat is episodic, site-specific, and strongly related to the meteorological conditions, induration and stability of rock formations within a given area; and to a combination of erosive agents, both natural and man-induced, that may act upon them.

An early attempt to estimate the rate of retreat of the cliffs was based upon the range of dates given by inscriptions and graffiti cut into mainly steep Cretaceous sandstone surfaces above sand beaches.[5] More recently, methods have involved placing nails in the rock face, followed by periodic measurement of retreat of the rock surface relative to the markers.[18] On the basis of 25-cm nails driven about 1 m above the base of the sea cliffs along a 30-km length of coast in San Diego County, Pinckney and Lee[18] indicated an average erosion rate of about 3 cm/year between 1971 and 1973. An extension of the measurements through 1976 by Lee et al.[17] yielded a range of 0 to 4 cm/year and a median of 0.6 cm/year of cliff retreat at 31 sites. In both studies some nails were lost due to small slides; thus the method is useful only for measuring the rate of surface abrasion rather than of slides.[8] Since episodic failures are rather large-scale and would never be indicated by the nails, it is felt that this method vastly underestimates the rates of erosion as they actually occur.

SUMMARY

Developments along the bluff margins north of San Diego have considerably increased cliff erosion. The many paths down the cliffs, along with stairways to the beach, have increased runoff. Attempts to grow nonnative vegetation such as ice-plant has seriously weighted the slopes. Watering has rapidly raised the water table levels to lubricate planes of slippage along the cliff front, and greatly weakened the bluffs. Grading the bluff tops has led to immediate and sizable retreats.

A

FIGURE 11. (A) (Before.) A 1977 view of storm drain designed to carry surface water down and away from slope face at Carlsbad State Park, Calif. Photo: Sea Grant, December 17, 1977. (B) (After.) A 1979 view at same location following collapse of storm drain pipe and severe gullying of bluff face. Photo: Sea Grant, February 11, 1979.

It has previously been believed that most beach sand along the shore in San Diego County was supplied by rivers in flood, with only minor contributions from the cliffs. However, both canyon-head erosion and landslides are presently contributing significant quantities of sand to beaches in northern San Diego County north of Oceanside.

A combination of continuing detailed observations in selected areas, detailed mapping on a large scale, careful review of historical records, and consideration of rock structures and competence along with hydrological considerations gives us the opinion that much of the coastal bluff area of San Diego County is a hazardous zone for building, particularly near bluff edges.

The past few decades, with little rain and small waves, have given builders a false confidence concerning the safety of the bluff edge, a confidence that has been shattered by recent rainstorms and increased wave activity. The rates of cliff retreat are proving to be much faster than previously estimated by rather ineffective studies using nails driven into the cliffs. The retreat is clearly episodic rather than steady, being controlled by the occurrence of slides. There are some indications that the southern California region may be returning to weather conditions that prevailed during the 19th century, with greater rainfall and storms. If so, this would result in greatly increased flood damage to low-lying coastal areas, and would further endanger developments placed too near eroding sea cliffs and bluffs.

FIGURE 11B.

REFERENCES

1. **Berggreen, R.,** Geology of the proposed Camp Pendleton LNG Site, San Diego County, California, in *Geologic Guide to San Onofre Nuclear Generating Station and Adjacent Regions of Southern California, Pacific Sections,* Fife, D. L., Ed., AAPG, SEPM, SEG Guidebook No. 461, American Association of Petroleum Geologists, Tulsa, Okla., 1979, A49.
2. California State Mining Bureau, 17th Rep. State Mineralogist, San Diego County Section, 1920.
3. **Douglas, A. V.,** Past Air-Sea Interactions off Southern California as Revealed by Coastal Tree-Ring Chronologies, M.S. thesis, University of Arizona, Tucson, 1973.
4. **Douglas, A. V.,** Past Air-Sea Interactions Over the Eastern North Pacific Ocean as Revealed by Tree-Ring Data, Ph.D. thesis, University of Arizona, Tucson, 1976.
5. **Emery, K. O.,** Rate of surface retreat of sea cliffs based on dated inscriptions, *Science,* 92, 617, 1941.
6. **Emery, K. O.,** Grain size of marine beach gravels, *J. Geol.,* 63, 39, 1955.
7. **Emery, K. O.,** *The Sea Off Southern California,* John Wiley & Sons, New York, 1960, 23.
8. **Emery, K. O. and Kuhn, G. G.,** Erosion of rock shores at La Jolla, California, *Mar. Geol.,* 37, 197, 1980.
9. **Emery, K. O. and Tschudy, R. H.,** Transportation of rock by kelp, *Geol. Soc. Am. Bull.,* 52, 855, 1941.
10. **Emery, K. O. and Kuhn, G. G.,** Seacliffs: their processes: profiles and classification, *Geol. Soc. Am. Bull.,* in press.
11. **Hannan, D. L.,** *Sea Cliff Stability, West of the Marine Biology Building,* Benton Eng. Proj. No. 75-9-18FG, Scripps Institute of Oceanography, La Jolla, Calif., 1975.
12. **Inman, D. L.,** Man's Impact on the California Coastal Zone, State Dept. Nav. Ocean Dev., State of California, Sacramento, 1976.
13. **Johnson, D. W.,** *Shore Processes and Shoreline Development,* John Wiley & Sons, New York, 1938, 65.
14. **Kuhn, G. G. and Shepard, F. P.,** Coastal erosion in San Diego County, California, in *Earthquakes and Other Perils, San Diego Region,* Abbott, P. L. and Elliott, W. J., Eds., Geological Society of America, Boulder, Colo., 1979.
15. **Kuhn, G. G. and Shepard, F. P.,** Should southern California build defenses against violent storms resulting in lowland flooding as discovered in records of past century? *Shore Beach,* 49(4), 2, 1981.
16. **Kuhn, G. G., Baker, E. D., and Campen, C.,** Greatly accelerated man-induced coastal erosion and new sources of beach sand, San Onofre State Park and Camp Pendleton, San Diego County, California, *Shore Beach,* 48(4), 9, 1980.
17. **Lee, L., Pinckney, C. J., and Bemis, C.,** Sea cliff base erosion, San Diego, California, Proc. Am. Soc. Civ. Eng. Natl. Water Resour. Ocean Eng. Conf., April 5 to 8, preprint no. 2708 American Society of Civil Engineers, New York, 1976.
18. **Pinckney, C. J. and Lee, L.,** Sea-cliff recession study, southern one-half of San Diego County, in Field Trip Guidebook: Eng. Probl. Areas, southwest San Diego Co., Calif., 1973, 14.
19. **Playfair, J.,** *Illustrations of the Huttonian Theory of the Earth,* Edinburgh, 1802.
20. **Shaw, M. J.,** Artificial sediment transport and structures in coastal southern California, Ref. No. 80-41, Scripps Institution of Oceanography, La Jolla, Calif., 1980.
21. **Shepard, F. P. and Grant, U. S., IV,** Wave erosion along the southern California coast, *Geol. Soc. Am. Bull.,* 58, 919, 1947.
22. **Sorben, D. R. and Sherrod, K. L.,** Groundwater occurrence in the urban environment: San Diego, California, in *Geology of Southwestern San Diego County and Northwestern Baja, California,* Farrand, G. T., Ed., San Diego Association of Geologists, Calif., 1977, 67.
23. **Soutar, A. and Crill, P. A.,** Sedimentation and climatic patterns in the Santa Barbara basin during the 19th and 20th centuries, *Geol. Soc. Am. Bull.,* 88, 1166, 1977.
24. U.S. Coast and Geodetic Survey, Descriptive report to accompany original field sheet entitled *Topography Pacific Coast Southward from San Dieguito Valley, California,* topogr. map no. 2014, National Ocean Survey Archives, Washington, D.C.
25. **Weber, F. H., Jr.,** Geology and Mineral Resources of San Diego County, California, Cty Rep. No. 3, California Division of Mines and Geol., Sacramento, 1963.

Chapter 14

EROSION OF THE U.S. SHORELINES

R. Dolan, B. Hayden, and S. May

INTRODUCTION

The River and Harbor Act of 1968 (33 USC 426i) required the U.S. Army Corps of Engineers to compile an inventory of the coastlines of the nation, including the Great Lakes. Their final summary, *A Report on the National Shoreline Study,* indicates that of the 135,500 km of coastline, 33,000 km were categorized as "seriously eroding" at the time (1971) the study was initiated.[9] Of the 12 reports in the study, 9 provide specific regional information and proposals for corrective measures. The General Accounting Office evaluated the National Shoreline Study and determined that inadequate funding and personnel, and inconsistencies in methodology and assessment criteria prevented the study from satisfying any but the broadest requirements of the 1968 Act.[2] However, to date, the National Shoreline Study has remained the only comprehensive assessment of coastal erosion on a nationwide scale.

Under sponsorship of the U.S. Geological Survey, we have assembled existing data collections on shoreline changes for the U.S. including Chesapeake Bay, the Great Lakes, Hawaii, and Alaska. The information will be presented as a series of 1:2,000,000 scale multicolor U.S. Geological Survey maps and as a 1:7,500,000 scale map in the *National Atlas.* The data bank of shoreline rates of change is accessible in a computer-based Coastal Erosion Information System (CEIS) at the University of Virginia. In this chapter we present a summary of the shoreline rates of change for various geographic regions of the U.S. based upon the CEIS data base.

MEASUREMENT OF SHORELINE CHANGE

CEIS contains shoreline change data originally determined by a variety of methods. These range from highly accurate engineering surveys to very general patterns detected by comparisons of old photographs. The paragraphs that follow summarize the advantages and disadvantages of the "standard" methods.

Repeated Shoreline Surveys

Cost: Data acquisition based upon field survey is expensive because of the high man-hours required.

Accuracy: Survey data are the most accurate available, with resolution on the order of 0.01 m or less.

Advantages: Measurements are direct. Individual measurements may be updated at relatively moderate cost. Measurement method commonly understood by the public. Measurement commensurate with local property survey records.

Disadvantages: Historical time line of data usually not available. Accuracy of measurement greatly exceeds the resolution of shoreline definition. Along-the-coast density is poor. Systematic updating of extensive coastal reaches is expensive and time consuming. Because surveying is time consuming, measurements for extensive reaches of the coast vary with time in the tidal cycle and with synoptic weather conditions.

Aerial Photography

Cost: Given the along-the-coast density of available data, the cost is rated as low.

Accuracy: Resolution of data varies with scale of photography and methods employed. Errors in the 2.5-m range are typical.

Advantages: High along-the-coast resolution. Historical data for the last 40 years usually available. Data is highly time-specific. Shoreline definition is within the resolution error of systems being analyzed. Repeated coverage is inexpensive if extensive coastal areas are included. Frequent coverage in time generally available.

Disadvantages: Longer time lines than the last 40 years are generally not available. Photo-interpretation skills needed to reduce data. Variation in photography type may result in errors. Some historical photo series are classified and not available for general use.

Historical Maps and Charts

Cost: Existing maps and charts are generally inexpensive but costs associated with new mapping are high.

Accuracy: Map accuracy is generally unavailable, but may be estimated in tens of meters.

Advantages: Maps and charts from the mid-1800s are available in the U.S. thus providing an unusually long time frame for determination of mean shoreline erosion.

Disadvantages: Irregular availability. Unstable map bases. Low accuracy and resolution. No correction for sea state or tide level.

Nonmetric Photography (Hand-Held Cameras)

Cost: Data is generally inexpensive.

Accuracy: At best, accuracy is poor with little information about either sea state or tide cycle.

Advantages: Historical information is possible.

Disadvantages: No systematic archives of such information are available. Photos are almost always oblique views requiring extensive correction. Scale is usually difficult to estimate.

In summary, ground-level surveys provide the highest quality shoreline data in terms of accuracy. Reliable historical survey records are lacking, however, for most coastal areas and new surveys are expensive and time consuming. With the exception of a few scattered sites, ground information is generally unavailable. Maps and National Ocean Survey (U.S. Coast and Geodetic Survey) nautical charts are available for numerous coastal locations, and frequently extend back to the mid-1800s. Although such charts are very useful, they are of questionable accuracy[3,5,6] and are frequently restricted in temporal coverage and to areas immediately adjacent to major shipping lanes and port facilities. Maps and charts best serve as supplemental information in determining historical trends in shoreline erosion.

Aerial photographs, taken with low-distortion mapping cameras, are available for most coastal locations in the U.S. The earliest photographs date to the 1930s, and photographs for subsequent decades are generally available. While aerial photographs taken with high-resolution mapping cameras provide considerable detail, they are not the photogrammetric equivalent of maps. This lack of equivalence results from scale variance within and between aerial photographs taken sequentially. Scale variations are generally of three types: (1) differences caused by changes in the altitude of the camera platform, (2) variations due to camera tilt, and (3) radial distortions due to relief variations of the surface photographed.[7,8] The scale variation due to land relief is not a source of error along low sedimentary coasts.[4] In the photography which we have used, mapping only the center section of each photo, we have not detected any measurable amount of radial distortion. Tilt error within images and scale error between images is usually minimal and is easily rectified using photogrammetric projectors.[7]

CEIS DATA BASE

The CEIS data bank was created from the best and most comprehensive of the existing information sets contributed by coastal scientists and engineers nationwide. It includes rate-of-change data from ground surveys, maps, charts, and aerial photos. Initially, the approach was to obtain maximum coverage using the most readily available data sets. Preference was given to studies with a wide range, both temporal and spatial, and which were already available in graphic form. Once the broad reaches were covered, high-resolution information was introduced. District and division offices of the U.S. Army Corps of Engineers were contacted for information. Because of its involvement with the National Shoreline Study, the Corps was able to provide up-to-date lists of coastal scientists and other contacts who could give us access to data. Requests for data were also made to state offices with responsibility for coastal affairs, Sea Grant offices, and university departments. Most contacts provided reports, data sets, and/or further contacts. It was quickly established that the mid-Atlantic, southeast, Gulf, and Great Lakes areas are data rich. In contrast, the Pacific, Alaskan, and northeast coasts have not been surveyed as extensively or quantitatively. Continuous shoreline monitoring is not a high priority in these areas because shoreline change is low in magnitude and/or there are low population densities in the hazard zone.

CEIS is designed to provide the user with original rate-of-change data, primary sources, statistical summaries, and notes regarding special interest features for any given reach of coast. The information is stored by grid cells which are 3′ latitude or longitude on a side.

A request can be initiated for any length of shoreline on the Atlantic, Gulf, Pacific, Alaskan, and Hawaiian coasts, the Great Lakes, and Chesapeake and Delaware Bays. The only information which the user must know prior to entering the system is the latitude and longitude limits of the area of interest. Rate-of-change information is printed out for each 3′ segment. Sources are provided and statistical summaries are calculated over the entire reach requested (Figure 1).

CHANGES OF THE U.S. SHORELINES — SUMMARY

We used CEIS to generate statistical summaries of rate-of-change data for each coastal state and for reaches with similar geology and shoreline geomorphology. These statistics are summarized in Tables 1 and 2. At the end of this chapter we have included a set of 1:7,500,000 scale maps that show the erosion rates for the U.S.

In Table 1 the mean rates of change for each coastal state, the within-state standard deviations of these rates, and the extreme range of rates in any one 3′ grid cell are listed. Along the Atlantic coast the average erosion rate is −0.8 m/year, with Virginia (−4.2 m/year), Maryland (−1.5 m/year), and New Jersey (−1.0 m/year) having the most pronounced erosion rates. The rapid erosion of Virginia's shoreline is largely confined to the Virginia barrier islands. Georgia, New York, and Delaware are, on the whole, accreting coasts. The Delaware Bay shoreline, however, is receding at a rate (−1.6 m/year) in excess of the

```
INPUT SEARCH MODE
? 2

ENTER LATITUDE OF LOCATION 1
? 360300N
ENTER LONGITUDE OF LOCATION 1
? 0753900W

ENTER LATITUDE OF LOCATION 2
? 355700N
ENTER LONGITUDE OF LOCATION 2
? 0753900W

--------------------------------------------------------------------

LATITUDE:    36- 3- 0 N
LONGITUDE:   75-39- 0 W

LOCATION:    BODIE ISLAND: CROATAN SHORES
             DARE
             NORTH CAROLINA

QUAD NAME: KITTY HAWK
QUAD YEAR: 1940
GRID CELL SIZE: 3 X 3      MINUTES

SHORELINE LENGTH: NOT LISTED IN RECORD
RATE OF CHANGE:      -.1
RECORD TYPE: AERIAL PHOTOGRAPHY
PERIOD OF RECORD: 1940-1977

MORE PRECISE LONG-TERM AVERAGE RATES OF SHORELINE CHANGE NOT
DETERMINED DUE TO LACK OF QUANTITATIVE DATA.

COMMERCIAL-RESIDENTIAL DEVELOPMENT ON/NEAR SHORELINE.

--------------------------------------------------------------------

LATITUDE:    36- 0- 0 N
LONGITUDE:   75-39- 0 W

LOCATION:    BODIE ISLAND: NAGS HEAD
             DARE
             NORTH CAROLINA

QUAD NAME: MANTEO
QUAD YEAR: 1975
GRID CELL SIZE: 3 X 3      MINUTES

SHORELINE LENGTH: NOT LISTED IN RECORD
RATE OF CHANGE:      -.9
RECORD TYPE: AERIAL PHOTOGRAPHY
PERIOD OF RECORD: 1940-1977

COMMERCIAL-RESIDENTIAL DEVELOPMENT ON/NEAR SHORELINE.

SHORELINE PROTECTION STRUCTURES AND/OR OTHER ENGINEERING STRUCTURE
(FISHING PIERS, OIL TERMINALS, BRIDGES, ETC.) PRESENT.

--------------------------------------------------------------------
```

A

FIGURE 1. (A, B). Representative output from the Coastal Erosion Information System (CEIS).

Atlantic coast average. Erosion rates along the Chesapeake Bay shoreline are of the same order as the Atlantic coast average. Overall, the Atlantic coast can be characterized as a receding system, with 403 of the 510 grid cells (79%) showing some measure of erosion.

The Gulf Coast states have the distinction of the most rapid average erosion rates (−1.8 m/year), more than twice the national average of −0.8 m/year. Within the Gulf Coast region the deltaic coast of Louisiana is by far the most dynamic (−4.2 m/year). Like the Atlantic coast, the Gulf region can be described as eroding. However, the percentage of receding areas is somewhat less (63%). This is due, in part, to the large area covered by the active deltaic system of the Mississippi River.

```
LATITUDE:   35-57- 0 N
LONGITUDE:  75-39- 0 W

LOCATION:  BODIE ISLAND: WHALEBONE
           DARE
           NORTH CAROLINA

QUAD NAME: ROANOKE ISLAND NE
QUAD YEAR: 1975
GRID CELL SIZE: 3 X 3      MINUTES

SHORELINE LENGTH: NOT LISTED IN RECORD
RATE OF CHANGE:     -1.2
RECORD TYPE: AERIAL PHOTOGRAPHY
PERIOD OF RECORD: 1940-1977

COMMERCIAL-RESIDENTIAL DEVELOPMENT ON/NEAR SHORELINE.

SHORELINE PROTECTION STRUCTURES AND/OR OTHER ENGINEERING STRUCTURE
(FISHING PIERS, OIL TERMINALS, BRIDGES, ETC.) PRESENT.

--------------------------------------------------------------------

LATITUDE:   35-57- 0 N
LONGITUDE:  75-38- 0 W

LOCATION:  BODIE ISLAND: WHALEBONE, JEANETTES PIER
           DARE
           NORTH CAROLINA

QUAD NAME: ROANOKE ISLAND NE
QUAD YEAR: 1975
GRID CELL SIZE: 3 X 3      MINUTES

SHORELINE LENGTH: NOT LISTED IN RECORD
RATE OF CHANGE:     -.8
RECORD TYPE: AERIAL PHOTOGRAPHY
PERIOD OF RECORD: 1945-1974

COMMERCIAL-RESIDENTIAL DEVELOPMENT ON/NEAR SHORELINE.

SHORELINE PROTECTION STRUCTURES AND/OR OTHER ENGINEERING STRUCTURE
(FISHING PIERS, OIL TERMINALS, BRIDGES, ETC.) PRESENT.

--------------------------------------------------------------------
```

```
PRIMARY SOURCES:

DOLAN, R. AND HAYDEN, R., PROJECT FOR COASTAL ENVIRONMENTS, DEPT. OF
ENVIRONMENTAL SCIENCES, CLARK HALL, RM. 101, UNIVERSITY OF VIRGINIA,
CHARLOTTESVILLE, VA.  22903.

USACE, 1971, NATIONAL SHORELINE STUDY, REGIONAL INVENTORY REPORT,
SOUTH ATLANTIC-GULF REGION:  U.S. ARMY ENG. DIV., SOUTH ATLANTIC.

--------------------------------------------------------------------

NUMBER OF GRID CELLS:          4
NUMBER OF CELLS WITH DATA :    4

MEAN RATE OF CHANGE:    -.750      MAXIMUM VALUE:     -.1
VARIANCE:                .217      MINIMUM VALUE:    -1.2
STANDARD DEVIATION:      .465
STANDARD ERROR:          .233
```

FIGURE 1(B).

In general, the Great Lakes shorelines are receding at rates (-0.8 m/year) commensurate with the Atlantic coast. The highest erosion rates are found along the shorelines of Lake Superior (-1.3 m/year) and Lake Erie (-0.7 m/year). Both these lakes have long east-west dimensions and, thus, large fetches during the eastward passage of storms. The Pacific

Table 1
RATE OF CHANGE STATISTICS FOR STATES AND REGIONS

Region	x(m/year)[a]	δ	Total range[a] Maximum	Minimum	N[b]
Oceanic Coastlines					
Atlantic Coast	−0.8	3.2	25.5	−24.6	510
Maine	−0.4	0.6	1.9	−0.5	16
New Hampshire	−0.5	—	−0.5	−0.5	4
Massachusetts	−0.9	1.9	4.5	−4.5	48
Rhode Island	−0.5	0.1	−0.3	−0.7	17
New York	0.1	3.2	18.8	−2.2	42
New Jersey	−1.0	5.4	25.5	−15.0	39
Delaware	0.1	2.4	5.0	−2.3	7
Maryland	−1.5	3.0	1.3	−8.8	9
Virginia	−4.2	5.5	0.9	−24.6	34
N. Carolina	−0.6	2.1	9.4	−6.0	101
S. Carolina	−2.0	3.8	5.9	−17.7	57
Georgia	0.7	2.8	5.0	−4.0	31
Florida	−0.1	1.2	5.0	−2.9	105
Gulf of Mexico	−1.8	2.7	8.8	−15.3	358
Florida	−0.4	1.6	8.8	−4.5	118
Alabama	−1.1	0.6	−0.8	−3.1	16
Mississippi	−0.6	2.0	0.6	−6.4	12
Louisiana	−4.2	3.3	3.4	−15.3	106
Texas	−1.2	1.4	0.8	−5.0	106
Pacific Coast	−0.0	1.5	10.0	−5.0	305
California	−0.1	1.3	10.0	−4.2	164
Oregon	−0.1	1.4	5.0	−5.0	86
Washington	0.5	2.2	5.0	−3.9	46
Alaska	−2.4	2.0	2.9	−6.0	69
Bays and Lakes					
Delaware Bay					
New Jersey	−1.9	1.3	0.3	−3.0	13
Delaware	−1.3	2.1	5.0	−3.0	12
Chesapeake Bay	−0.7	0.7	1.5	−4.2	136
Western Shore	−0.7	0.5	1.5	−1.9	67
Maryland	−0.7	0.3	−0.1	−1.3	35
Virginia	−0.8	0.7	1.5	−1.9	32
Eastern Shore	−0.7	0.8	0.1	−4.2	69
Maryland	−0.8	0.9	−0.3	−4.2	47
Virginia	−0.5	0.4	0.1	−1.2	22
Great Lakes	−0.7	0.5	0.6	−2.7	327
Lake Erie	−0.7	0.6	−0.2	−2.4	98
Ohio	−0.6	0.6	−0.2	−2.2	68
Pennsylvania	−0.3	0.1	−0.2	−0.4	14
New York	−1.4	0.6	−0.5	−2.4	20
Lake Ontario	−0.5	0.2	−0.2	−1.2	58
Lake Huron	−0.4	0.3	−0.3	−1.3	28

Table 1 (continued)
RATE OF CHANGE STATISTICS FOR STATES AND REGIONS

Region	x(m/year)[a]	δ	Total range[a] Maximum	Minimum	N[b]
Lake Michigan	−0.6	0.8	0.6	−9.9	184
Western Shore	−0.6	0.4	0.6	−1.5	62
Eastern Shore	−0.7	0.9	0.3	−9.9	122
Wisconsin	−0.7	0.3	−0.3	−1.5	46
Illinois	−0.2	0.4	0.6	−0.9	16
Indiana	−0.4	0.5	−0.3	−0.9	12
Michigan	−0.7	0.9	−0.3	−9.9	110
Lake Superior	−1.3	0.7	−0.3	−2.7	35
Minnesota	−0.8	0.4	−0.3	−1.5	16
Wisconsin	−1.8	0.6	−0.9	−2.7	19

[a] (−) values indicate erosion; (+) values indicate accretion.
[b] Total number of 3′ grid cells over which the statistics are calculated.

Table 2
RATE OF CHANGE STATISTICS FOR COASTAL LANDFORM TYPES

Region	x(m/year)[a]	δ	Total range[a] Maximum	Minimum	N[b]
Mud flats					
Florida	−0.3	0.9	1.5	−1.5	9
La.—Tex.	−2.1	2.2	3.4	−8.1	84
All Gulf	−1.9	2.2	3.4	−8.1	93
Rock shorelines					
Atlantic	1.0	1.2	1.9	−4.5	36
Pacific	−0.5	—	−0.5	−0.5	7
Pocket beaches					
Atlantic	−0.5	—	−0.5	−0.5	9
Pacific	−0.2	1.1	5.0	−1.1	144
Sand beaches					
Maine—Mass.	−0.7	0.5	−0.5	−2.5	17
Mass.—N.J.	−1.3	1.3	2.0	−4.5	22
Atlantic	−1.0	1.0	2.0	−4.5	39
Gulf	−0.4	1.6	8.8	−4.5	121
Pacific	−0.3	1.0	0.7	−4.2	19
Sand beaches with rock headland	0.3	1.9	10.0	−5.0	134
Deltas	−2.5	3.5	8.8	−15.3	155
Barrier Islands					
La.—Tex.	−0.8	1.2	0.8	−3.5	76
Fla.—La.	−0.5	1.7	8.8	−4.5	82
Gulf	−0.6	1.5	8.8	−4.3	158
Maine—N.Y.	0.3	2.6	4.5	−1.5	12
N.Y.—N.C.	−1.5	4.5	25.5	−24.6	153
N.C.—Fla.	−0.4	2.6	9.4	17.7	256
Atlantic	−0.8	3.4	25.5	−24.6	421

[a] (−) values indicate erosion; (+) values indicate accretion.
[b] Total number of 3′ grid cells over which the statistics are calculated.

coast, including Alaska, has the lowest erosion rates (-0.005 m/year), as indicated by the available sources, as well as the lowest overall percentage of eroding areas (30%). Washington State has a net accretion while Alaska, especially its Arctic shoreline, is eroding rapidly (-2.4 m/year). While erosion rates for the bays and lakes are of the same order as ocean coasts, the along-the-coast variability of the rates of erosion are much larger for ocean coasts. For example, Chesapeake Bay (-0.7 m/year) and the Atlantic coast (-0.8 m/year) have about the same rate of change, but the along-the-coast variation for the Atlantic coast is ± 3.2 m/year while Chesapeake Bay is only ± 0.7 m/year. The disparity in the range of rates (Table 1) between lake and bay coasts and ocean coasts is even more extreme.

When the national erosion rates are grouped on a basis of the shorezone geology the resulting patterns are, for the most part, as one would expect (Table 2). We used the four rock-type categories listed on the Coastal Landforms Map published in the U.S. Geological Survey *National Atlas* (1971). We further subdivided each rock type into the geographic regions. Coasts with fine grain sediments, deltas, and mudflats have the highest mean erosion rates (-2 m/year) while most sandy beaches and barrier islands are eroding at -0.8 m/year. The rock shorelines of the Atlantic coast (Maine) in the glacial rebound areas are accreting at 1.0 m/year. Table 2 also provides a tabulation of the erosion rates for the 295 Gulf and Atlantic Coast barrier islands. The Gulf Coast barrier islands are eroding at a mean rate of -1.6 m/year, but some islands have rates as high as -15.3 m/year. The Atlantic Coast barriers are eroding on an average of -0.8 m/year. The noticeably higher rates for the New York to North Carolina reach (-1.5 m/year) are due to the very high rate for the Virginia barrier islands (-4.2 m/year).

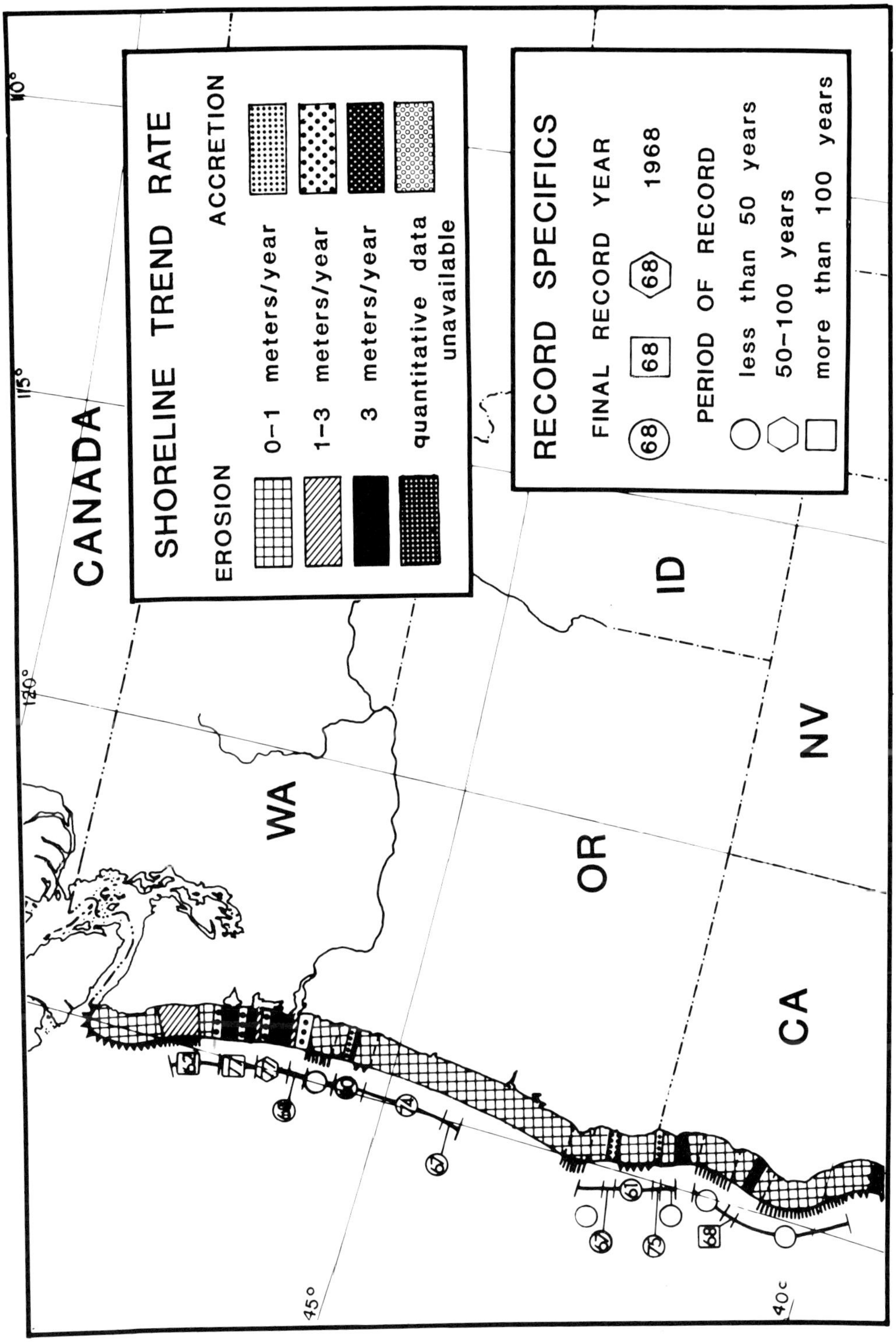
SHORELINE TREND RATE
EROSION
ACCRETION
0–1 meters/year
1–3 meters/year
3 meters/year
quantitative data unavailable
RECORD SPECIFICS
FINAL RECORD YEAR
68 68 68 1968
PERIOD OF RECORD
less than 50 years
50–100 years
more than 100 years
CANADA
WA
OR
ID
NV
CA
45°
40°
120°
115°
110°

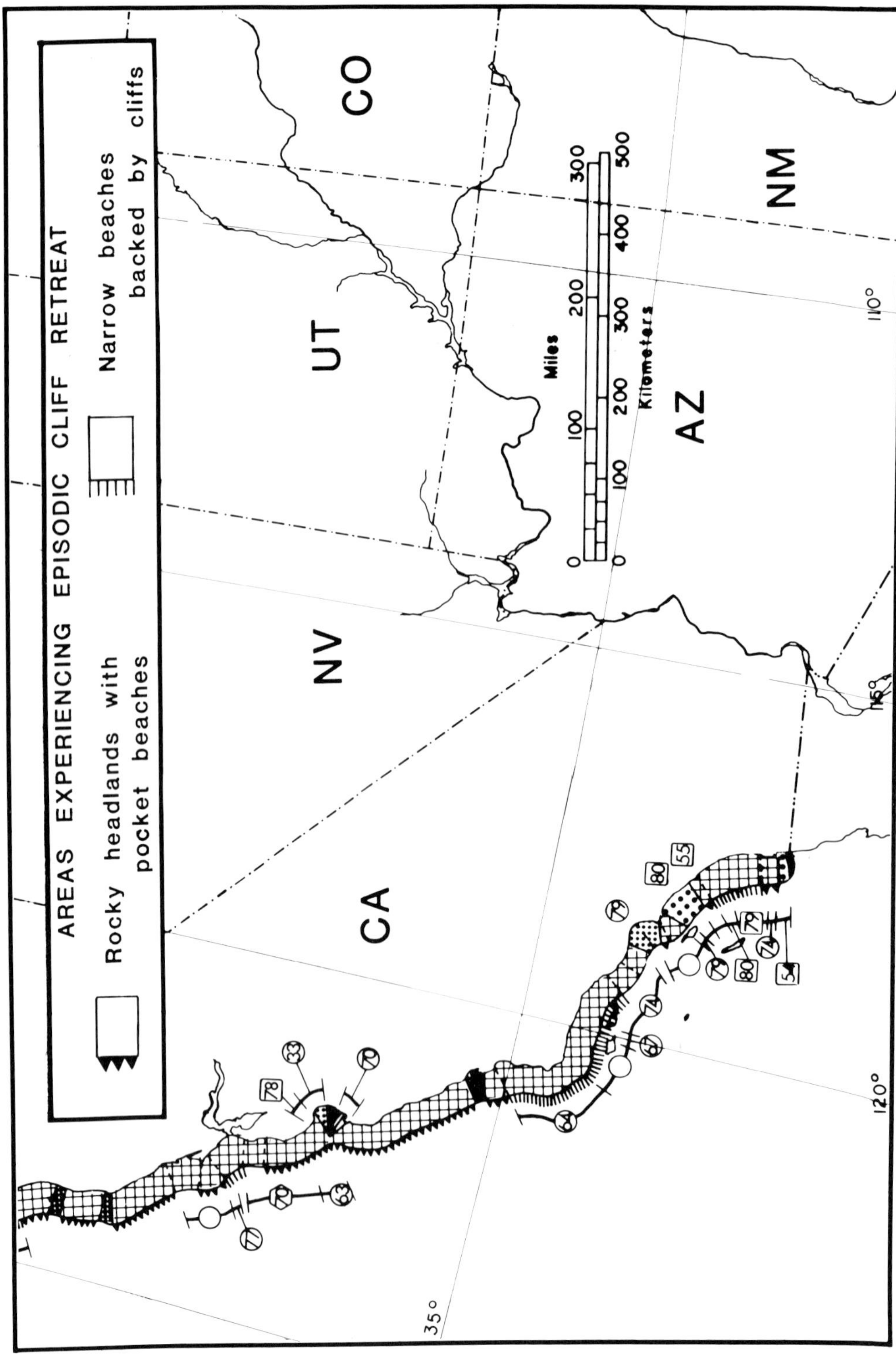
AREAS EXPERIENCING EPISODIC CLIFF RETREAT
Rocky headlands with pocket beaches
Narrow beaches backed by cliffs
CO
UT
NV
CA
AZ
NM
Miles
Kilometers
35°
120°
110°

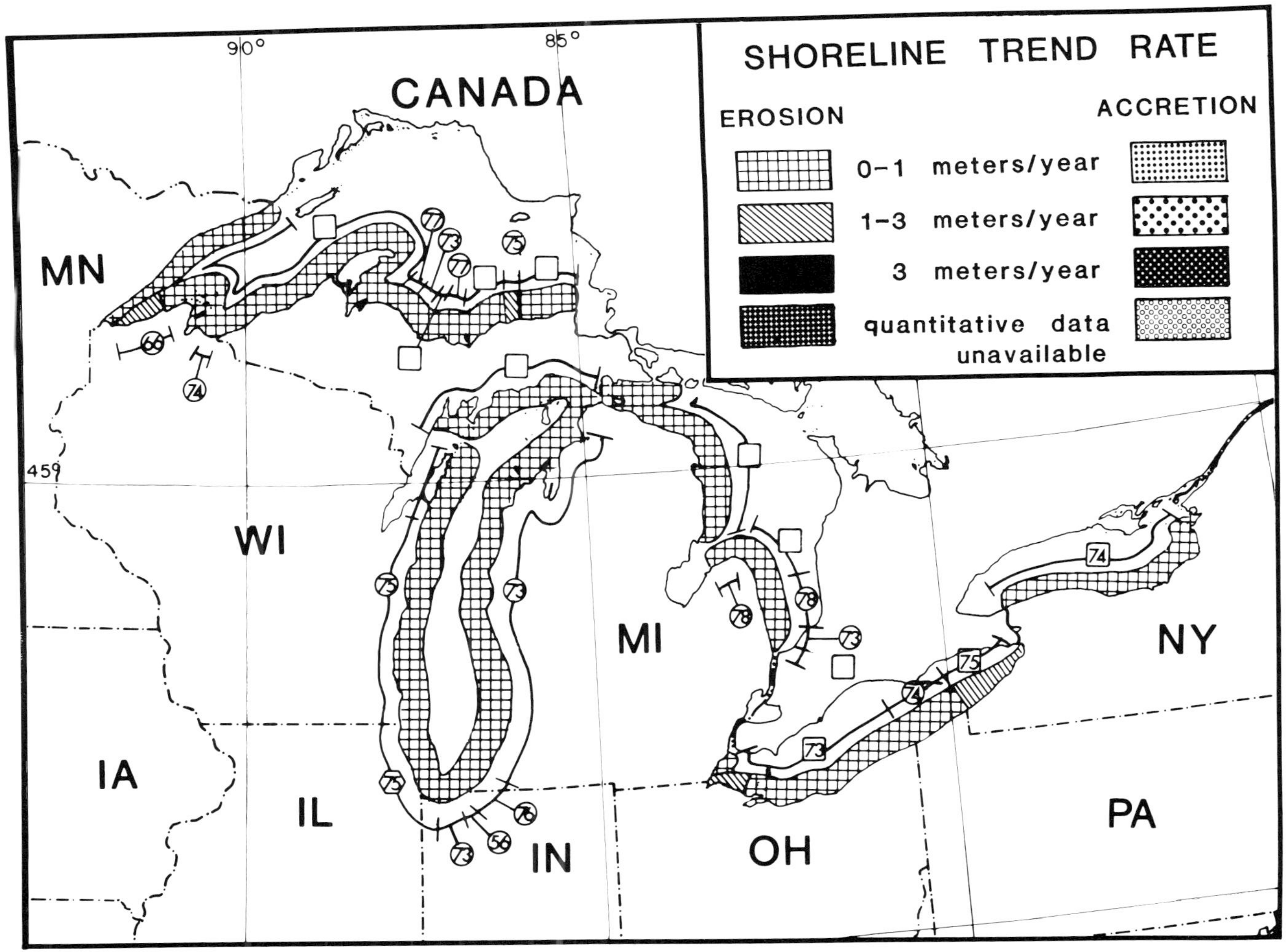
SHORELINE TREND RATE
EROSION
ACCRETION
0–1 meters/year
1–3 meters/year
3 meters/year
quantitative data unavailable
CANADA
MN
WI
MI
IL
IN
OH
NY
PA
IA
90°
85°
45°

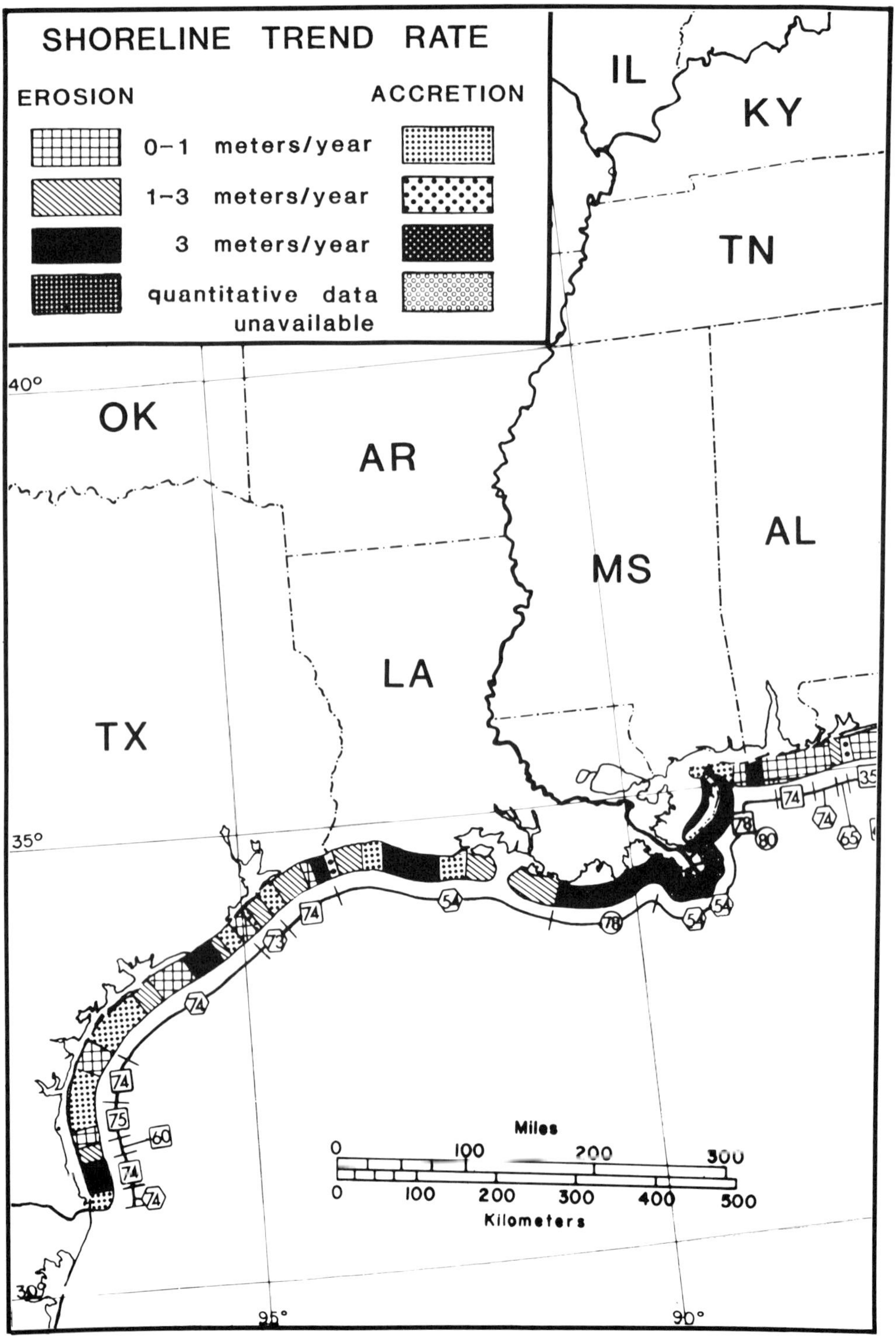
SHORELINE TREND RATE
EROSION
ACCRETION
0–1 meters/year
1–3 meters/year
3 meters/year
quantitative data unavailable
IL
KY
TN
OK
AR
AL
MS
LA
TX
40°
35°
30°
95°
90°
Miles
0
100
200
300
0
100
200
300
400
500
Kilometers

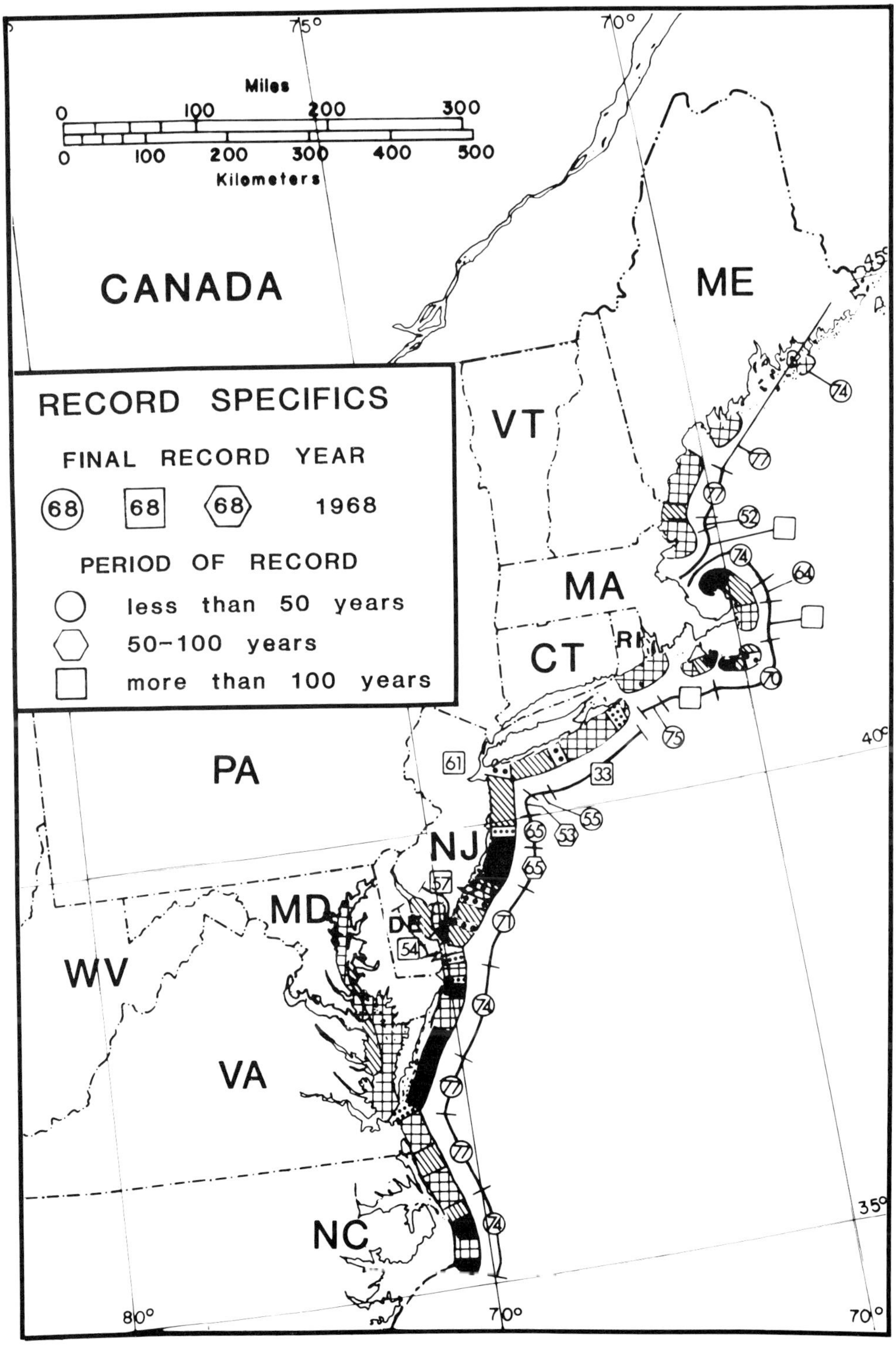
Miles
0 100 200 300
0 100 200 300 400 500
Kilometers
75°
70°
45°
40°
35°
80°
CANADA
ME
VT
MA
CT
RI
PA
NJ
MD
DE
WV
VA
NC
RECORD SPECIFICS
FINAL RECORD YEAR
68 68 68 1968
PERIOD OF RECORD
less than 50 years
50–100 years
more than 100 years
74
77
77
52
74
64
70
75
61
33
55
53
65
65
57
54
71
74
77
77
74

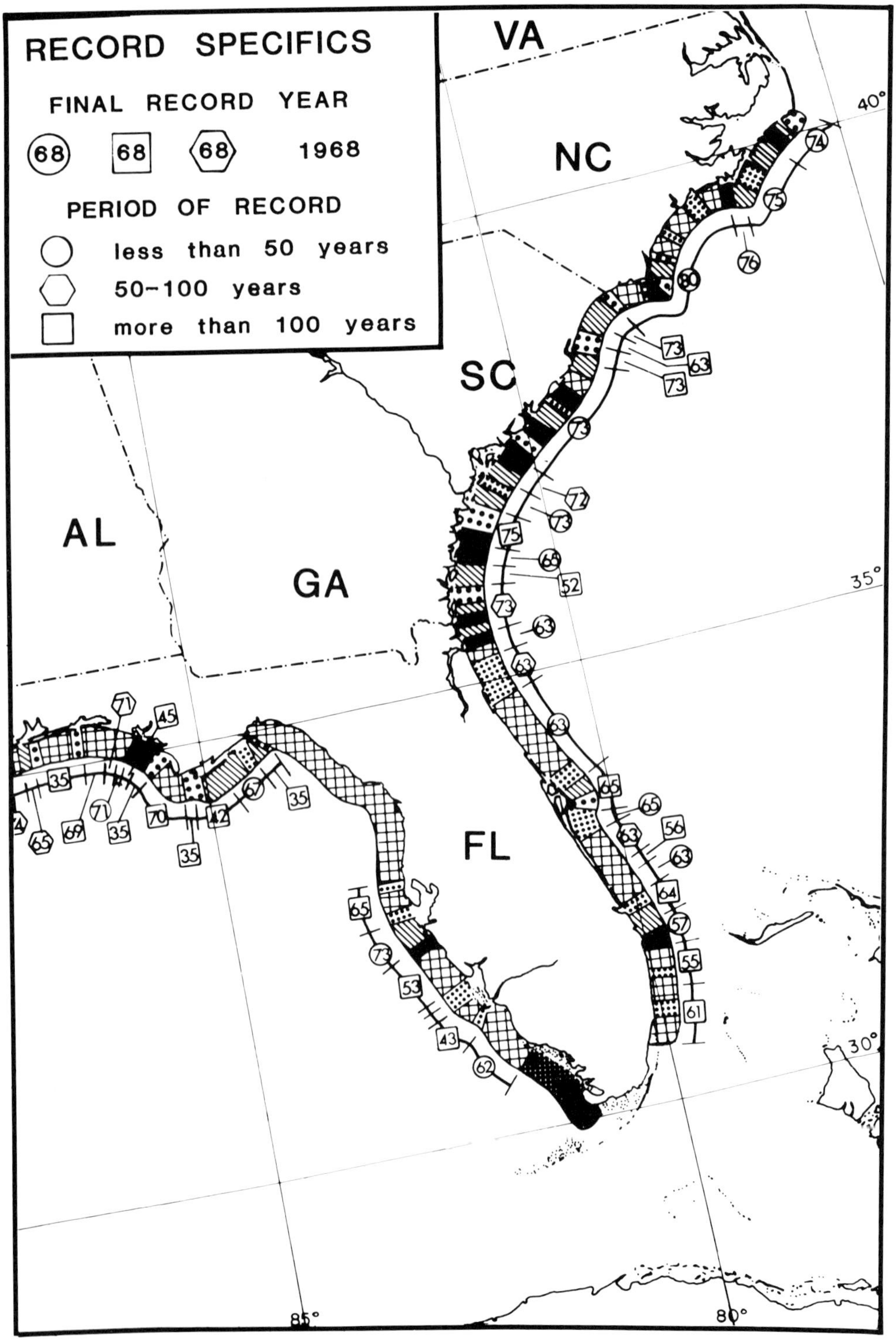
RECORD SPECIFICS
FINAL RECORD YEAR
68 68 68 1968
PERIOD OF RECORD
less than 50 years
50–100 years
more than 100 years
VA
NC
SC
GA
AL
FL
40°
35°
30°
85°
80°

REFERENCES

1. **Dolan, R. et al.,** Shoreline erosion along the middle Atlantic coast of the United States, *Geology,* 7, 602, 1979.
2. **Anon.,** Report to the Congress — National Effort to Preserve the Nation's Beaches and Shorelines — a Continuing Problem, United States General Accounting Office, Washington, D.C., 1975.
3. **Karo, H. A.,** One hundred and twenty year record of coastal changes, *Shore Beach,* 29(2), 33, 1961.
4. **Kelley, R. A.,** Applications of Terrestrial Photogrammetry for Determination of Shore Movements, Master of Science thesis, Ohio State University, Columbus, 1956.
5. **McCowan, J. H. and Brewton, J. L.,** *Historical Changes and Related Coastal Processes, Gulf and Mainland Shorelines, Matagorda Bay Area, Texas,* Bur. Econ. Geol., University of Texas, Austin, 1975.
6. **Shalowitz, A. L.,** Shore and Sea Boundaries, Publ. 10-1, U.S. Department of Commerce, Washington, D.C., 1964.
7. **Stafford, D. B. and Langfelder, J.,** Air photo survey of coastal erosion, *Photo Eng.,* 37, 565, 1971.
8. **Tanner, W. F., Ed.,** *Standards for Measuring Shoreline Changes, Report of a Workshop,* Dept Geol., Florida State University, Tallahassee, 1978.
9. USACE, National Shoreline Study, U.S. Army Corps of Engineers, Washington, D.C., 1971.
10. U.S. Geological Survey, National Atlas, U.S. Government Printing Office, Washington, D.C., 1971.

INDEX

A

B

C

R

S

T

U

V

W